DEPLETED URANIUM

Properties, Uses, and Health Consequences

DEPLETED URANIUM

Properties, Uses, and Health Consequences

Edited by
Alexandra C. Miller, Ph.D.

CRC Press
Taylor & Francis Group
Boca Raton London New York

CRC Press is an imprint of the
Taylor & Francis Group, an informa business

CRC Press
Taylor & Francis Group
6000 Broken Sound Parkway NW, Suite 300
Boca Raton, FL 33487-2742

First issued in paperback 2019

ISBN-13: 978-0-367-45328-2 (pbk)
ISBN-13: 978-0-8493-3047-6 (hbk)

Library of Congress Cataloging-in-Publication Data

Depleted uranium : properties, uses, and health consequences / edited by
Alexandra C. Miller.
p. cm.
Includes bibliographical references and index.
ISBN 0-8493-3047-5 (alk. paper)
1. Depleted uranium--Toxicology. 2. Depleted uraniumm--Health aspects. I.
Miller, Alexandra C.

RA1231.U7D48 2006
615.9'25431--dc22 2006045504

Visit the Taylor & Francis Web site at
http://www.taylorandfrancis.com

and the CRC Press Web site at
http://www.crcpress.com

Preface

The use of depleted uranium (DU) in armor-penetrating munitions remains a source of controversy because of the numerous unanswered questions about its long-term health effects. Although there are no conclusive epidemiological data correlating DU exposure to specific health effects, studies using cultured cells and laboratory rodents continue to suggest the possibility of leukemogenic, genetic, reproductive, and neurological effects from chronic exposure. On the other hand medical surveillance studies of U.S. soldiers wounded by DU shrapnel demonstrate that despite persistent urine uranium elevations more than 12 years since first exposure, renal and other clinical abnormalities have not been observed. Continuing surveillance is indicated, however, due to the ongoing nature of the exposure. Until issues of concern are resolved with further research, the use of DU by the military will continue to be controversial.

One of the reasons that complete information on DU exposure health hazards is not available is because there was little to no research done on DU biological effects before my laboratory at the U.S. Armed Forces Radiobiology Research Institute began its studies in 1995. Initially, it may have been assumed by the designers of DU applications that health hazards of DU would be similar to those of natural uranium. But significant radiological differences between DU and natural uranium make a direct comparison of biological and health effects difficult. Second, there is a great deal of controversy regarding the occupational health hazards of natural uranium as well. There are epidemiological studies demonstrating that natural uranium exposure is a health hazard, as well as reports showing that there is little significant effect of natural uranium exposure on health in general. Therefore, it was concluded that studies directly examining DU biological and health effects were warranted. Despite limited funding available, many scientists answered the call to study aspects of DU ranging from toxicology, carcinogenesis, and neurotoxicity, to human medical surveillance studies and risk modeling.

The goal of this book is to comprehensively describe the research progress made during the last 11 years by the leading radiobiology and toxicology scientists to answer the question of DU biological and health effects. Given the absence of any other books addressing DU health questions, there is a definite need for a book that includes the broadest collection of DU-studied topics. However, as this field of DU research is still in its earliest stages, each topic addressed is still in the process of evolving toward finalizing DU health questions, and many research topics cannot be fully answered yet. In fact, the briefness of several chapters reflects the situation of having limited DU information at this time. The field of DU research has produced many important scientific publications that benefit from being brought together to provide a current comprehensive resource. Interest in DU has greatly increased and broadened in recent years due to military use of DU, as evidenced by a growing interest by both scientists and the general public. The topics addressed in this book should provide the information so often sought by these two diverse groups.

This volume reviews key findings on DU biological and health effects ranging from cellular malignant transformation and carcinogenesis to animal toxicity and neurotoxicity, and concludes with human medical surveillance studies, uranium measurement methodologies, risk modeling, and environmental monitoring. The chapters are authored by clinical and basic researchers who are at the forefront of toxicology, carcinogenesis, and human epidemiology research.

We anticipate that this book will be a valuable resource for clinical and basic researchers, as well as media writers, legislators, and research advocates, and will provide a comprehensive collection of current research on DU. The knowledge about DU is expected to grow and in the future may include information on exposure assessment and hereditary effects. It is our hope that this book will stimulate further interest and research in DU health effects.

Acknowledgments

The editor would like to thank the following scientists for their helpful discussion regarding DU and related topics: Dr. Terry Pellmar, Dr. William Blakely, Dr. David Livengood, Dr. Gayle Woloschak, Dr. Dudley Goodhead, and Dr. Munira Kadhim.

I would like to express my gratitude to the contributors for their diligence and commitment to producing this book. I would also like to thank the staff of Taylor & Francis Group and CRC Press, especially Dilys Alam, Steve Zollo, and Barbara Norwitz. I am especially grateful to David Fausel, Kim Youngman, and Rachael Panthier for their special assistance and patience helping this first-time editor.

On a personal note, I would like to thank my mother for her continued belief in me and my endeavors. I would like to thank my entire family for their support as well.

Editor

Alexandra C. Miller is currently a senior scientist and principal investigator at the U.S. Armed Forces Radiobiology Research Institute and the Uniformed Services University of the Health Sciences in Bethesda, Maryland. She received her Bachelor of Science in chemistry from the University of Maryland, College Park, in 1981 and in 1986 she received her doctorate in radiation biology and experimental pathology from Roswell Park Cancer Institute Division at the State University of New York in Buffalo. Dr. Miller completed two postdoctoral projects: the first at the Armed Forces Radiobiology Research Institute in the area of cellular radiobiology and the second at the Uniformed Services University of the Health Sciences in molecular biology and carcinogenesis. She has received training awards from the American Society of Therapeutic Radiation Oncology and the Photo-Radiation Society. Dr. Miller has been a visiting scientist at the National Cancer Institute Division of Cancer Treatment and in France at the University of Paris. Dr. Miller has been nominated for the Radiation Research Society Michael J. Fry Award three times since 1996 and has over 45 peer-reviewed publications. Dr. Miller has received extramural funding from the Henry M. Jackson Foundation for Medical Research, NASA, the U.S. Army Medical Research and Material Command (USAMRMC), and the USAMRMC congressionally directed medical research program. The author has been the principal investigator for more than nine research projects since 1992. She has been an *ad hoc* member of panels for the World Health Organization, U.S. Army Medical Research Command, U.S. Congressionally Directed Medical Research Programs (CDMRP), U.S. Agency on International Development Committee, and the U.S. Department of Energy.

Dr. Miller's research interests include carcinogenesis, transgenerational genotoxicity, chemoprevention, and biodosimetry. Current research projects include heavy metal carcinogenicity, transgenerational genotoxicity, radiation protection against internal radiation emitters and external radiation exposure, chemoprevention of radiation- and heavy metal–induced leukemias, and biomarkers of exposure and disease development. Past research focused on molecular genetics of radiation resistance and photodynamic therapy of cancer.

List of Contributors

M.R. Bailey, Ph.D.
Centre for Radiation, Chemical and Environmental Hazards
Chilton, Didcot, Oxfordshire
United Kingdom

Wayne Briner, Ph.D.
University of Nebraska at Kearney
Kearney, Nebraska

Mario Burger, Ph.D.
United Nations Environmental Programme
Geneva, Switzerland

José A. Centeno, Ph.D.
Armed Forces Institute of Pathology
Washington, District of Columbia

Valérie Chazel, Ph.D.
IRSN/DRPH/SRBE/LRTOX
Pierrelatte, France

John W. Ejnik, Ph.D.
Northern Michigan University
Marquette, Michigan

Patrik Gerasimo, Ph.D.
Service de Protection Radiologique des Armées
Paris, France

Fletcher F. Hahn, D.V.M., Ph.D.
Scientist Emeritus
Lovelace Respiratory Research Institute
Albuquerque, New Mexico

Pascale Houpert, Ph.D.
IRSN/DRPH/SRBE/LRTOX
Pierrelatte, France

John F. Kalinich, Ph.D.
Armed Forces Radiobiology Research Institute
Bethesda, Maryland

Pierre Laroche, Ph.D.
Service de Protection Radiologiquc des Armées
Paris, France

R.W. Leggett, Ph.D.
Oak Ridge National Laboratory
Oak Ridge, Tennessee

David E. McClain, Ph.D.
Armed Forces Radiobiology Research Institute
Bethesda, Maryland

Melissa A. McDiarmid, M.D., M.P.H.
University of Maryland School of Medicine
Baltimore, Maryland

Alexandra C. Miller, Ph.D.
Armed Forces Radiobiology Research Institute
Bethesda, Maryland

Florabel G. Mullick, M.D.
Armed Forces Institute of Pathology
Washington, District of Columbia

Edward A. Ough, Ph.D.
Royal Military College of Canada
Kingston, Ontario, Canada

François Paquet, Ph.D.
IRSN/DRPH/SRBE/LRTOX
Pierrelatte, France

A.W. Phipps, Ph.D.
Centre for Radiation, Chemical
and Environmental Hazards
Chilton, Didcot, Oxfordshire
United Kingdom

Gérard Romet, M.D.
Service de Protection Radiologique
des Armées
Paris, France

Henri Slotte, Ph.D.
United Nations Environmental Programme
Geneva, Switzerland

Brian G. Spratt, Ph.D.
St. Mary's Hospital Campus
London, United Kingdom

Katherine S. Squibb, Ph.D.
University of Maryland School
of Medicine
Baltimore, Maryland

Todor I. Todorov, Ph.D.
Armed Forces Institute
of Pathology
Washington, District
of Columbia

Abstract

There is limited medical and research information available to enable a complete understanding of the potential health hazards of depleted uranium exposure. This book provides the first compilation of the most current published research results regarding the potential health hazards of depleted uranium. Although full and complete knowledge of depleted uranium health effects is limited due to the paucity of research data, the chapters contained in this book provide the most current and comprehensive collection of information regarding depleted uranium health hazards. The chapters include information on cellular and animal studies, *in vivo* carcinogenesis, risk modeling, uranium measurement methodologies, medical surveillance programs, and environmental monitoring. This book will be of value to those who are relatively unfamiliar with depleted uranium scientific and medical study methods and to those who are science savvy but want to know what is the most current information available on depleted uranium research. There is no other single source for this type of information on depleted uranium.

Contents

1 Depleted Uranium Biological Effects: Introduction and Early *In Vitro* and *In Vivo* Studies

David E. McClain and Alexandra C. Miller

CONTENTS

INTRODUCTION

The use of depleted uranium in armor-penetrating munitions remains a source of controversy because of the numerous unanswered questions about its long-term health effects. Although there are no conclusive epidemiological data correlating depleted uranium exposure to specific health effects, studies using cultured cells and laboratory rodents continue to suggest the possibility of leukemogenic, genetic, reproductive, and neurological effects from chronic exposure. Until issues of concern are resolved with further research, the use of depleted uranium by the military will continue to be controversial.

Advances in metallurgy and weapons design in the past several decades have led to new munitions whose effectiveness has provided tactical advantages on the battlefield and, consequently, saved lives of personnel. However, decisions to deploy these munitions have sometimes outpaced our knowledge of how they impact the health of those exposed to them.

Depleted uranium (DU) kinetic energy penetrators are perhaps the best-known example of these advanced munitions, primarily because of their outstanding, well-publicized performance against enemy armor in the 1991 Persian Gulf War. DU munitions were again used in the NATO military actions in Bosnia-Herzegovina (1995) and Kosovo (1999) and, more recently, coalition actions in Iraq.

DU munitions were used only by coalition forces during the 1991 Gulf War, but their use led to DU fragment injuries among coalition forces as a result of friendly fire incidents. Other personnel were exposed via inhalation/ ingestion after working around vehicles struck by DU munitions. Such exposures were not considered especially dangerous at the time, because numerous epidemiological studies of uranium miners and millers working with natural uranium had shown few concrete health effects from exposure, and DU has 40% less radioactivity than natural uranium. However, the exposure of wounded personnel to uranium as embedded fragments had no medical precedent, so the earlier studies dealing primarily with inhalation or ingestion exposures in miners were of uncertain utility. As a result, questions were soon raised as to whether it was wise to leave in place fragments possessing the unique radiological and toxicological properties of DU, especially when considering that exposures might extend as long as the 40–50 years remaining in a person's life span. As these treatment questions were being addressed, a growing public concern about the long-term health and environmental impact of using a radioactive metal like DU on the battlefield fueled forceful national and international efforts to ban the use of DU in munitions.

This chapter aims to summarize the current status of knowledge about the potential health effects of DU based on cellular and animal studies. Studies have been conducted using cultured cells and animal models and have attempted to answer questions relating to toxicity, carcinogenicity, and involvement of radioactivity.

BACKGROUND

Uranium was discovered in the mineral pitchblende in 1789 by the German chemist Martin Heinrich Klaproth. Uranium does not exist in pure metallic form in nature because it is quickly oxidized in air. It occurs most commonly as U_3O_8, uranium oxide, in ores such as pitchblende. Refined uranium metal used in reactors is in the form of UO_2, uranium dioxide.

A sample of uranium was used by the French physicist Henri Becquerel in his discovery of the concept of radioactivity in 1896. Natural uranium has three predominant natural isotopes: ^{234}U, ^{235}U, and ^{238}U, all of which are radioactive; other uranium isotopes can be produced artificially in a reactor. The half lives of the natural isotopes are 2.44×10^5 years for ^{234}U, 7.10×10^8 years for ^{235}U, and 4.5×10^9 years for ^{238}U, and their composition in natural uranium by mass is 0.005% ^{234}U, 0.711% ^{235}U, and 99.284% ^{238}U. Considering the isotope half lives and their mass percentages, it can be calculated that 48.9% of the radioactivity of natural uranium is derived from the isotope ^{234}U, 2.2% from ^{235}U, and 48.9% from ^{238}U (ATSDR 1999). Thus, ^{234}U contributes as much to the radioactivity of natural uranium as does ^{238}U, despite the fact it is 20,000 times less abundant. Natural uranium has a low specific radioactivity of about 0.68 μCi/g or 1.8×10^7 Bq, which means natural uranium is considered only a weakly radioactive element. A comparison of depleted uranium to natural uranium is shown in Table 1.1. This chart quite simply shows that DU has an approximately 50% lower specific activity than natural uranium. This chart also shows that DU has a different radioisotope decay chain than natural uranium, making DU very different from natural uranium.

All isotopes of uranium, natural or manmade, decay by emission of alpha particles of various energies, a process by which the uranium is transformed into another element that is also radioactive. The decay series continues until reaching a nonradioactive isotope of lead. Alpha particles have very low penetrating power but deposit large amounts of energy during penetration. Thus, alpha particles represent little hazard when on the surface of the skin but are potentially a significant hazard if inhaled or ingested, whereby they come in close contact with sensitive tissues (Hartmann et al., 2000). Beta and gamma radiation are also emitted during certain transformations, but those radiation levels are lower. Workers exposed to natural uranium could receive radiation exposures to all of the isotopes in the transformation series.

The use of uranium as nuclear fuel or in nuclear weapons requires enrichment of the fissionable isotope ^{235}U. The enrichment process concentrates ^{235}U in the metal to specific activities required to sustain nuclear reactions. The by-product of enrichment is uranium with reduced levels of the ^{235}U isotope, or "depleted" uranium. The Nuclear Regulatory Commission considers the specific activity of DU to be no more than 0.36 μCi/g, but more aggressive enrichment processes can drive this value slightly lower (~0.33 μCi/g) (ATSDR 1999). This means DU has roughly 50% of the radioactivity of natural uranium. Even though DU has less specific radioactivity than natural uranium, it retains all of its chemical properties. The large-scale

TABLE 1.1
Comparison of the Relative Contribution of Uranium Isotopes[a] (Natural and Depleted)

Isotope	Specific Activity (Ci/g)μ	DU SA by Wt% (Ci/g)μ	Natural Uranium SA by Wt% (Ci/g)μ
^{238}U	0.333	0.332	0.331
^{236}U (not naturally occurring)	63.6	0.0001	0
^{235}U	2.2	0.0044	0.051
^{234}U	6200	0.093	0.310
Total		**0.4295**	**0.692**

[a] Contribution of the daughter products is not included.

production of enriched uranium for nuclear weapons and fuel over the decades has resulted in an abundance of cheap DU, a factor that has played a role in its use in a wide variety of applications (e.g., radiation shielding, compact counterweights, armor, kinetic energy weapons). The properties of DU that make it useful as an armor-penetrating munition are its density (1.68 times that of lead) and the ability to engineer into it a molecular structure that facilitates entry into a hardened target by "shedding" outer layers of the metal during penetration.

URANIUM TOXICITY AND HEALTH EFFECTS: COMPILED REPORTS

Toxicology studies of natural uranium partially relevant to understanding DU health effects are numerous, beginning with the first reported observations of uranium-induced kidney abnormalities in the mid-1800s (see Hodge 1973b). Most of our detailed knowledge of uranium toxicity is derived from studies in the 1940s and early 1950s as the Manhattan Project, and the need for enriched uranium for reactors led to a requirement to understand better the occupational hazards presented to uranium workers. Much of that original work is described in the classic multivolume monographs of Voeglen and Hodge (Voeglen 1949, 1953) and in Hodge et al. (Hodge et al. 1973a), which are often together considered the definitive compilations of toxicology and pharmacokinetic data for uranium in animals and humans. It is important to keep in mind, however, that these studies are about natural uranium.

Additional biological information about uranium has accumulated since then that has reinforced our understanding of both uranium and DU health effects. The Agency for Toxic Substances and Disease Registry (ATSDR) has produced a very thorough reference that summarizes what is known and not known about the toxic effects of natural uranium exposure (ATSDR 1999). The controversy surrounding the use of DU during and after the 1991 Gulf War led to a number of other excellent literature reviews of uranium and DU health effects (Institute of Medicine 2000; The Royal Society 2001, 2002). These literature searches were not able, however, to answer many questions as DU studies had not been completed and the significant differences between natural uranium and DU in terms of total radioactivity and decay chain, continue to make it difficult to make claims about the potential hazards of DU exposure. Additionally, DU exposure conditions, chronic, wounding, etc., are considerably different than that for natural uranium (intermittent, inhalation). Therefore, it is important to keep in mind that these reports are about natural uranium which is considerably different from DU. Secondly, the type of exposure to natural uranium (intermittent, short term) considered in these reports is considerably different from the type of exposure to DU (chronic) that could occur following wounding by DU shrapnel.

RESEARCH APPROACHES: *IN VITRO* STUDIES

The earliest studies on DU effects involved using cellular model systems and were conducted in the mid-1990s at the U.S Armed Forces Radiobiology Research Institute. A strategic research approach involving the progression from cellular studies to

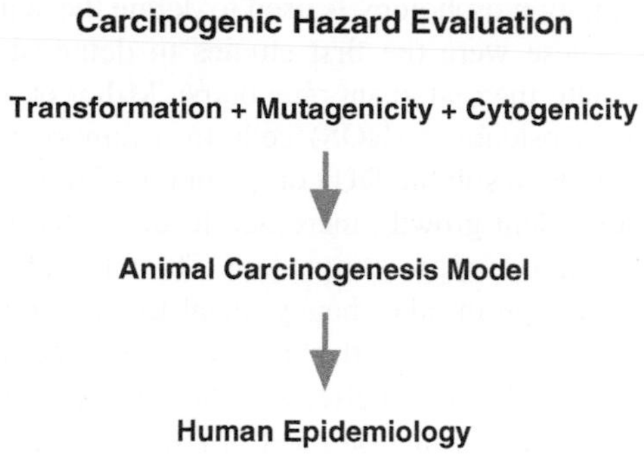

FIGURE 1.1 Strategic research approach. Carcinogenic hazard evaluation. This type of hazard evaluation is a multistep process involving *in vitro* and animal models. Human epidemiology is used as the last step in the process.

animal model and finally to human epidemiology considerations was the approach used by Dr. Miller and colleagues (Figure 1.1). This research group used an immortalized human cell line to study neoplastic transformation, genomic instability, genotoxicity, and radiation-induced damage. The neoplastic transformation model approach was used to evaluate carcinogenic potential of DU (Figure 1.2). In this approach, human cells are exposed to the test material and then plated for colony formation.

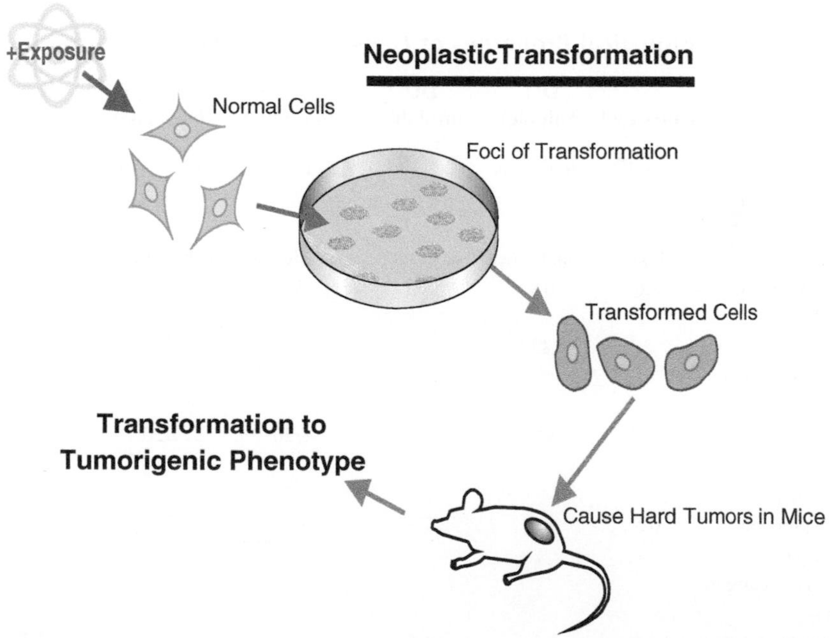

FIGURE 1.2 (See color insert following page 80) Neoplastic transformation. The Figure illustrates the process of neoplastic transformation which is a multistep process.

An evaluation of the colony morphology is used to define the state of transformation of the exposed cells. These were the first studies to demonstrate that DU could transform human cells into the malignant phenotype. Miller et al. (1998b) observed transformation of human osteoblast (HOS) cells to a tumorigenic phenotype after exposure to uranyl chloride, a soluble DU compound. DU-treated cells also demonstrated anchorage-independent growth, increased levels of the *k*-ras oncogene, and decreased levels of Rb tumor suppressor protein. The latter changes are associated with the malignant phenotype in other heavy metal treated cells. A comparison to insoluble DU (uranium dioxide), and other nonradioactive heavy metals was done to enable a comparison of DU-induced effects to those of better known carcinogenic heavy metals like nickel. Transformation rates of DU-exposed cells were 9.6 times that of untreated controls, and transformed cells formed tumors in nude mice. The results from all these transformation studies are compiled in Table 1.2. A comparison of the transformation values indicates that DU can neoplastically transform human cells similar to other better-known heavy metals and at a similar magnitude. These results were the first indication that DU could be carcinogenic; however, as is the case for any determination of carcinogenicity, *in vivo* studies would be necessary to more fully understand DUs potential as a carcinogen. The studies done to assess DU carcinogenicity *in vivo* are discussed in detail in a later chapter. Not only was the human cell model excellent as a means to study DU, but the same model was used to evaluate additional carcinogenic, mutagenic, and genotoxic endpoints.

TABLE 1.2
Transformation Biological Properties of Transformed HOS Cells

	Untreated	DU (Soluble)	DU (Insoluble)	rWNiCo (Insoluble)	Beryllium (Insoluble)	Nickel (Insoluble)
Transformation Frequency (per survivor × 10^{-4})	4.4 ± 1.1	49.6 ± 4.8	110.2 ± 10.1	37.6 ± 5.1	31.4 ± 4.4	39.5 ± 3.1
Morphology	Flat	Transformed	Transformed	Transformed	Transformed	Transformed
Saturation density (× 10^5 cells)	2.6	6.9	6.6	6.1	6.3	7.1
Soft-agar colony formation (plating efficiency %)	2	47	61	29	31	28
Tumorigenicity (mice tumors/mice inoculated)	0/82	19/20	14/20	8/20	7/20	6/20
Kras expression (densitometric analysis; relative to actin)	0.5	8.1	7.3	6.9	7.0	5.7
pRb phosphorylation (densitometric analysis)	6.1	.1	1.5	2.0	1.2	1.1

Experiments with cultured cells have also demonstrated the capacity of uranium to induce genotoxic changes. (Lin 1991) showed that uranyl nitrate increased frequencies of micronuclei, sister chromatid exchange, and chromosomal aberrations in Chinese hamster ovary (CHO) cells. Assessments using a human cell model were again conducted by Miller et al., who used *in vitro* studies to demonstrate that DU was genotoxic by measuring the induction of sister chromatid exchanges, micronuclei, and dicentric chromosomes in the same human cell model used to study neoplastic transformation. Three examples of chromosomal damage measured in these studies are shown in Figure 1.3, including micronuclei, dicentric, and sister chromatid exchange. Studies examining the effect of DU exposure on the induction of these chromosomal changes revealed that DU (in both soluble and insoluble) form could induce an increase in genotoxic damage in comparison to control levels (Table 1.3). Furthermore, these were the first studies to demonstrate that DU could induce a radiation-specific marker, a dicentric chromosome. This finding led to additional studies to address the role of alpha particle radiation damage caused by DU exposure. The nonradioactive heavy metals like nickel which were genotoxic, causing sister chromatid exchanges and micronuclei formation, did not induce dicentric chromosomes; therefore, these data also suggest that DU causes genotoxic damage via a radioactive mechanism.

To address the question of the role of alpha particle radiation and DU, Miller et al. (2002a) showed that incubating HOS cells with uranyl nitrate solutions at a fixed uranium concentration but increasing specific radioactivity resulted in increasing transformation rates (Figure 1.4). These results demonstrated that uranium toxicity can result from both chemical and radiological toxicity. Previously, it was thought that DU caused it effects through its chemical effects alone as the radioactive contribution was calculated to be low. The results from Miller et al. have indicated that the role of radiation in DUs cellular effects cannot be discounted.

As evidence has shown that radiation can induce a transmissible persistent destabilization of the genome, studies were done to evaluate the involvement of genomic

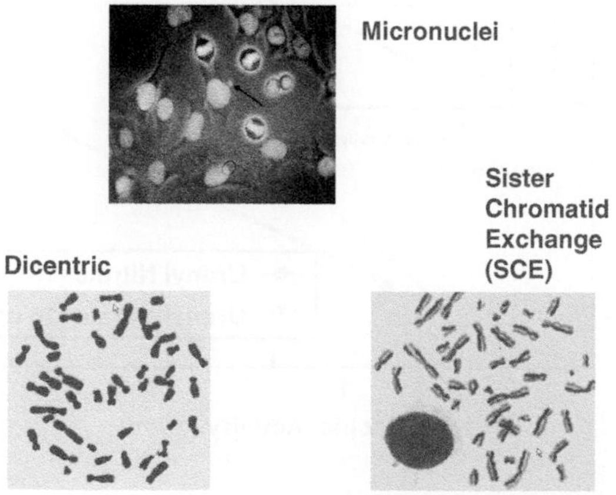

FIGURE 1.3 **(See color insert following page 80)** Examples of chromosomal damage: measures of genotoxicity.

TABLE 1.3
Induction of Genotoxic Damage

	DU (Soluble)	DU (Insoluble)	rWNiCo	Be	Ni
Micronuclei Induction	↑	↑	↑	↑	↑
Sister Chromatid Exchange	↑	↑	↑	↑	↑
DNA Filter Elution (DNA strand break)	↑	↑	↑	↑	↑
Dicentric Formation	↑	↑	no change	ND	no change

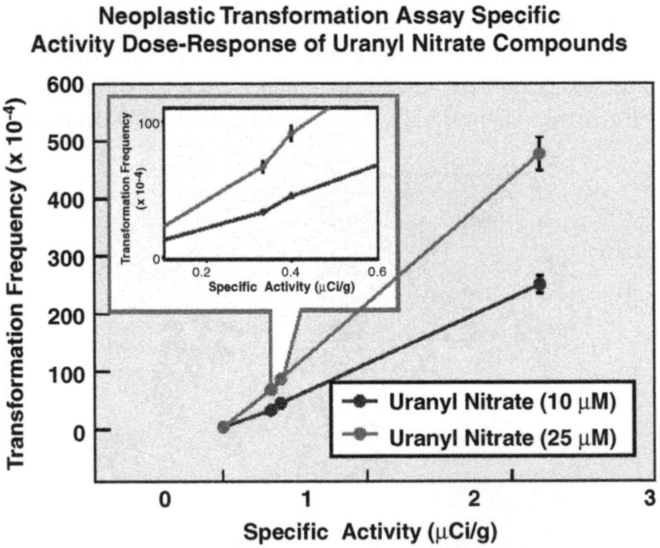

FIGURE 1.4 Radiation Specific Effects Human cells were exposed to equal concentrations of uranyl nitrate that had different specific activities based on their isotopic content.

instability in the mechanism of DU-induced effects (Miller et al. 2000, Miller et al. 2003, Miller et al. 2004). The yields of cell lethality and micronucleus formation measured immediately, or at delayed times after DU-, Ni, and Gamma radiation-exposure were determined in HOS cells. Micronuclei yields were also measured in clonal progeny of surviving DU-exposed cells. Following the initial concentration-dependent acute response for both endpoints data demonstrated that there was *de novo* genomic instability in the surviving progeny. Delayed reproductive death was observed for many generations (36 d; 30 population doublings) following cellular exposure to DU-, Ni, or gamma radiation. Delayed production of micronuclei was observed at times up to 36 d post-DU exposure. In contrast to DU, both gamma irradiated or Ni-exposed cells showed that the initial response at 3 d was the peak expression time for micronuclei damage. There was also a persistent increase in micronuclei in all clones isolated from individual cells which had been exposed to nontoxic concentrations of DU. Although clones isolated from gamma-irradiated individual cells (at doses equitoxic to the DU exposure) demonstrated an moderate increase in micronuclei, clonal progeny of Ni exposed cells did exhibit an increase in micronuclei frequency. It appears that DU exposure *in vitro* resulted in genomic instability manifested as delayed reproductive death and micronucleus formation. These studies confirm previous results demonstrating that heavy metals like cadmium and Ni could induce genomic instability *in vitro* similar to that observed with radiation (Coen, 2001). As there are also extensive data showing that alpha particles can induce a persistent instability in the genome of progeny of irradiated cells, it is difficult to determine whether the alpha particle- or metal-component of DU is responsible for the induced genomic instability.

The *in vitro* studies examining cellular effects show that DU is neoplastically transforming, genotoxic, mutagenic, and induces genomic instability. Cellular damage following DU exposure is caused in part by radiation effects but the contribution of chemical vs. radiation is still not known. These findings strongly supported the hypothesis that DU is carcinogenic. In the next section of this chapter we will review what the limited carcinogenesis *in vivo* studies have shown regarding the question of DU carcinogenicity.

RESEARCH APPROACHES IN *IN VIVO* STUDIES

Cancer in Laboratory Animals

Experiments with laboratory animals have expanded our understanding of the carcinogenic potential of uranium and DU. Not long after the 1991 Gulf War, in an effort to understand more about the potential health effects in personnel wounded in that conflict by DU shrapnel, Pellmar et al. (1999a) carried out a toxicological investigation using Sprague-Dawley rats implanted with various numbers of DU pellets (cylinders 1 mm in diameter and 2 mm long) to mimic shrapnel injuries in humans. Although cancer was not specifically designed as an endpoint in these studies, necropsies of subject rats showed no increased number of tumors in DU-implanted rats compared to tantalum pellet-implanted controls. The high levels of spontaneous tumor development typical of Sprague-Dawley rats confounded the interpretation of that data, however.

In similar studies, Hahn et al. (2002) implanted male Wistar rats with either pellets or thin foils (1 × 2 or 2 × 5 mm) of DU. They showed that there was a slightly elevated risk of cancer in the DU foil– vs. tantalum foil–implanted rats. Implanted DU pellets similar in size to those used in research by others (Pellmar et al. 1999a) did not produce tumors. Interpretation of these data were complicated by the fact that rats are prone to developing tumors around implanted foreign bodies shaped like the foils used, irrespective of the object's chemical composition. In experiments assessing the carcinogenic potential of DU and a tungsten alloy proposed as a surrogate for DU in armor-penetrating munitions, Kalinich et al. (2005) demonstrated DU pellets (1 × 2 mm cylinders) implanted into the leg muscles of Fisher 344 rats for 18 months caused no cancer development. While DU implants have shown varying solid tumor responses in the Hahn et al. and Kalinich et al. studies, the induction of nonsolid tumors was also examined in other studies using different rodent models.

Leukemia in Animals

Although studies using rats models showed that DU caused a solid-state induction of solid tumors, other results indicated that DU could induce leukemia in a mouse model. Recent studies from Miller's laboratory demonstrated that internalized DU could be a carcinogenic risk by causing leukemia in laboratory mice (Miller 2005). To better assess this risk, they developed an *in vivo* leukemogenesis model using murine hematopoietic cells (FDC-P1) that are dependent on stimulation by granulocyte macrophage–colony stimulating factor. Although immortalized, these cells are not tumorigenic on subcutaneous inoculation. As shown in Figure 1.5, intravenous injection of FDC-P1 cells into DBA/2 mice was followed by the development of leukemias in 76% of all mice implanted with DU pellets. In contrast, only 10%

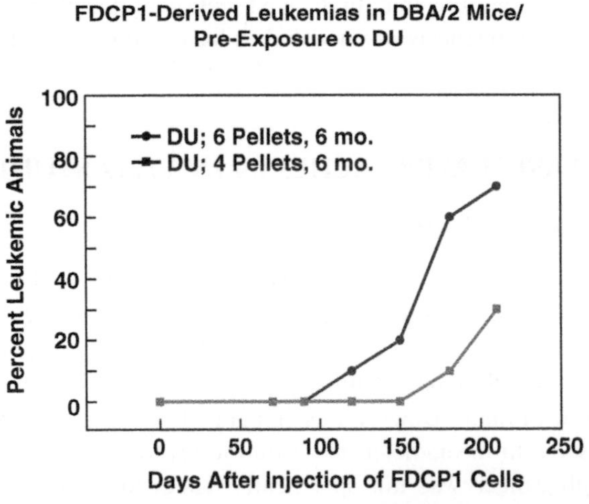

FIGURE 1.5 DU-induced leukemia. Rodents were chronically exposed to embedded DU for 3 months. They were injected with hematopoietic progenitor cells and monitored for leukemia development.

of control mice developed leukemia. Karyotypic analysis confirmed that the leukemias originated from FDC-P1 cells. The full impact of these results is not understood as human studies have not shown the development of leukemia in DU-exposed individuals; however, it appears that a DU-altered *in vivo* environment may be involved in the pathogenesis of a DU-induced leukemia in an animal model. Further studies are necessary to determine how the DU-altered environment enhanced the development of leukemia in mice.

Uranium Genotoxicity

There have been very few studies comparing uranium exposure in humans to genotoxic endpoints. Such studies are relevant because destabilization of the genome can indicate an increased susceptibility to cancer development. Martin et al. (1991) reported that levels of chromosomal aberration, sister chromatid exchange, and dicentrics measured in nuclear fuel workers increase proportionally with uranium exposure. McDiarmid et al. (2004), in their 10-year follow-up of 39 veterans exposed to DU in friendly fire incidents during the 1991 Gulf War, reported that study participants exposed to the highest levels of DU showed a statistically significant increase in chromosomal aberrations compared to the low-exposure groups. HPRT mutation frequencies were also significantly higher in the high-DU groups, but sister chromatid exchanges were not.

The consistent observation of uranium-induced genetic changes remains a cause for concern because they are known to precede cancer development. Hu and Zhu (Hu and Zhu 1990) injected uranyl fluoride into the testes of mice and showed that chromosomal aberrations in spermatogonia and primary spermatocytes are dependent on the amount of injected uranium. In experiments in which rats were implanted with pellets of DU and/or the biologically inert metal tantalum, urine mutagenicity levels increased in a DU dose-dependent manner (Miller et al. 1998a). Figure 1.6 demonstrates that DU internalization results in the excretion of a mutagen which is assumed to be uranium. Rodents were chronically exposed to DU for 18 months

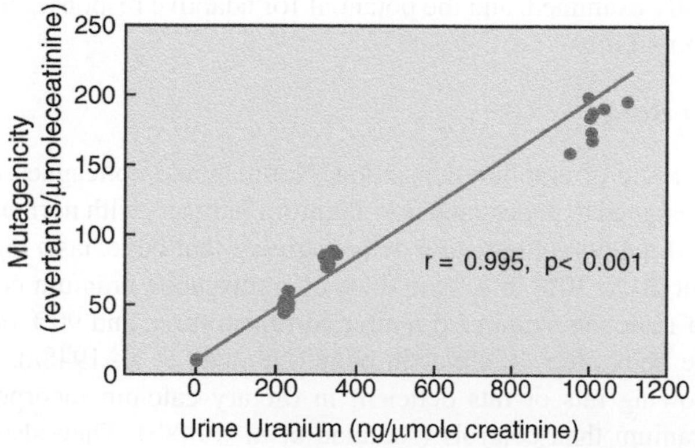

FIGURE 1.6 DU-induced mutagenicity. Rodents were chronically exposed to embedded DU for 3 to 18 months. Urine was collected and assayed for mutagenicity using the Ames test.

and the mutagenicity was measured using the Ames test. Mutagenicity versus urine uranium is shown in Figure 1.6.

Nephrotoxicity

The review by Hodge (1973b) of uranium toxicity prior to the Manhattan Project (1824–1942) shows that it has long been known that uranium is toxic to humans, animals, and other living things. Kidney toxicity of uranium was first recognized in animals around the middle of the 19th century, and kidney toxicity remains the primary basis for the regulation of uranium exposure. Limits for inhalation and ingestion of uranium are aimed at not allowing uranium content in the kidney to exceed a set value which, for most countries, is set at a maximum of 3 μg of uranium per gram of kidney tissue; effects caused by exposure of the kidneys at these levels are considered to be minor and transient.

The pharmacokinetics and pharmacodynamics of uranium and, therefore, DU are well established (Wrenn et al. 1985; Leggett 1994; Taylor 1997; ICRP 1995; Leggett and Pellmar, 2003). There have been many studies that have investigated the results of uranium exposure in laboratory animals (Morrow et al. 1982; La Touche et al. 1987; Ortega et al. 1989; Wrenn et al., 1989). Once absorbed it circulates in the blood primarily as the stable uranyl ion UO_2^{2+} bound to bicarbonate, albumin, or proteins (Diamond 1989; Kocher 1989; Leggett 1989). At the kidney, uranium is filtered through the glomerulus and most is excreted within 24 h. Renal kidney toxicity occurs when residual uranium is subsequently taken up by the proximal tubules and causes damage by forming complexes with phosphate ligands and proteins in the tubular walls, thereby impairing kidney function (Blantz 1975).

Pellmar et al. (1999a) showed that DU from pellets implanted in muscle of rats can be measured in their urine within one day after pellet implantation. Over the course of the 18-month experiment, kidney uranium content reached levels well above 5 μg/g of kidney tissue, a concentration that, if reached in an acute exposure, would normally prove lethal to the animal. The findings suggested that the kidney adapted to the high levels during the chronic exposure. This aspect of DU exposure has not been fully examined, and the potential for adaptive responses to DU remains an intriguing possibility.

Uranium and Bone

Bone is a major site of uranium deposition. Neuman and colleagues, in a series of early articles designed to understand how uranium interacts with normal bone metabolism, published the first observation demonstrating that bone has a high affinity for uranium. About 20 to 30% of a toxic does of intravenous uranium could be found in the bones of male rats within 2.5 h after administration, and 90% of the uranium retained by the body after 40 d was in bone (Neuman et al. 1948a). They showed that young growing rats or rats deficient in dietary calcium incorporated greater amounts of uranium than controls (Neuman et al. 1948b). They also showed that uranium is preferentially incorporated in areas of active calcification and becomes more refractory to resorption as new calcification covers areas of uranium deposition (Neuman and Neuman, 1948c).

Uranium incorporates itself into the bone matrix by displacing calcium to form complexes with phosphate groups in the matrix (Domingo et al. 1992; Guglielmotti et al. 1989). Bone-bound uranium establishes an equilibrium with uranium in the blood, and as the circulating uranium is excreted by the kidneys, bone-bound uranium slowly returns it to the circulation over time (Wrenn et al. 1985). Pellmar et al. (Pellmar et al. 1999a) demonstrated that DU from implanted pellets rapidly distributes throughout the body and accumulates at high levels in the bone, though histological examination showed no bone lesions as a result.

Uranium Neurological Effects

The neurophysiological effects of uranium exposure have been investigated for many decades. Among the early findings was the observation that uranyl ions potentiate the twitch response of frog sartorious muscles by prolonging the active state of contraction. The fact this effect was reversed by administration of phosphate ions suggested that uranium prolongs the action potential (Sandow and Isaacson, 1996). In a study of uranium workers, Kathren and Moore (1986) showed individuals excreting up to 200 µg U per liter urine manifested abnormal mental function. High doses of oral (210 mg/kg) or subcutaneous (10 mg/kg) uranyl acetate caused tremors in rats (Domingo et al. 1987). It was also shown that uranium applied at high concentrations to the ileal longitudinal muscle of guinea pig (Fu 1985) and mouse phrenic nerve-diaphragm preparation (Lin 1988) enhanced muscle contraction.

In a study investigating the toxicology of embedded pellets of DU in rats to mimic shrapnel wounds in wounded 1991 Gulf War veterans, Pellmar et al. (1999b) demonstrated that DU crosses the blood brain barrier and accumulates in the hippocampus, where electrophysiological changes were observed. Briner and Murray (Briner 2005) tested behavioral effects and brain lipid peroxidation in rats exposed to various concentrations of uranyl acetate in drinking water for 2 weeks or 6 months. Open-field behavior was altered in male rats receiving the highest dose of DU after two weeks of exposure; female rats demonstrated behavioral changes after 6 months of exposure. Lipid peroxidation levels increased in the brain and correlated with some of the behavioral changes, but the correlation was ambiguous. Barber et al. (2005) sought to determine the kinetics of uranium content in the brains of rats following a single intraperitoneal injection of uranyl acetate (1mg/kg). They found that uranium content in all areas of the brain tested increased rapidly after injection and remained elevated for 7 d postinjection. Interestingly, rats stressed by daily forced swimming before uranium injection accumulated less uranium in their brains and had lower levels than unstressed animals 7 d after exposure.

Uranium Reproductive/Developmental Effects

Despite nearly a century of studies of uranium toxicity, there were few detailed studies of uranium reproductive and developmental toxicity until the late 1980s (Domingo 1995). In most exposure scenarios, including exposure to DU, the chemical toxic effects from uranium compounds appear to occur at lower exposure levels than radiological toxicity (Hartmann 2000), and this is thought to be the case for reproductive effects as well (Domingo 1995). In the early 1980s, Domingo and his colleagues began extensive investigations of uranium reproductive toxicity, and they have provided most of our current knowledge on the subject to date. There are only a few and preliminary studies investigating the reproductive and developmental health effects of DU specifically. Given the likelihood that uranium chemical toxicity plays the major role in any reproductive and developmental toxicity, it is reasonable to assume that uranium and DU reproductive health effects are similar.

Early studies (Maynard and Hodge 1949) identified uranium as a possible reproductive toxicant in rats. Male and female rats fed diets containing 2% uranyl nitrate hexahydrate for 7 months, followed by normal diets for 5 months, produced fewer litters with fewer pups per litter than control rats (Maynard 1949). However, it was difficult to determine in these experiments whether uranium toxicity or nutritional effects arising from retarded weight gain in the uranium-fed rats caused the decreased reproductive success. In follow-up studies, rats fed diets containing 2% uranyl nitrate hexahydrate for a single 24-h period after weaning also produced fewer litters with fewer pups per litter than control rats with no signs of maternal toxicity (Maynard 1953), an observation that strengthened the connection between uranium exposure and reproductive toxicity.

Llobet et al. (Llobet 1991) showed that male mice continuously receiving water containing uranyl acetate dihydrate and mated with untreated females resulted in a significantly decreased, but dose-unrelated, pregnancy rate; but testicular function and spermatogenesis were unaffected. Domingo et al. (1989) showed that pregnant female mice given uranyl acetate dihydrate (0.05-50 mg/kg per day) by oral gavage from gestational day 13 through postnatal day 21 demonstrated no significant decrease in litter size, pup litter size, and pup viability except at the highest dose (50 mg/kg; about 1/5 of the oral LD_{50}). On the other hand, injection of 1/40 to 1/10 the subcutaneous LD_{50} dose (20 mg/kg) into pregnant female mice between gestational days 6–15 produced both maternal and fetal toxicity (Bosque 1993). Some of the malformations noted in pups could have occurred as a result of maternal toxicity, but defects such as cleft palate and certain other variations are not known to be associated with maternal toxicity and were interpreted to be the result of uranium developmental toxicity (Domingo 2001).

Reports of the health status of military veterans the 1991 Gulf War have provided certain insight into the possible health effects of DU. A follow-up examination of DU-exposed individuals (via embedded DU fragments and/or inhaled DU dusts) showed that there were no significant differences in semen and sperm characteristics among veterans with high or low DU urine concentrations (McDiarmid et al. 2000; McDiarmid et al. 2004). As of 1999, 50 of the Gulf War veterans in the McDiarmid studies had fathered 35 children since the conflict, and none had birth defects (McDiarmid et al. 2001). The relatively small number of individuals involved in these studies and the endpoints that were possible limit their contribution toward understanding reproductive health effects.

EPIDEMIOLOGICAL STUDIES

A series of significant epidemiological studies of nuclear industry workers and uranium miners and millers carried out since the mid-1960s have added a wealth of data to the uranium health effects database. Several investigations of uranium millers (Wagoner 1965; Archer et al. 1973; Waxweiler 1983), workers whose occupation exposes them to uranium dust inhalation in the workplace, used death certificates and in some cases health records to investigate cancers and other diseases (e.g., renal) as a cause of death. These studies failed to clearly identify a link between uranium exposure and any specific health effects, including cancer. Studies have also been carried out of workers at the Y-12 nuclear processing plant in Oak Ridge, TN (Dupree et al. 1995; Loomis 1996) The studies, which often included controls for age, race, gender, radiation dose, other chemical exposures, and medical history (when available) showed no association between cancer and occupational exposure to radiation from external and internal sources. The relatively small sizes of these epidemiological studies, uncertainties about the amount of uranium workers were exposed to, and the impact of confounding factors such as parallel exposures to agents such as radon, silicates, and other toxic metals (e.g., arsenic) lead to large statistical errors in the results, so caution should be exercised in overinterpreting the results of such studies.

In the 1991 Gulf War, an unknown number of personnel were exposed to DU aerosols (primarily uranium oxides) after being in vehicles that were struck by DU munitions, rescuing personnel in struck vehicles, reclaiming or investigating struck vehicles, or moving through areas where DU dusts were left in the environment. Even though satisfactory exposure model exists for such personnel, it is generally considered that the brief exposures to DU dust experienced by personnel would have been far below exposures experienced by uranium miners and millers in earlier studies, so no cancers would be expected by any route of exposure. McDiarmid et al. (2004) calculated radiation dose estimates of personnel carrying DU shrapnel in their bodies as a result of fragment injuries. Whole body radiation counting using the ICRP 30 biokinetic model for uranium yielded an upper dose limit of 0.1 rem/year, a dose is not considered particularly dangerous.

Although human uranium epidemiology studies continue to remain controversial as to whether uranium exposure by any route is associated with cancer (ATSDR 1999), the *in vivo* results are significant and only serve to strengthen the relevance of *in vitro* experiments showing genetic changes consistent with cancer development. The BEIR IV report (BIER IV 1988) on radon and other alpha particle emitters states that large statistical uncertainties in most of the epidemiological studies looking for cancer in uranium workers may be hiding small populations of adversely affected individuals; it cautions against minimizing the risk until more studies are available.

SUMMARY AND CONCLUSIONS OF DEPLETED URANIUM HEALTH EFFECTS

As internalization of uranium in any form will result in a combined chemical and radiation exposure, there exists the potential for subtle differences in health effects between DU and uranium. Recent developments in cell biology technology and understanding are providing more sensitive approaches towards understanding those differences (see (Miller et al. 2001)). Furthermore, because there are conflicting results regarding the carcinogenicity of chronic DU exposure (negative solid tumor results versus positive leukemia results), the question still remains whether chronic DU exposure is a significant toxicological threat.

HEAVY METAL ALTERNATIVES

In many ways the development of substitutes for DU in munitions has followed a pattern similar to that for DU deployment, in that incomplete toxicological information was available prior to their release. In terms of the new tungsten alloys, it was assumed that many years of industrial use of tungsten and alloys such as tungsten carbide (which showed common exposures to the metals [e.g., inhalation, ingestion, or skin contact] represents a manageable risk) meant they could be used as safely in armaments. However, until recently, limited toxicological studies had never been carried out with many of the alloys of specific military interest, and there has been no meaningful prior experience with exposures of special interest to the military, such as via embedded fragments. Recent research into the health effects of embedded pellets of a tungsten/nickel/cobalt alloy have led to those assumptions being questioned.

The health effects of tungsten metal exposure is receiving a new look in other circumstances as well. Environmental testing of the leukemia cluster around Fallon, NV, in the western U.S. showed slightly elevated levels of several heavy metals including uranium and cobalt, but significantly elevated levels of tungsten (CDC 2003). Although no definitive link between elevated tungsten levels and cancer has been established, because of the uncertainty surrounding this issue, the U.S. National Toxicology Program recently added tungsten to their list of compounds to be assessed for adverse health effects. Further study of the health effect of tungsten and tungsten alloys is clearly indicated.

Our present understanding and experience reinforces the advisability of including more effective health effects testing early in weapons development programs. The relatively insignificant cost of such testing would be paid back many times over by helping to redirect expensive engineering programs to more acceptable alternatives earlier in the development process.

REFERENCES

Archer, V.E., Wagoner, J.K., Lundin, F.E., Cancer mortality among uranium mill workers, *J. Occup. Med.* 15: 11–14, 1973.

ATSDR, Toxicological Profile for Uranium, U.S. Department of Health and Human Services, Public Health Service, Agency for Toxic Substances and Disease Registry, Atlanta, GA, 1999.

Barber, D.S., Ehrlich, M.F., Jortner, B.S., The effects of stress on the temporal and regional distribution of uranium in rat brain after acute uranyl acetate exposure, *J. Toxicol. Environ. Health*, Part A, 68: 99–111, 2005.

BEIR IV: Health risks of radon and other internally deposited alpha emitters, Committee on the Biological Effects of Ionizing Radiations, National Research Council, National Academy Press, Washington, DC, 1988.

Blantz, R.C., The mechanism of acute renal failure after uranyl nitrate, *J. Clin. Invest.* 55: 621–635. 1975.

Bosque, M.A., Domingo, J.L., Llobet, J.M., Corbella, J., Embryotoxicity and teratogenicity of uranium in mice following SC administration of uranyl acetate, *Biol. Trace Element Res.* 36: 109–118, 1993.

Brand, K.G., Buoen, L.C., Johnson, K.H., Brand, T., Etiological factors, stages, and the role of the foreign body in foreign-body tumorigenesis: a review, *Cancer Res.* 35: 279–286, 1975.

Briner, W., Murray, J., Effects of short-term and long-term uranium exposure on open-field behavior and brain lipid peroxidation in rats, *Neurotoxicol. Teratol.* 27: 135–143, 2005.

CDC, Cross-Sectional Exposure Assessment of Environmental Contaminants in Churchill County, Nevada, Centers for Disease Control and Prevention, Atlanta, GA, 2003.

Coen, N., Mothersill, C., Kadhim, M., Wright, E.G., Heavy metals of relevance to human health induce Genomic Instability. *J. Patholo.* Oct. 195(3) 293–299.

Diamond, G.L., Biological consequences of exposure to soluble forms of natural uranium, *Radiat. Prot. Dosimetry* 26: 23–33, 1989.

Domingo, J.L., Llobet, J.M., Tomas, J.M., Corbella, J., Acute toxicity of uranium in rats and mice, *Bull. Environ. Contam. Toxicol.* 39: 168–174, 1987.

Domingo, J.L., Paternain, J.L., Llobet, J.M., Corbella J., Evaluation of the perinatal and postnatal effects of uranium in mice upon oral administration, *Arch. Environ. Health* 44: 395–398, 1989.

Domingo, J.L., Colomina, M.T., Llobet, J.M., Jones, M.M., Singh, P.K., The action of chelating agents in experimental uranium intoxication in mice: variations with structure and time of administrations, *Fundam. Appl. Toxicol.* 19: 350–357, 1992.

Domingo, J.L., Chemical toxicity of uranium, *Toxicol. Ecotoxicol. News* 2: 74–78, 1995.

Domingo, J.L., Reproductive and developmental toxicity of uranium and depleted uranium: a review, *Reprod. Toxicol.* 15: 603–609, 2001.

Dupree, E.A., Watkins, J.P., Ingle, J.N., Wallace, P.W., West, C.M., Tankersley, W.G., Uranium dust exposure and lung cancer risk in four processing operations, *Epidemiology* 6: 370–375, 1995.

Guglielmotti, M.B., Ubios, A.M., Larumbe, J., Cabrini, R.L., Tetracycline in uranyl nitrate intoxication: its action on renal damage and U retention in bone, *Health Phys.* 57: 403–405, 1989.

Hahn, F.F., Guilmette, R.A., Hoover, M.D., Implanted depleted uranium fragments cause soft tissue sarcomas in the muscles of rats, *Environ. Health Perspect.* 110: 51–59, 2002.

Hartmann, H.M., Monette, F.A., Avci, H.I., Overview of toxicity data and risk assessment methods for evaluating the chemical effects of depleted uranium compounds, *Hum. Ecol. Risk Assessment* 6: 851–874, 2000.

Heath, J.C., Cobalt as a carcinogen, *Nature* 173: 822–823, 1954.

Heath, J.C., The production of malignant tumors by cobalt in the rat, *Br. J. Cancer* 10: 668–673. 1956.

Heath, J.C., Daniel, M.R., The production of malignant tumors by nickel in the rat, *Br. J. Cancer* 18: 261–264. 1964.

Hodge, H.C., A history of uranium poisoning (1824–1942), In: Uranium, plutonium, transplutonic elements, *Handbook of Experimental Pharmacology*, Vol. 36, H.C. Hodge, J.N. Stannard, J.B. Hursh (Eds.), Springer-Verlag, New York, 1973b, pp. 5–69.

Hu, Q., Zhu, S., Induction of chromosomal aberrations in male mouse germ cells by uranyl fluoride containing enriched uranium, *Mutation Res.* 244: 209–214, 1999.

Huang, X., Iron overload and its association with cancer risk in humans: evidence for iron as a carcinogenic metal, *Mutation Res.* 533: 153–171, 2003.

IARC, Overall Evaluation of Carcinogenicity, IARC monographs on the evaluation of carcinogenic risk of chemicals to humans, Supplement 7, Lyon, France, International Agency for Research on Cancer, 1987, 440 pp.

ICRP, Age-dependent doses to members of the public from the intake of radionuclides, Part 3. Ingestion dose coefficients, Publication 69 of the International Commission on Radiological Protection, Pergamon Press, Oxford, 1995.

Institute of Medicine, *Gulf War and Health, Vol. 1. Depleted Uranium, Sarin, Pyridostigmine Bromide, Vaccines*, Committee on Health Effects Associated with Exposures During the Gulf War, Institute of Medicine, National Academy Press, Washington, DC, 2000.

Kalinich, J.F., Emond, C.A., Dalton, T.K., Mog, S.R., Coleman, G.D., Kordell, J.E., Miller, A.C., McClain, D.E., Embedded weapons-grade tungsten alloy shrapnel rapidly induces metastatic high-grade rhabdomyosarcomas in F344 rats, *Environ. Health Perspect.* 113: 729–734, 2005.

Kasprzak, K.S., Gabryel, P., Jarczewska, K., Carcinogenicity of nickel (II) hydroxides and nickel (II) sulfate in Wistar rats and its relation to the in vitro dissolution rates, *Carcinogenesis* 4: 275–279, 1983.

Kathren, R.L., Moore, R.H., Acute accidental inhalation of U: a 38-year follow-up, *Health Phys.* 51: 609–619, 1986.

Kocher, D.C., Relationship between kidney burden and radiation dose from chronic ingestion of U: implications for radiation standards for the public, *Health Phys.* 57: 9–15, 1989.

La Touche, Y.D., Willis, D.L., Dawydiak, O.I., Absorption and biokinetics of U in rats following oral administration of uranyl nitrate solution, *Health Phys.* 53: 147–162, 1987.

Leggett, R.W., Basis for the ICRP's age-specific biokinetic model for uranium, *Health Phys.* 67: 589–610, 1994.

Leggett, R.W., The behavior and chemical toxicity of U in the kidney: a reassessment, *Health Phys.* 57: 365–383, 1989.

Leggett, R.W., Pellmar, T.C., The biokinetics of uranium migrating from embedded DU fragments, *J. Environ. Radioact.* 64: 205–225, 2003.

Lin, R.H., Fu, W.M., Lin-Shiau, S.Y., Presynaptic action of uranyl nitrate on the phrenic nerve-diaphragm preparation of the mouse, *Neuropharmacology* 27: 857–863, 1988.

Lin, R.H., Wu, L.J., Lee, C.H., Lin-Shiau, S.Y., Cytogenetic toxicity of uranyl nitrate in Chinese hamster ovary cells, *Mutation Res.* 319: 197–203, 1991.

Lison, D., Lauwerys, R., Study of the mechanism responsible for the elective toxicity of tungsten-carbide-cobalt powder toward macrophages, *Toxicol. Lett.* 60: 203–210, 1997.

Loomis, D.P., Wolf, S.H., Mortality of workers at a nuclear materials production plant at Oak Ridge, Tennessee, 1947–1990, *Am. J. Ind. Med.* 29: 131–141, 1996.

Maynard, E.A., Hodge, H.C., Studies of the toxicity of various uranium compounds when fed to experimental animals, in C. Voeglen (Ed.), *Pharmacology and Toxicology of Uranium Compounds*, Vol. I, McGraw-Hill, New York, 1949, pp. 309–376.

Maynard, E.A., Downs, W.L., Hodge, H.C., Oral toxicity of uranium compounds, in C. Voeglen, H.C. Hodge (Eds.), *Pharmacology and Toxicology of Uranium Compounds*, Vol. III., McGraw-Hill, New York, 1953, 1221–1369.

Martin, F., Earl, R., Tawn, E.J., A cytogenetic study of men occupationally exposed to uranium, *Br. J. Ind. Med.* 48: 98–102, 1991.

McDiarmid, M.A., Keogh, J.P., Hooper, F.J., McPhaul, K., Squibb, K.S., Kane, R., DiPino, R., Kabat, M., Kaup, B., Anderson, L., Hoover, D., Brown, L., Hamilton, M., Jacobson-Kram, D., Burrows, B., Walsh, M., Health Effects of depleted uranium on exposed Gulf War veterans, *Environ. Res.*, Section A 82: 168–180. 2000.

McDiarmid, M.A., Squibb, K.S., Engelhardt, S., Oliver, M., Gucer, P., Wilson, P.D., Kane, R., Kabat, M., Kaup, B., Anderson, L., Hoover, D., Brown, L., Jacobson-Kram, D., Surveillance of depleted uranium exposed Gulf War veterans: health effects observed in an enlarged "friendly fire" cohort, *J. Occup. Environ. Med.* 43: 991–1000, 2001.

McDiarmid, M.A., Engelhardt, S., Oliver, M., Gucer, P., Wilson, P.D., Kane, R., Kabat, M., Kaup, B., Anderson, L., Hoover, D., Brown, L., Handwerger, B., Albertini, R.J., Jacobson-Kram, D., Thorne, C.D., Squibb, K.S., Health effects of depleted uranium on exposed Gulf War veterans: a 10-year follow-up, *J. Toxicol. Environ. Health*, Part A 67: 277–296, 2004.

Miller, A.C., Fuciarelli, A.F., Jackson, W.E., Ejnik, J.W., Emond, C.A., Strocko, S., Hogan, J., Page, N., Pellmar, T., Urinary and serum mutagenicity studies with rats implanted with depleted uranium or tantalum pellets, *Mutagenesis* 13: 643–648, 1998a.

Miller, A.C., Blakely, W.F., Livengood, D., Whittaker, T., Xu, J., Ejnik, J.W., Hamilton, M.M., Parlette, E., John, T.S., Gerstenberg, H.M., Hsu, H., Transformation of human osteoblast cells to the tumorigenic phenotype by depleted uranium-uranyl chloride, *Environ. Health Perspect.* 106: 465–471, 1998b.

Miller, A.C., Xu, J., Stewart, M., Emond, C., Hodge, S., Matthews, C., Kalinich, J., McClain, D., Potential health effects of the heavy metals, depleted uranium and tungsten, used in armor-piercing munitions: comparison of neoplastic transformation, mutagenicity, genomic instability, and oncogenesis, *Metal Ions* 6: 209–211, 2000.

Miller, A.C., Mog, S., McKinney, L., Luo, L., Allen, J., Xu, J., Page, N., Neoplastic transformation of human osteoblast cells to the tumorigenic phenotype by heavy-metal tungsten-alloy metals: induction of genotoxic effects, *Carcinogenesis* 22: 115–125, 2001.

Miller, A.C., Xu, J., Prasanna, P.G.S., Page, N., Potential late health effects of the heavy metals, depleted uranium and tungsten, used in armor piercing munitions: comparison of neoplastic transformation and genotoxicity using the known carcinogen nickel, *Mil. Med.* 167: 120–122, 2002a.

Miller, A.C., Xu, J., Stewart, M., Brooks, K., Hodge, S., Shi, L., Page, N., McClain, D., Observation of radiation-specific damage in human cells exposed to depleted uranium: dicentric frequency and neoplastic transformation as endpoints, *Radiat. Prot. Dosimetry* 99: 275–278, 2002b.

Miller, A.C., Brooks, K., Smith, J., Page, N., Genomic instability in human osteoblast cells after exposure to depleted uranium: delayed lethality and micronuclei formation. *J. Environ. Radioact.*, 64(2–3): 247–259, 2003.

Miller, A.C., Brooks, K., Smith, J., Page, N., Effect of the militarily-relevant heavy metals, depleted uranium and heavy metal tungsten-alloy on gene expression in human liver carcinoma cells (HepG2), *Mol Cell Biochem* 255: 247–256, 2004.

Miller, A.C., Leukemic transformation of hematopoietic cells in mice internally exposed to depleted uranium. *Mol Cell Biochem* 279(1–2): 97–104, 2005.

Morrow, P., Gelein, R., Beiter, H., Scott, J., Picano, J., Yuile, C., Inhalation and intravenous studies of UF6/UO2F in dogs, *Health Phys.* 43: 859–873, 1982.

Neuman, W.F., Fleming, R.W., Dounce, A.L., Carlson, A.B., O'Leary, J., Mulryan, B.J. Distribution and secretion of injected uranium, *J. Biol. Chem.* 173: 737–748, 1948a.

Neuman, W.F., Neuman, M.W., Mulryan B.J., The deposition of uranium in bone: I Animal studies, *J. Biol. Chem.* 175: 705–709, 1948b.

Neuman, W.F., Neuman, M.W., The deposition of uranium in bone: II Radioautographic studies, *J. Biol. Chem.* 175: 711–714, 1948c.

Ortega, A., Domingo, J.L., Llobet, J.M., Tomas, J.M., Paternain, J.L., Evaluation of the oral toxicity of uranium in a 4-week drinking-water study in rats, *Bull. Environ. Contam. Toxicol.* 42: 935–941, 1989.

Pellmar, T.C., Fuciarelli, A.F., Ejnik, J.W., Hamilton, M., Hogan, J., Strocko, S., Emond, C., Mottaz, H.M., Landauer, M.R., Distribution of uranium in rats implanted with depleted uranium pellets, *Toxicol. Sci.* 49: 29–39, 1999a.

Pellmar, T.C., Keyser, D.O., Emery, C., Hogan, J.B., Electrophysiological changes in hippocampal slices isolated from rats embedded with depleted uranium fragments, *Neurotoxicology* 20: 785–792, 1999b.

Sandow, A., Isaacson, A., Topochemical factors in potentiation of contraction by heavy metal cations, *J. Gen. Physiol.* 49: 937–961, 1996.

Sunderman, F.W., Jr., Maenza, R.M., Comparisons of carcinogenicities of nickel compounds in rats, *Res. Commun. Chem. Pathol. Pharmacol.* 14: 319–330, 1976.

Sunderman, F.W., Jr., Maenza, R.M., Alpass, P.R., Mitchell, J.M., Damjanov, L., Goldbalatt, P.J., Carcinogenicity of nickel subsulfide in Fischer rats and Syrian hamsters after administration by various routes, *Adv. Exp. Med. Biol.* 91: 57–67, 1977.

Sunderman, F.W., Jr., Carcinogenicity of metal alloys in orthopedic prostheses: clinical and experimental studies, *Fundam. Appl. Toxicol.* 13: 205–216, 1989.

The Royal Society, The health hazards of depleted uranium munitions. Part I. Radiological effects, The Royal Society, London, 2001.

The Royal Society, The health hazards of depleted uranium munitions. Part II. Non-radiological effects and environmental impact, The Royal Society, London, 2002.

Voeglen, C., Pharmacology and toxicology of uranium compounds, Vol. I, C. Voeglen (Ed.), McGraw-Hill, New York, pp. 309–376, 1949.

Voeglen, C., Pharmacology and toxicology of uranium compounds, Vol. III, C. Voeglen, H.C. Hodge (Eds.), McGraw-Hill, New York, pp. 1221–1369, 1953.

Wagoner, J.K., Archer, V.E., Lundin, F.E., Jr., Holaday, D.A., Lloyd, J.W., Radiation as the cause of lung cancer among uranium miners, *New Engl. J. Med.* 273: 181–188.

Wrenn, M.E., Durbin, P.W., Howard, B., Lipszten, J., Rundo, J., Still, E.T., Willis, D.L., Metabolism of ingested U and Ra, *Health Phys.* 48: 601–633.

2 Characteristics, Biokinetics, and Biological Effects of Depleted Uranium Used in Weapons and the French Nuclear Industry

Valérie Chazel, Pascale Houpert, and François Paquet

CONTENTS

INTRODUCTION

A metallic form of depleted uranium (DU) was used in weapons during the 1991 Gulf War and more recently in Bosnia and Kosovo [1]. This slightly radioactive metal is particularly toxic when it penetrates the body through inhalation, ingestion, or injury. When a depleted uranium penetrating round hits a tank, for example, substantial quantities of uranium-bearing particles may be produced and therefore inhaled by soldiers near or inside the tank. Bursts of fragments may also cause injuries. Several reports have addressed the risks to health of accidents of this nature [2]. The conclusions are normally that there is very little risk of developing cancer from internal radiological exposure to depleted uranium. Numerous uncertainties do, however, exist, particularly on the estimation of the amounts of depleted uranium likely to be inhaled. In addition, little is known on the properties of depleted uranium aerosols generated when firing on armored vehicles. A nonexhaustive, state-of-the-art study carried out on depleted uranium in weapons, presented later on in this chapter, is followed by a summary of experiments performed within the Institute for Radiological Protection and Nuclear Safety (IRSN) in France.

To remove uncertainties previously linked to aerosols produced by the use of depleted uranium in weapons, realistic experiments reproducing the firing of depleted uranium weapons at an armored target have been performed by IRSN and the French armament procurement agency (DGA). Atmospheric samples were taken of aerosols generated during impacts of arrow shells containing depleted uranium on tank structures. These aerosols were then characterized to determine the chemical nature of the products formed, their granulometric distribution, and the volume concentrations near the impact. *In vitro* chemical dissolution tests were performed on these particles in various media representing acid rain, mineral water, and a pulmonary fluid simulant to determine their type of solubility and, especially, to predict the doses likely to be received from inhaling these particles.

The results are compared with those obtained for depleted uranium oxides with a similar chemical nature but produced in different plants in the French nuclear industry.

Other studies are also presented in this chapter that aim to provide additional understanding of the biokinetics of depleted uranium after inhalation of uranium oxide aerosols with varying degrees of solubility. Repeated, acute inhalations are compared to determine the legitimacy of using the chronic model described by the ICRP which postulates that chronic contamination can be modeled by iteration of successive and independent acute intakes. Two different approaches are used to test this hypothesis. In a first study, a chronic inhalation model for rats was developed based on acute exposure to UO_2, then used to compare the theoretical results with experiments of repeated inhalations of the same UO_2. The second approach compared the biokinetics of a soluble DU oxide, UO_4, after a single, acute inhalation or after prior exposure to insoluble UO_2 inhaled acutely or repeatedly. Both approaches were used to determine the influence of prior exposure on the biokinetics of a DU compound inhaled subsequently.

In the final section in this chapter, the biological effects of acute and repeated inhalations of DU oxides are studied in rats and emphasize the effects on behavior, genotoxicity, and pulmonary inflammation.

DU IN WEAPONS: STATE OF THE ART

Depleted uranium is a by-product of the nuclear industry, with two possible sources: the enrichment of natural uranium to manufacture nuclear fuel and the reprocessing of irradiated fuels.

The percentages by mass and by activity of the uranium isotopes 234, 235, and 238 may vary in DU according to the technological process used (gaseous diffusion, ultracentrifugation). Isotope 236 is only found in fuel recovered after passing through the nuclear reactors. Table 2.1 summarizes these percentages in natural (Nat. U), depleted (DU). and reprocessed (Reproc. U) uranium and gives the specific activity of each.

^{236}U is a sign of reprocessed uranium, also containing traces of fission products and activation and transuranic products.

TABLE 2.1
Percentages by Mass and by Activity of the Various Uranium Isotopes Found in Natural, Depleted, and Reprocessed Uranium

	Mass (%)				Activity (%)				Specific Activity
	^{234}U	^{235}U	^{236}U	^{238}U	^{234}U	^{235}U	^{236}U	^{238}U	*(Bq.mg⁻¹)*
Nat. U	0.0055	0.72	—	99.274	49.5	2.3	—	48.2	25,2
DU	0.0008	0.202	—	99.797	12.8	1.1	—	86.1	14,4
Reproc. U	0.014	0.92	0.26	98.806	61.8	1.1	14.9	20.3	60,0

Given its abundance, low cost, metallurgical properties, and high density, metallic forms of DU are widely used in such fields as aeronautics (wings and rudders), navigation (boat keels) and radiological protection (protection screens, collimators). It is used in military applications to manufacture armor-piercing shells and armor-plating for tanks. When the shell hits a heavy target like a tank, a cloud of DU dust is released inside and outside the vehicle which bursts spontaneously into flames, thereby producing a fire which increases the damage already caused. Some significant data from published works on the characteristics of the aerosols produced during the impact of a DU shell are summarized in Table 2.2. [2]. The experiments listed in this table were all performed on DU from weapons fired at targets.

Hanson et al. described concentrations of 500–1700 $mg.m^{-3}$ at the weapon exit chamber and 70 to 600 $mg.m^{-3}$ at the entrance [3]. A rapid buildup of ultrafine particles is also noted in the first 30 sec, thereby increasing the median aerodynamic diameter by mass and reducing the standard geometric deviation, σ_g. Fragments were also recovered, mainly at the exit chamber, that could have a significant impact on the environment given their large size.

Patrick et al. observed the morphological characteristics of the particles formed after impact of the DU missiles (antitank ammunition 105 mm XM 744) on steel-armored targets [4]. As indicated in Table 2.2, the authors observe a wide range of particle sizes, from submicronic to diameters of over 50 µm. Most of them are spherical or ellipsoidal in shape and have very "tortuous" surfaces, suggesting that the particles melted. Most of them are U/Fe alloys caused by the impact. One unexpected result is the presence of ultrafine particles (< than 0.1 µm) present overall on the surface of the large particles, but also agglomerated.

Similarly, the Capstone study demonstrates that the majority of aerosols generated, at least during the first minutes after impact, may be inhaled with an AMAD of around 1µm [5]. Solubility tests on aerosols from the cyclone stage were used to calculate the dissolution parameters, where the rapidly-dissolved fraction varies between 1–28%, and the slow dissolution rate varies between 0.0004–0.0095 d^{-1}.

XM 774 ammunition was also used in experiments by Glissmeyer et al., fired from a weapon 200 m from a target of three sloping plates so that the fragments were deflected [6]. The weapon was fired five times. The measured concentrations were in

TABLE 2.2
Granulometric Distribution, Composition, and Dissolution Parameter of the Various DU Aerosols Described in Published Works

Reference	MMAD (μm)	σ_g (= $d_{84\%}/d_{50\%}$)	f_r	s_r (d^{-1})	s_s (d^{-1})	Composition
Hanson, 1974	2.1–3.3 (entrance chamber) 2.4–4.2 (exit chamber)	1.8–3.3 1.8–3.1	—	—	—	—
Patrick, 1978	Very large size range: submicronic to fragments > 50 μm		—	—	—	Majority of U and Fe
Glissmeyer, 1979	0.8–3.1	1.6–18	0.43 respirable 0.15 total	—	< 0.01	25% UO_2 + 75% U_3O_8
Chambers, 1982	1.6 (1.4–2.0)	13 (12–17)	—	—	—	—
Brown, 2000	3.7 (1.1–7.5) inside 1.8 (1.3–2.7) outside	3.5 (2.8–4.2) 4.1 (3.9–4.5)	—	—	—	—
Scripsick, 1985a	—	—	0.25	1.7	0.0014	60% UO_2+ 40% U_3O_8 (air filter, total) 18% UO_2 + 62% U_3O_8 + 20% amorphous (respiable)
Scripsick, 1985b	—	—	0.04	4.7	0.004	97% UO_2 + 3% U_3O_8 (core sample, total) 54% UO_2 + 46 % U_3O_8 (respirable)
Capstone; Parkhurst, 2004	BFV first 10s: 0.6–4 BFV after 10s: 0.4–4 Abrams first 10s: 0.2–8 Abrams after 10s: 0.3–7 Abrams DU armor first 10s: 0.8–8 After 10s: 0.1–5		1–28	0.1–30	0.0004–0.0095	

the order of 8–35 $mg.m^{-3}$. The volume of the dust cloud was 1000 to 2000 m^3. About 70% of the mass of the penetrating round was found in the form of particles, including 50% respirable (< 3.3 μm according to the authors). Scanning electron microscopy revealed that the particles were small in size (0.03 to 0.1 μm). Lastly, the UO_2/U_3O_8 ratio was 25/75.

Experiments by Chambers et al., nine in total, used 75 mm bullets fired at a target made up of one or three plates [7]. Atmospheric samples started to be taken 2 min after each shot and lasted 1–4 min. The measured concentration was 130 $mg{\cdot}m^{-3}$ at target and 6, 25, and 36 $mg{\cdot}m^{-3}$ 1.5, 7.6, and 15 m, respectively, from the target.

The mass median aerodynamic diameters (MMAD) are given in Table 2.2; the authors concluded that about 70% of the particles were less than 7 μm in diameter and could be considered respirable.

Brown et al. describe the results of atmospheric samples taken during English tests of firing with DU ammunition [8]. The MMAD data are given in Table 2.2. The measured concentrations are 13–60 $mg.m^{-3}$ 3 m from the target and 7–17 $mg{\cdot}m^{-3}$ 7 m away. The CRACUK computer code indicates that 20% of the DU is dispersed in the atmosphere, 10% in a plume and that all the particles are nearly 1 μm in diameter.

Scripsick et al. studied the dissolution of uranium oxides collected on filters on the battlefields produced by impacts from DU penetrating rounds, compared with aerosols prepared in the laboratory [9,10].

The *in vitro* tests were performed using the technique of a suspension deposit sandwiched between two filters of 0.1 μm for 30 d with simulated pulmonary interstitial fluids (SUF without DTPA = Gamble). X-ray diffraction revealed the presence of UO_2 and U_3O_8 and 20% of an amorphous phase on the samples taken from a filter.

The authors find that the rapidly dissolved fraction is equal to 25% (for aerosols on filters), unlike the fraction relating to the oxides synthesized in the laboratory which is 6–10%. The rapid dissolution rate, s_r, is 4.7 d^{-1} and the slow dissolution rate is 0.004 d^{-1}. The high rapid dissolution rate may be attributed, in the authors' view, to a larger specific surface area than for the synthesized compounds.

Spectrometry analyses of energy dispersion coupled to a scanning electron microscope on fragments of penetrating rounds recovered on the battlefields in Kosovo revealed the presence of metallic uranium containing impurities or small uranium oxide phases. The impurities were aluminum, silicon, iron, copper, chromium, and titanium. The authors conclude that these impurities come from the ground.

Secondly, ICP-MS analyses of these fragments after dissolution have shown the presence of ^{236}U, ^{239}Pu, and ^{240}Pu, thus proving that the uranium used in ammunition comes either from depleted uranium after reprocessing (in other words, after passing through a nuclear reactor) or from depleted uranium contaminated by reprocessing uranium in the enrichment facilities [11].

Another article indicates lastly that DU shells are made up of metallic uranium and titanium to bolster or improve the hardness of the penetrating round and other components that could add to the toxicity of this ammunition. The author provides no further information on these components to avoid revealing trade secrets. The article also indicates that the metallic form of DU is supplied by an American company, Nuclear Metal, Inc., to the French company SICN (1000 t exported in 1993) to machine and manufacture penetrating rounds [12].

DU IN WEAPONS: FRENCH STUDY

The aerosols produced by impacts on a tank turret and glacis from depleted uranium arrow shells were sampled on filters and impactors to characterize them from a physicochemical viewpoint, determine their solubility type and the quantities likely to be inhaled, and generate a dose in the organism [13]. Two shots were fired with shells containing 99% of depleted uranium. Ten aerosol samplers, including an impactor, were placed around the tank structures 1, 2.5, and 4 m away, as indicated in Figure 2.1. An additional sampler was placed inside each structure. The sampling started 20 min before the shot was fired and stopped 15 min afterward. The temperature and pressure inside the structures were also measured.

The particles collected by filters were characterized by x-ray diffraction to identify the phases present and the related crystalline structures. The form and element composition were determined by scanning electron microscopy (SEM) coupled to an X analyzer. Thermo ionization mass spectrometer (TIMS) was used to measure the isotopic composition. The chemical dissolution kinetics of sampled particles were assessed *in vitro* (static dissolution test in sandwich) in solutions of bicarbonate, HCl, and Gamble, simulating mineral water, acid rain, and pulmonary fluids, respectively. Kinetic phosphorescence analysis (KPA) was used to analyze the uranium. The dissolution rates were calculated using the Gigafit software program (NRPB), and the predictive doses were calculated using the LUDEP software program [14].

The results showed that the uranium used, with a specific activity equal to 14.06 $kBq \cdot g^{-1}$, contained the isotope ^{236}U, characteristic of a reprocessed uranium. The average concentrations measured around the turret and glacis were 122 *(± 29)* $Bq \cdot m^{-3}$ and 118 *(± 26)* $Bq \cdot m^{-3}$, respectively, which corresponds to 8.7 and 8.4 mg of $DU \cdot m^{-3}$. Differences in concentration were insignificant despite the impact of the shell being different on the turret and the glacis. The highest concentration (although underestimated due to the filter being partially destroyed by fire) of 135 $Bq \cdot m^{-3}$ (9.6 $mg \cdot m^{-3}$) was measured inside the glacis, with the temperature and pressure reaching 600°C and 15 bars, respectively. These concentration values agree with published works that

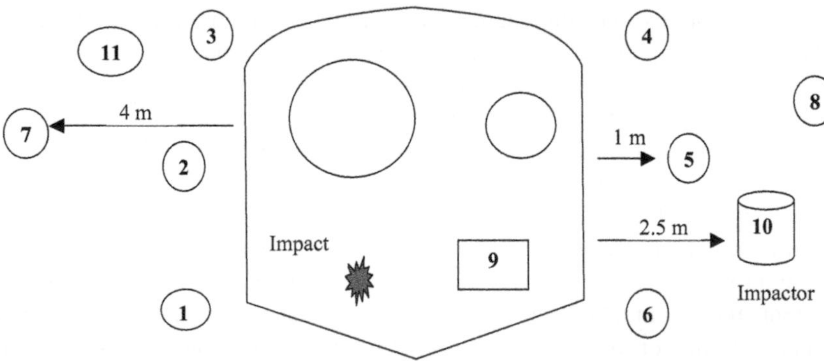

FIGURE 2.1 Location of filters (numbers 1–9 and 11) and impactor (number 10) around the turret and impact point.

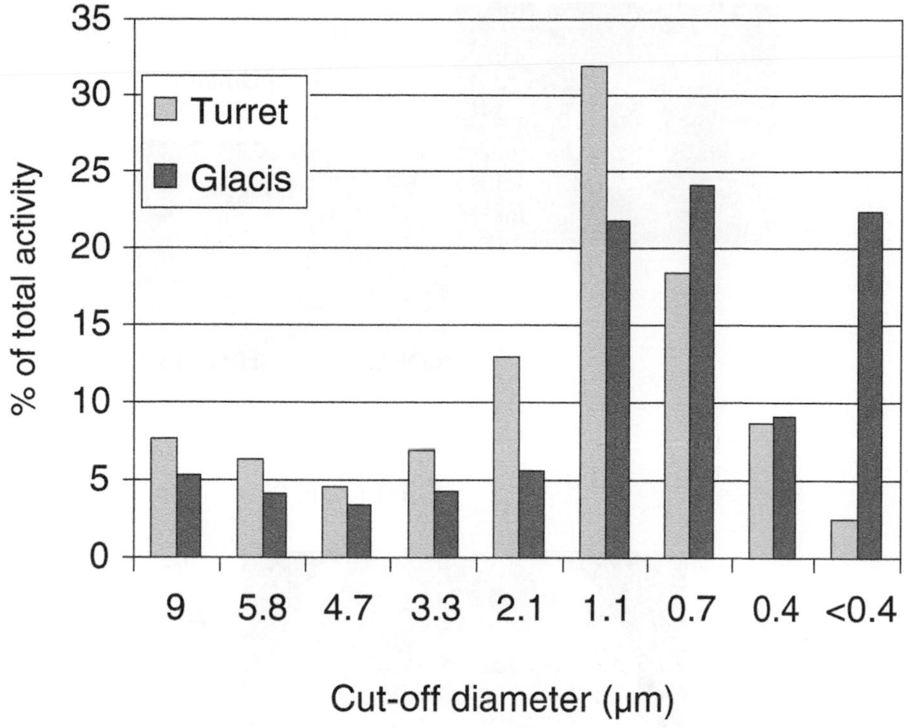

FIGURE 2.2 Granulometric distribution of DU aerosols from turret and glacis.

find concentrations varying between 6 and 1700 $mg \cdot m^{-3}$, according to the type of ammunition and sampling location.

Particles generated during the impact on the glacis had an AMAD (aerodynamic median activity diameter) of 1.05 µm ($\sigma_g = 3.7$) with 22% of the activity carried by aerosols with an aerodynamic diameter of less than 0.4 µm (Figure 2.2). The granulometric distribution obtained for the turret was very significantly different, with an AMAD of 2 µm (($\sigma_g = 2.5$) and a majority of particles (32%) carried by the filters with cut-off diameters between 0.7 and 2.1 µm. Overall, the majority of particles generated during firing of the shells were fine particles ($d_{ae} \leq 2$ µm). Published works mention values between 0.8 µm and 7.5 µm and sometimes the presence of ultrafine particles (< 0.1 µm) [4].

The SEM observation (Figure 2.3) showed several types of particles: (1) fine particles (geometric diameter, d_g = 0.5 to 2 µm), mainly composed of a uranium-aluminum mix, (2) particles composed entirely of uranium (d_g = 30 µm), (3) larger particles (d_g > 50 µm) of melted Ni seemingly covered with a porous surface containing uranium and aluminum, and (4) much larger fragments (d_g = 40 to 170 µm) containing Al, U, Si (from the sand on which the tank structures were parked), Fe, Zn, P, Cu, and Ni.

These alloys may contribute to the toxicity of the depleted uranium. The toxicity of all these elements after inhalation is well known. Aluminum in particular is

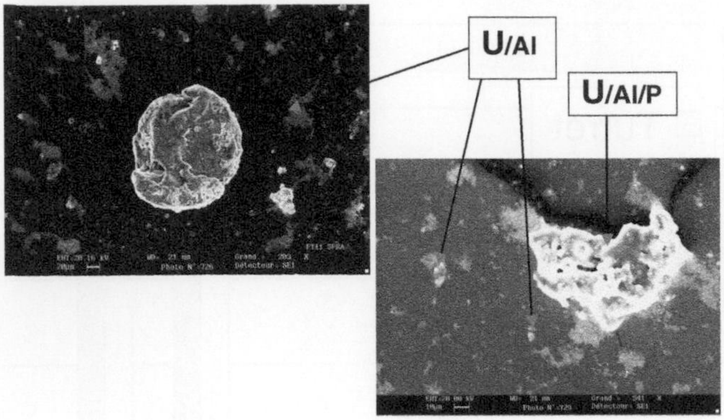

(a)

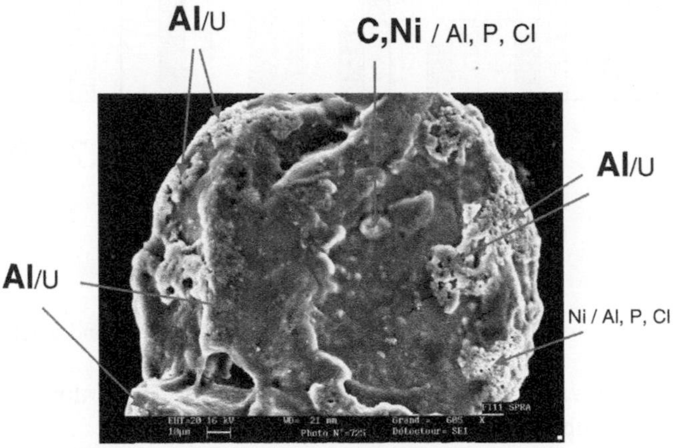

(b)

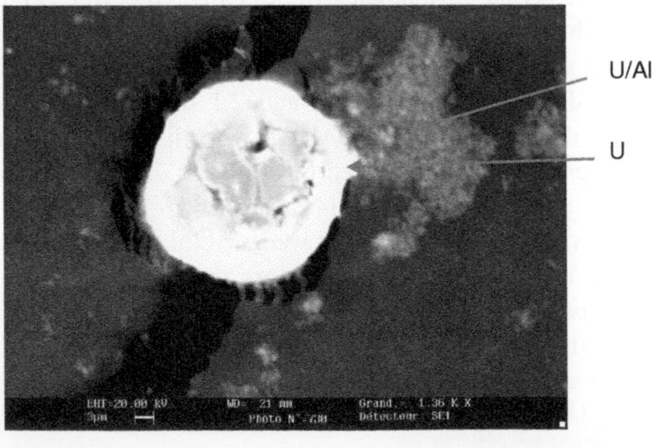

(c)

FIGURE 2.3 SEM micrographs of particles sampled when firing the DU shell at the tank.

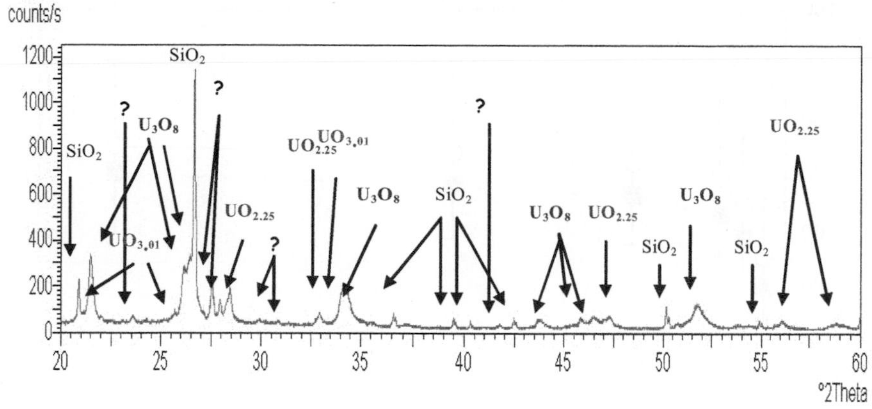

FIGURE 2.4 X-ray diffractogram of aerosols sampled during impact.

genotoxic, and uranium combined with metallic alloys containing nickel and iron causes human cells to transform into osteoblasts [15].

The phase analysis revealed the presence of U_3O_8 (30–40%) in alpha orthorhombic form, $UO_{2.25}$ (= U_4O_9, 25–40%) in cubic form, $UO_{3.01}$ (20%) in orthorhombic form, and a phase that could be attributed to a U/Al mix (where the crystalline structure was not referred in the database used) (Figure 2.4). The published works on compounds produced during impact of DU weapons mainly report UO_2 and U_3O_8 oxides, although UO_3 is also reported occasionally [6, 9,10].

The solubility of particles collected proved highest in the HCl medium than in the Gamble medium (Figure 2.5) and lowest in the bicarbonate medium. Significant differences between the results obtained from aerosols generated on the turret, and the glacis were revealed in the Gamble and bicarbonate media with a higher solubility for the turret aerosols.

In an acid medium, the high solubility (35% dissolved in 6 d) could be the cause of soil contamination after the deposit of particles and leaching by acid rains. On the other hand, the low particle solubility in a medium simulating mineral water (bicarbonate, pH = 7.9) could be both the reason for the particles being transported by the water toward the organic or mineral matter, and also the cause of contamination.

The dissolution parameters calculated in the Gamble medium (simulating the pulmonary fluids) from the dissolution kinetics (Figure 2.5) were used to classify the uranium oxides generated during impact into a type of intermediate solubility between M and F, with rapid dissolution rate values (f_r) higher than those described in the published works for U_3O_8 and UO_2.

This may be explained, first, by the specific surface area of the aerosols ($m^2 \cdot g^{-1}$) generated during firing. Their production at 600°C must cause the formation of high porosity and, therefore, considerable solubility by virtue of a high contact surface with the fluids. This has already been described for these industrial UO_2 et U_3O_8 oxides [16,17]. Second, the small diameter of the particles and the very nature of the oxides present (M to F type UO_3 present, based on the hydration rate) could also explain these high dissolution rates.

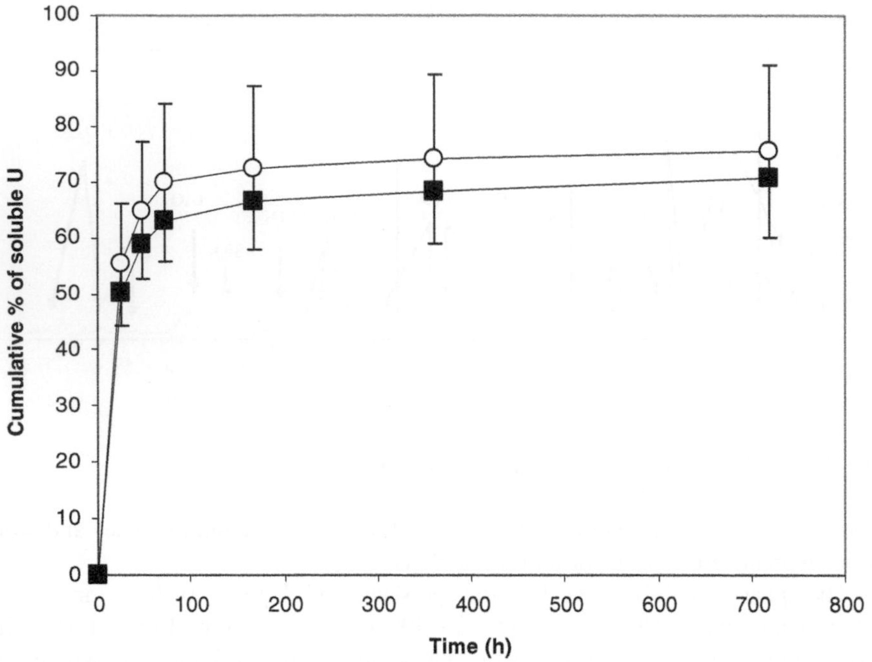

FIGURE 2.5 Kinetic of dissolution *(Mean ± SD, n = 3)* of aerosols from turret (white circle) and glacis (black square) in a Gamble solution.

The risk of inhalation for individuals exposed to this type of aerosol has been established in terms of the dose likely to be received during exposure to an average concentration measured in this study, i.e., 120 $Bq{\cdot}m^{-3}$. An exposure scenario lasting 15 min with a ventilation rate of 3 $m^{-3}{\cdot}h^{-1}$ was used, which corresponded to a unique inhalation of 6.4 mg of depleted uranium. These hypotheses suggest a committed effective dose over 1 year of 0.5 mSv for the entire body. By way of comparison, the annual dose limits for workers in the nuclear industry are 20 mSv, and 1 mSv for members of the public [18].

For the distribution of the dose through the various organs, the highest doses were calculated for the lungs (about 6 mSv after 50 years) and the lymph nodes (14 to 36 mSv). A 0.04 mSv dose was calculated for the bone marrow, which is already far beyond the annual value of 1.2 mSv for "normal" exposure to natural radiation sources as defined by UNSCEAR [19]. Lastly, based on the dissolution results in a medium simulating the pulmonary fluids (3–4% dissolved per day) and a fraction of 12% of uranium dissolved in the blood, reaching the kidneys, a concentration of 0.07 to 0.1 $\mu g\ U{\cdot}\ g^{-1}$ kidney could be reached after inhalation of 6.4 mg of DU. Although controversial, the nephrotoxic limit in humans of 3 $\mu g{\cdot}g^{-1}$ kidney is far from being achieved in the exposure hypothesis used.

The IRSN study has thus rounded out the data on depleted uranium used in weapons with experimental results from a very realistic context of firing with DU shells. The results have shown that the particles generated, mainly composed of a mix of U_3O_8, $UO_{2.25}$, UO_3 oxides, and a U/Al alloy, were small in size (AMAD of 1 to 2 μm), with high solubility in an acid medium and in a medium simulating the pulmonary fluids, and with low solubility in a bicarbonate medium. An exposure hypothesis unique to the characterized aerosols was produced; it showed that the doses likely to be received after inhalation were lower than those recommended by the ICRP. Lastly, the risk of nephrotoxicity can apparently be excluded in this scenario, as the calculated kidney concentrations proved to be far lower than the known limits.

DU IN THE FRENCH NUCLEAR INDUSTRY

In the French nuclear industry, uranium oxides similar to those produced during impact of a DU shell are commonly encountered in various plants. During the uranium enrichment process for the manufacture of fuel, depleted uranium is also produced in the form of UF_6. This is then defluorinated, resulting in the formation of depleted U_3O_8. This DU is also used in the form of UO_2 combined with plutonium oxide, PuO_2, to manufacture MOX (mixed oxide) fuel. In France, Compagnie Générale des Matières Nucléaires (General Nuclear Materials Company) (COGEMA) is responsible for transforming DU produced by enrichment into depleted uranium. Similarly, COGEMA conditions the uranium after it has passed through the nuclear plant. The nuclear fuel is recovered after use and taken to the production unit for depleted uranium oxides such as U_3O_8, in particular, to be stored and reused later in the fuel cycle.

Studies at IRSN on the various uranium-bearing compounds in a variety of isotopic compounds (DU, reprocessed U, enriched U, natural U) have been summarized in a review article [20]; the DU data have been extracted below for comparison with those obtained during the study on DU in weapons described previously.

The purpose of the studies performed in the fuel cycle plants in France was to determine the risk of internal contamination of agents subjected to inhalation of uranium aerosols at their workplaces. This involved characterizing the particles sampled at the workplaces from a physicochemical viewpoint and to determine the type of absorption into the blood (fast, moderate, or slow), either from *in vitro* studies in media simulating pulmonary fluids or via *in vivo* experiments by instillation in rats. The dose coefficients (DPUI in $Sv.Bq^{-1}$) were also calculated from experimental results and compared with the default values recommended by ICRP for the compounds studied.

The methods used were similar to those described for the previous study, namely aerosol sampling on impactors to determine the granulometric distribution, then their characterization by x-ray diffraction and measurement of the density and specific surface area ($m^2 \cdot g^{-1}$). The chemical dissolution tests used the same static technique (sandwiched) as for the previous study, with a commercial chemical medium (GIBCO). The animal experiments consisted of intratrachea instillations of particle suspension in groups of 10–20 rats per batch. The retention kinetics in the organs of interest, especially lungs and kidneys, and excretion in urine and feces, were monitored over periods of between 1–60 d to determine the blood absorption parameters used in the dose calculation.

Sticking to depleted uranium, the compounds characterized in the French nuclear industry were made up of a mix of oxides of varying proportions, except for UO_2 used to manufacture MOX fuel and its by-product U_3O_8, which were pure.

For the defluorination workshop for depleted UF_6, the U_3O_8 manufactured contained traces of UO_2F_2 in variable quantities. Similarly, the U_3O_8 and UO_4 from reprocessing included UO_3.

The aerodynamic median activity diameter (AMAD) of the DU particles collected in the COGEMA workshops varied between 4–7 μm, with standard geometric deviations of 2 to 3. The densities and specific surface areas were also variable and depended on the nature of the uranium-bearing compound. Thus, UO_2 showed a density of 11 and a specific surface area of 3.1 $m^2 \cdot g^{-1}$, the U_3O_8 (alone or mixed) had densities of between 7.3 and 8.3 and specific surface areas of 1.2 to 4.5 $m^2 \cdot g^{-1}$ and, last, the reprocessed UO_4 had a density of 5.2 and a specific surface area of 5.8 $m^2 \cdot g^{-1}$.

The f_r (rapidly dissolved fraction), s_r (rapid dissolution rate, in d^{-1}), and s_s (slow dissolution rate, in d^{-1}) parameters were calculated from *in vitro* chemical dissolution results and *in vivo* absorption into the blood. These *in vitro* and *in vivo* values were comparable, as they were also for two compounds of the same nature. However, the absorption and dissolution parameters obtained were different from those for the same pure compounds. The differences in the particle AMAD, the presence of traces of soluble uranium oxides (like UO_3 and UO_2F_2), and the various specific surface areas could explain these differences. To compare these values given by default by ICRP involved classifying these compounds in a standard intermediate absorption between a type M and S.

The uranium oxides produced during the impact of a shell on a tank were of the same nature as those described above. Unlike the ICRP which considers U_3O_8 and UO_2 to be very insoluble oxides, the mix made up predominantly of U_3O_8, UO_2 (plus UO_3 and another U/Al phase) was classified as an intermediate type of solubility between F and M.

Both types of studies described previously on the depleted uranium-bearing compounds, therefore, indicate that the physicochemical characteristics, solubility, and absorption into the blood are major parameters in determining the risk of contamination. The prediction of risk of contamination should be based as far as possible on specific information rather than default values, which may be incomplete.

DU BIOKINETICS AFTER INHALATION

As we have just seen, the physicochemical characteristics of uranium-bearing compounds, as well as their chemical solubility and their type of absorption into the blood, are essential factors in assessing the risk of internal contamination.

The study described earlier involved *in vivo* experiments on rats through contamination via the intratrachea path of a suspension of uranium-bearing particles. Other studies involving rats were conducted recently to assess DU biokinetics after contamination by acute and repeated inhalations of soluble and insoluble uranium oxides commonly encountered in the nuclear industry, namely, UO_2 and UO_4.

The main objective of these studies was to test the ICRP hypothesis postulating that chronic contamination may be modeled by the sum of successive and independent acute intakes. This model is based on acute or short-term exposure, and little data are available after chronic inhalation. In some specific cases, particularly nickel, it has been shown that repeated inhalations of NiO reduced the pulmonary clearance of the same oxide inhaled subsequently.

Repeated or chronic exposure should not be excluded, be it in the industrial nuclear sector or in the context of DU used in weapons. In these exposure cases, the assessment of the risk of internal contamination and, in particular, the calculation of the dose likely to be received, are based on the chronic model described previously. This has therefore been verified by realistic experiments of exposure by repeated inhalation to an insoluble, depleted uranium oxide, UO_2 [21]. The experimental biokinetics obtained from the various organs of interest and excreta were compared with those obtained after iteration of the results from acute inhalation of the same compound, with an identical cumulated pulmonary intake.

To achieve this, two series of experiments were carried out with several groups of rats subjected to various concentrations and exposure frequencies of UO_2, maintaining the same total exposure time. Another group of rats was contaminated uniquely to obtain a final pulmonary intake of the same order of magnitude as both the previous experiments.

Sixteen-week-old, adult male rats weighing 552 g *(± 36 g)* were subjected to insoluble, type S UO_2 inhalations using the nose-only system (Figure 2.6). The DU used came from an industrial source. The UO_2 aerosols generated showed an AMAD of 2.5 μm ($\sigma_g = 1.9$).

Two types of repeated exposure were used: (1) a group of 16 rats was subjected to a concentration of 190 *(± 41)* mg·m^{-3} (about 200 μg of lung intake per inhalation), 1 hour a day, 2 d a week for 3 weeks, (2) a second group of 16 rats was subjected to

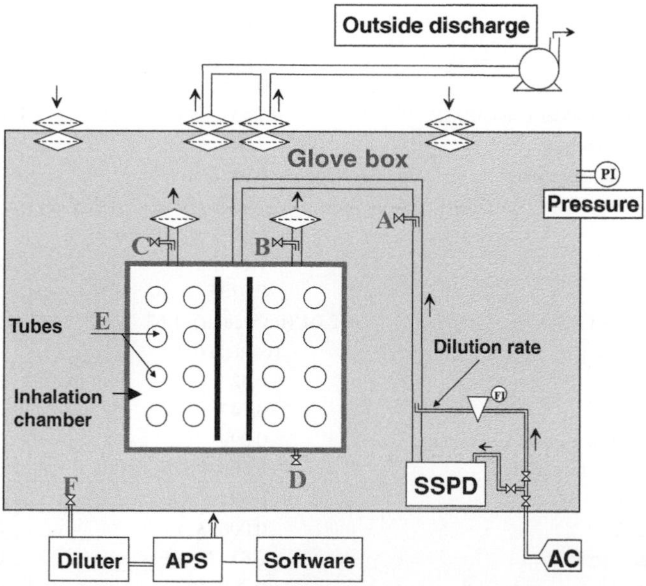

FIGURE 2.6 General layout of the inhalation system. SSPD: small-scale powder disperser, aerosol generation; APS: aerodynamic particle sizer, particle size distribution, and concentration.

TABLE 2.3
Parameter Values Used in the Biokinetic Model for Uranium in Rats

Transfer	Transfer Rate (d^{-1}) for Humans (ICRP)	Transfer Rate (d^{-1}) for Rats
s_p	0.1	0.5
s_t	0.0001	0.0005
s_{pt}	100	100
AI_1 to bb_1	0.02	0.02
AI_2 to bb_1	0.001	0.001
AI_3 to bb_1	0.0001	0.0001
AI_3 to LN_{TH}	0.00002	0.00002
bb_1 to BB_1	2	2
bb_2 to BB_1	0.03	0.03
bb_{seq} to LN_{TH}	0.01	0.01
BB_1 to ET_2	10	10
BB_2 to ET_2	0.03	0.03
BB_{seq} to LN_{TH}	0.01	0.01
ET_{seq} to LN_{ET}	0.001	0.001
Skin to ET_2	—	0.291
ET_2 to ST	100	100
ST to SI	24	26.4
SI to ULI	6	7.2
ULI to LLI	1.8	4.86
LLI to feces	1	5.5
SI to plasma	0.012	0.0271
Plasma to ST0	10.5	60
Plasma to RBC	0.245	1.4
Plasma to urinary bladder content	15.43	126
Plasma to urinary path	2.94	25.5
Plasma to other kidney tissue	0.0122	0.7
Plasma to ULI contents	0.122	7
Plasma to liver 1	0.367	5.94
Plasma to ST1	1.63	21
Plasma to ST2	0.0735	2.8
Plasma to bone surface	2.04 (trabecular)–1.63 (cortical)	17.5
ST0 to plasma	8.32	8.32
RBC to plasma	0.347	0.139
Urinary path to urinary bladder content	0.099	0.1083
Other kidney tissue to plasma	0.00038	0.01386
Liver 1 to plasma	0.092	0.147
Liver 1 to liver 2	0.00693	0.00693
ST1 to plasma	0.0347	0.0578
ST2 to plasma	0.000019	0.00347
Bone surface to plasma	0.0693	0.693
Bone surface to exch volume	0.0693	0.0693

(continued)

TABLE 2.3(CONTINUED)
Parameter Values Used in the Biokinetic Model for Uranium in Rats

Transfer	Transfer Rate (d^{-1}) for Humans (ICRP)	Transfer Rate (d^{-1}) for Rats
Liver 2 to plasma	0.00019	0.00693
Nonexchange. volume to plasma	0.000493 (trabecular)– 0.0000821 (cortical)	0.018
Exchange volume to bone surface	0.0173	0.0173
Exhange volume to non-exch volume	0.00578	0.03468
Urinary bladder content to urine	12	1

the same UO_2 concentration, 30 min a day, 4 d a week for 3 weeks (100 μg of lung intake per inhalation). A control group of a dozen rats was also exposed to air under the same conditions.

A final group of 18 rats exposed to air for 3 weeks was then contaminated by an acute inhalation of UO_2 at a concentration of 375 *(±120)* mg·m^{-3} for 2 h (about 1000 μg of lung intake), so that the acutely contaminated animals had the same age as the animals undergoing the repeated inhalation experiments.

The excreta of the contaminated animals was collected every day, and the organs of interest (lungs and trachea, kidneys, liver, 1g of blood, GIT, femurs, spleen and carcass with muscles) were sampled on the day the animal was put down, between 1 and 90 d postexposure (the precise times for putting the animals down depending on the experiment). Kinetic phosphorescence analysis (KPA) was used to calculate the uranium content.

The Cyclomod software program developed to predict the excretion and retention of radioactive elements in humans, based on the ICRP biokinetic models [18, 22, 23] and the algorithm described by Birchall and James [24], were altered to predict excretion and retention in rats. To achieve this, the biokinetic data obtained after acute inhalation of UO_2 were used to design this rat model. Some parameters, as indicated in Table 2.3, like the diagram of particle deposits in the respiratory tract, the transfer and absorption rates in the GIT, in the respiratory tract and the various organs, and the rate of urinary excretion were altered, so that the model adjusted as far as possible to the experimental retention and excretion curves after acute UO_2 inhalation. A new compartment representing the skin and the hair was also added to fine-tune the rat model. Organs that did not relate to a specific compartment, like the spleen, were included in the compartment "other soft tissues."

The results from all organs showed good correlation between the prediction of this new model and the experiment. In the lungs in particular, the divergence did not exceed 10% on average, with nevertheless an underestimation of 40% of the predicted retention 30 days after exposure. The best correlation was observed for the kidneys with only 7% divergence on average, although the experimental peak obtained 3 days post exposure could not be modeled correctly. In the other organs, like the femur, the divergence was 22% and for the GIT and the feces, a slight under-estimation by the model was noted. On the other hand, the model overestimated the urinary excretion for the first days.

This model was subsequently used to predict the biokinetics over 1 month of the two repeated inhalation experiments described above. The results presented in Figure 2.7 and Figure 2.8 showed, first, that the biokinetics of both experiments were comparable in the majority of organs except for GIT and that secondly, a good correlation existed between the two experiments and the model in the majority of organs except the femur. In this particular case, the model overestimated the retention by a factor of three.

For the majority of organs and excreta tested, both repeated inhalation models were therefore modeled correctly by the modified chronic "rat" model, which suggests that under the experimental conditions described, repeated inhalation is the clear equivalent of the sum of successive and independent acute intakes.

Several points in this study require attention, however. For a comparable final intake, the DU causes various effects on the general health parameters of the animals after acute and repeated inhalation. Compared with the control rats, a drop in food consumption and weight loss were observed in the rats subjected to acute inhalation, whereas no change was noted during the repeated inhalation experiments. In addition, the frequency of repeated exposure could influence the mucociliary movement mechanism or the accumulation in the GIT. The DU concentration in the GIT of rats exposed 4 times a week was higher than in the control rats whereas comparable to an exposure rate of twice a week. In addition, very high U concentrations were found in the feces on the first days post exposure and rapid excretion was noted.

Lastly, the model overestimating the uranium concentrations in the femur could be explained by a reduction in the absorption rate into the blood after repeated inhalation or a drop in the osseous loading. The first hypothesis seems unlikely, however, as the retention in the other organs, particularly the kidney, should also be affected should the absorption rate drop. Uranium is known to be toxic to bones. Exposure in rats, by injecting uranyl nitrate or by subcutaneous implants of UO_2 powder did, in fact, reduce the bone formation rate and increase the resorption rate [25–27]. In the event of repeated contamination, it is therefore probable that the uranium is fixed on the bone after a first exposure and causes osseous toxicity as described previously, which could have an effect on the biokinetics of the uranium introduced during subsequent inhalations and reduced the osseous retention.

Another recent study on DU contamination (in the form of uranyl nitrate) through chronic ingestion via drinking water has also shown different biokinetics from those obtained acutely, particularly for the kidneys and femurs [28]. The results from both these studies show that after chronic or repeated exposure, DU osseous retention cannot be predicted by extrapolating the acute exposure data.

Another biokinetics study after DU inhalation used the same batch of soluble UO_2 as previously and another uranium oxide, a more soluble, reprocessed UO_4 (containing 0.061% of ^{236}U mass). The biokinetics of the UO_4 inhaled acutely following acute or repeated inhalations of UO_2 were compared with those of the UO_4 inhaled alone, to determine the impact of prior exposure [29]. For sulfur, it showed that sulfur dioxide inhaled beforehand affected the clearance of the TiO_2 particles inhaled subsequently. This is also true for repeated inhalations of nickel

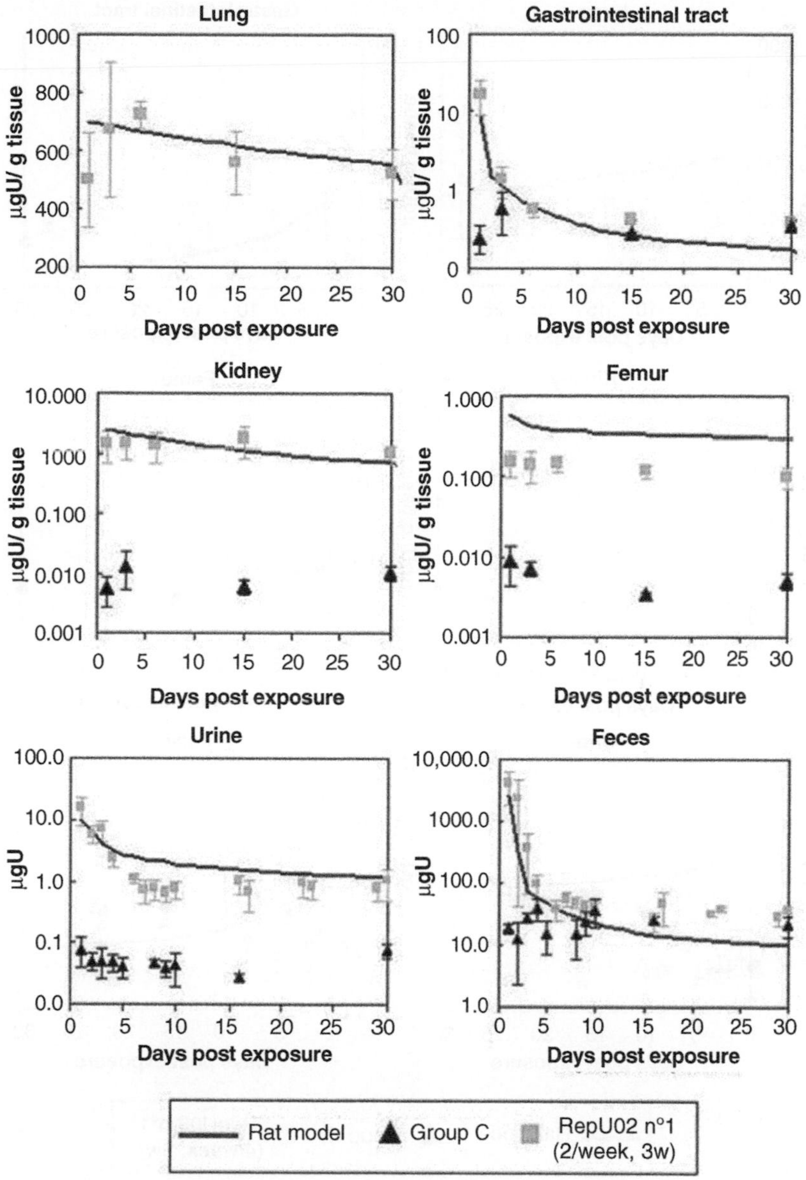

FIGURE 2.7 Uranium content µg/gram (lung, gastrointestinal tract, kidney, and femur) or in urine and feces over 24 h as a function of days post exposure for group RepUO2 no. 1 (repeated inhalation twice a week for 3 weeks); blue line shows theoretical biokinetics of repeated UO2 inhalation predicted with the iteration of the rat model; green square means ± SD of experimental content values of group RepUO2 no. 1 rats exposed 1 h twice a week; n = 3 for tissues; n = 6 – 15 for urine or feces up to 10 d, n = 3 for urine or feces after 10 d; black triangle means ± SD of the three experimental content values of control rats, group C.

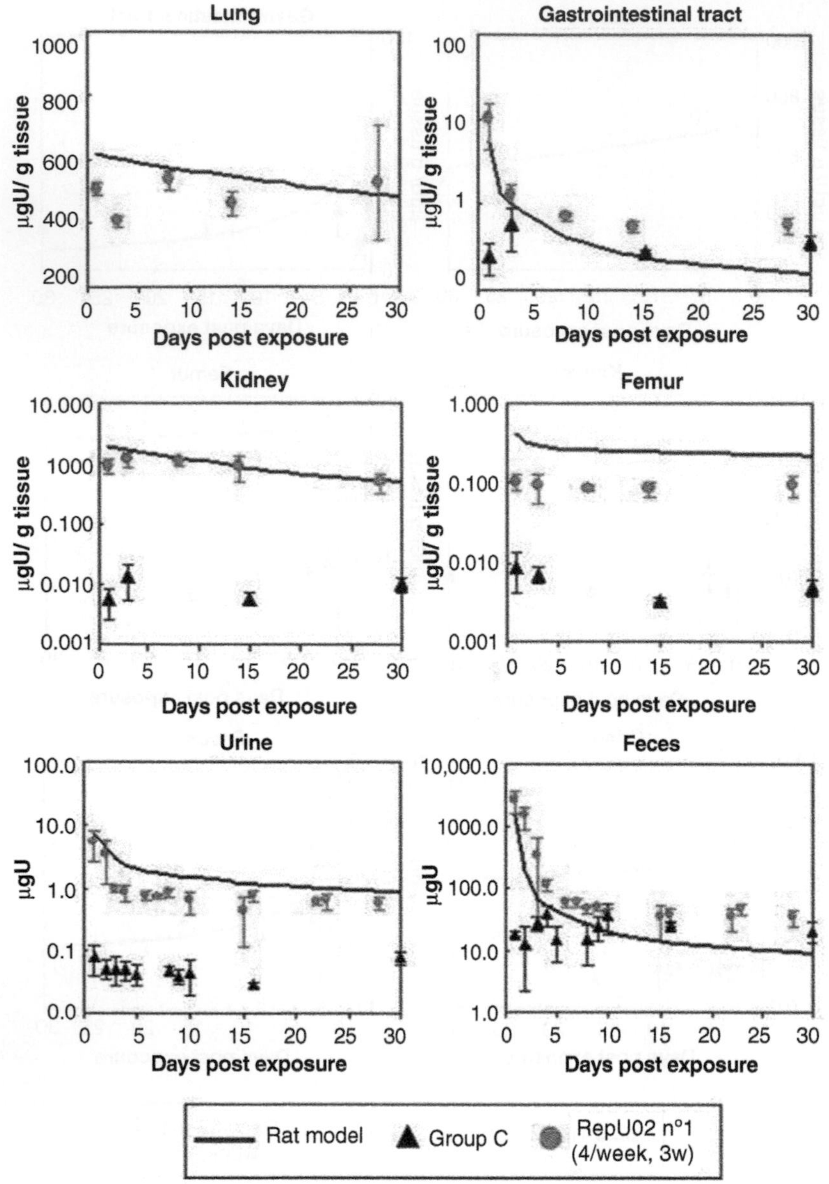

FIGURE 2.8 Uranium content μg/gram (lung, gastrointestinal tract, kidney, and femur) or in urine and feces over 24 h as a function of days post exposure for group RepUO_2 no. 2 (repeated inhalation 4 times a week for 3 weeks); blue line shows theoretical biokinetics of repeated UO_2 inhalation predicted with the iteration of the rat model; pink dot means ± SD of experimental content values of group RepUO_2 no. 2 rats exposed half an hour 4 times a week; n = 3 for tissues; n = 6 – 15 for urine or feces up to 10 d, n = 3 for urine or feces after 10 d; black triangle means ± SD of the three experimental content values of control rats, group C.

oxide in rats, which alters the clearance of NiO inhaled subsequently [30, 31]. This study also sought phenomena of this nature.

The UO_2 used was identical to the previous study. The reprocessed UO_4 had an AMAD of 2.3 μm (σ_g = 2.0). Several exposure series were performed: (1) acute inhalation of UO_4 for 30 min, of 116 *(± 60)* mg·m^{-3} (35 rats) (AcUO_4), (2) acute inhalation of UO_2 for 3 h at 375 *(± 70)* mg·m^{-3}, followed 11 d later by an identical acute inhalation of UO_4 as previously (30 rats) (AcUO_2 + UO_4), and (3) repeated inhalation of UO_2 for 30 min at 190 *(± 41)* mg·m^{-3}, 4 times a week for 3 weeks, followed 11 days later by acute inhalation of UO_4, identical to the first (40 rats) (RepUO_2 + UO_4). A last group of 15 control rats was exposed to air under similar conditions.

The lung intake of the rats subjected to acute inhalation of UO_2 was comparable with the total of the lung intakes at the end of repeated exposure; however, it was about 7 times higher than the lung intake after acute inhalation of the UO_4.

The organs of interest (lungs + trachea, kidneys, femurs, and GIT) were sampled 4 h, 1 d, 3 d, 8 d, and 16 d post exposure, and the excreta was collected for 4 d post exposure, then twice a week. ICP-MS was used to determine the uranium content.

Taking the general health parameters into consideration, exposure to UO_4 in the three groups caused a drop in the average weight of the rats and a reduction in the amount of water and food consumed, compared with the control rats. In addition, the exposure to UO_4 alone, plus the repeated double exposure to UO_2 followed by acute exposure to UO_4, caused an increase in the urinated volume in the 5 d following inhalation. The liver showed no signs of lesion, but for the three exposure series, the kidneys were seen to be affected through an increase in urea and creatinine.

The biokinetics of the DU in the three exposure series were compared by expressing the amount of UO_4 as a percentage of the initial lung deposit (ILD) taken as equal to the quantity of UO_4, 4 h after exposure. Figures 2.9 and 2.10 present the results of the biokinetics in the main organs and excreta.

No difference in lung retention was revealed in the three groups. Kidney retention, on the other hand, was about 4 times higher in the two groups exposed previously to UO_2 compared with the group which had only inhaled UO_4. Prior exposure to UO_2 therefore appears to modify the biokinetics of UO_4 in the kidneys.

The K/K+U ratio (where K is the amount of UO_4 in the kidneys and U the amount in the urine) representing the urinary excretion from the kidneys, was comparable in the first days (about 0.4) for the single inhalation group and the repeated inhalation of UO_2 + acute inhalation of UO_4 group, but higher (about 0.7) for the acute UO_2 + acute UO_4 group. The ratio was identical for all three groups in the days that followed. These results, therefore, suggest that repeated exposure to UO_2 does not modify the urinary excretion diagram of UO_4 (unlike a single prior exposure to UO_2), but rather increases the lung dissolution rate in the blood. This is confirmed by the amount of UO_4 in the femurs. Only the group with the double repeated contamination of UO_2 followed by the acute inhalation of UO_4 increased by a factor of 2 compared with the other two groups. Lastly, the amount of UO_4 in the GIT was lower in the double exposure groups compared with the single exposure group. This correlates with the quantity in the feces which follow the same diagram. These results could be explained by alteration in the mucociliary movement triggered by prior exposure to UO_2, of any kind.

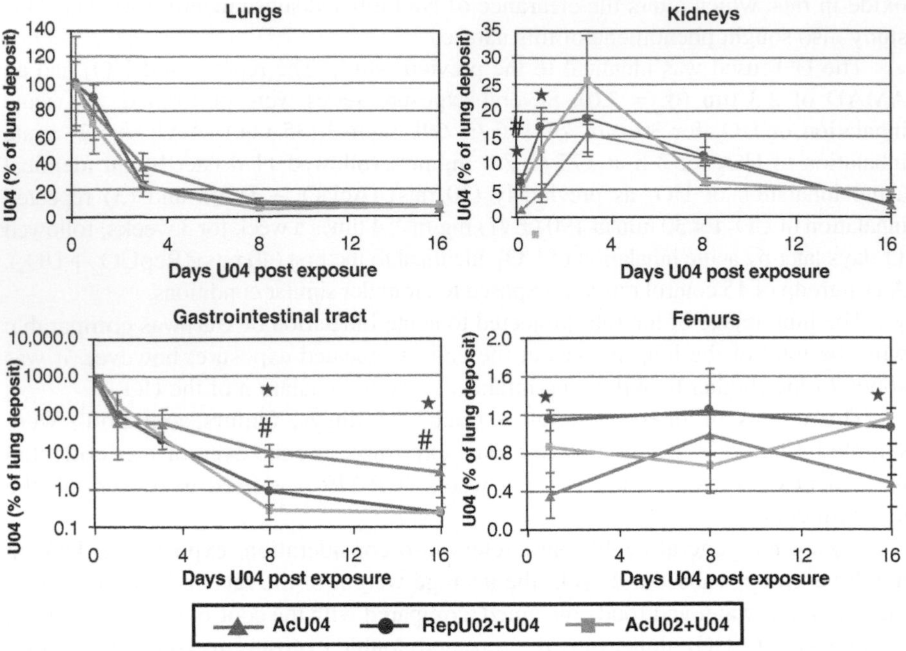

FIGURE 2.9 **(See color insert following page 80)** Biokinetics of UO_4 in the lungs, kidneys, gastrointestinal tract, and femurs as a function of days after UO_4 exposure. Percent of UO_4 lung deposit, mean ± SD. N = 5 for $RepUO_2 + UO_4$, N = 3 for $AcUO_2 + UO_4$ and N = 4 for $AcUO_4$. *: $p < 0.05$ between the $RepUO_2 + UO_4$ and $AcUO_4$ groups, #: $p < 0.05$ between the $AcUO_2 + UO_4$ and $AcUO_4$ groups.

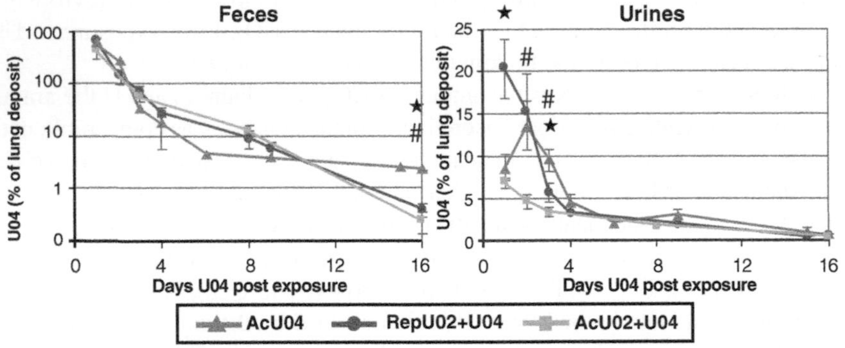

FIGURE 2.10 **(See color insert following page 80)** Daily excretion of UO_4 as a function of days after UO_4 exposure. % of UO_4 lung deposit, mean ± SD. N = 5 for $RepUO_2 + UO_4$, N = 3 for $AcUO_2 + UO_4$ and N = 4 for $AcUO_4$. *: $p < 0.05$ between the $RepUO_2+UO_4$ and $AcUO_4$ groups, #: $p < 0.05$ between the $AcUO_2 + UO_4$ and $AcUO_4$ groups.

Prior exposure to an insoluble depleted uranium oxide like UO_2 may therefore significantly disturb the biokinetics of a more soluble oxide like UO_4 subsequently inhaled acutely in the kidneys, GIT, and excreta of rats. The consequences of the double exposures could therefore result from the combined effects of UO_2 and UO_4. The inhalation of soluble UO_4 could reveal modifications induced by insoluble UO_2, for the solubility of a compound influences its biokinetics as well as how it acts.

All the results from both these DU inhalation studies, compared with the results of an ingestion study, therefore indicate that prior exposure may influence the biokinetics of DU for some organs in a case of subsequent contamination by an identical compound or with different solubility. Be it in the context of medical surveillance of workers in the nuclear industry or the use of DU in weapons, this should be taken into account to assess the risk of internal contamination which cannot therefore be predicted simply by using an iterative model.

BIOLOGICAL EFFECTS OF EXPOSURE TO DU AFTER INHALATION

Bioaccumulation and Effect on the Central Nervous System

As we have just seen, after inhalation the DU accumulates in some organs and is excreted with various biokinetics depending on the nature of the inhaled compound and the inhalation method.

Some studies suggest a link between exposure to uranium and neurological toxicity. Be it in the context of the nuclear industry or the use of DU in weapons, inhalation is without doubt the main exposure path for workers and soldiers [32–35].

Within the framework of the repeated inhalation experiments described previously, the biokinetics of a DU oxide in the brain were assessed and behavioral tests carried out in an attempt to correlate a possible effect of inhaled DU with the bioaccumulation in the central nervous system [36].

A group of 42 rats was subjected to repeated inhalations of the same of UO_2 as used in the previous studies, at 190 *(± 41)* $mg.m^{-3}$ (lung intake of around 40 $\mu g{\cdot}g^{-}1$ lung) for 30 min, 4 d a week for 3 weeks. A group of 33 control rats inhaled air only under the same conditions. 27 rats were used to determine DU biokinetics in the brain and were put down 1, 3, and 7 d postexposure. The hippocampus, olfactory bulb, front cortex, cerebellum, the rest of the brain, and the lungs were sampled to dose uranium by KPA (kinetic phosphorescence analyzer). Two types of behavioral tests were performed on a group of 48 rats based on 2 independent experiments noted A and B, each with 24 rats (including 12 control rats [Group C] and 12 rats contaminated with DU [Group U]). The spontaneous locomotion was studied in an open field fitted with two levels in infrared cells that recorded the horizontal (representing the locomotion behavior) and vertical (representing the rearing activity) movements by the animals over 20-min periods. This test was performed 1 d after the end of experiment A and 5 d after the end of experiment B. The working memory was assessed the following day on the same animals in a Y maze; the number of visits to each branch were counted and used to assess the overall exploratory activity, with the spontaneous alternation accounting more specifically for the spatial working memory. This test was performed 2 d post

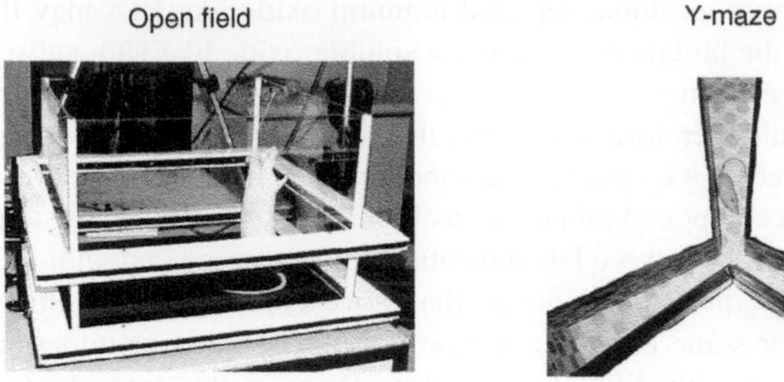

FIGURE 2.11 Behavioral tests.

exposure for experiment A and 6 d later for experiment B. Figure 2.11 illustrates the two systems used.

The results of the biokinetics presented in Figure 2.12 have shown that the accumulation of DU in the brain of contaminated rats was statistically greater and more heterogeneous than in the brains of the control rats, 1 d postrepeated exposure.

The DU concentrations found in the various structures of the central nervous system were 567 ng·g^{-1} in the olfactory bulb (16 ng.g^{-1} for the controls), 156 ng·g^{-1} in the hippocampus (20 ng·g^{-1} for the controls), 122 ng·g^{-1} in the front cortex

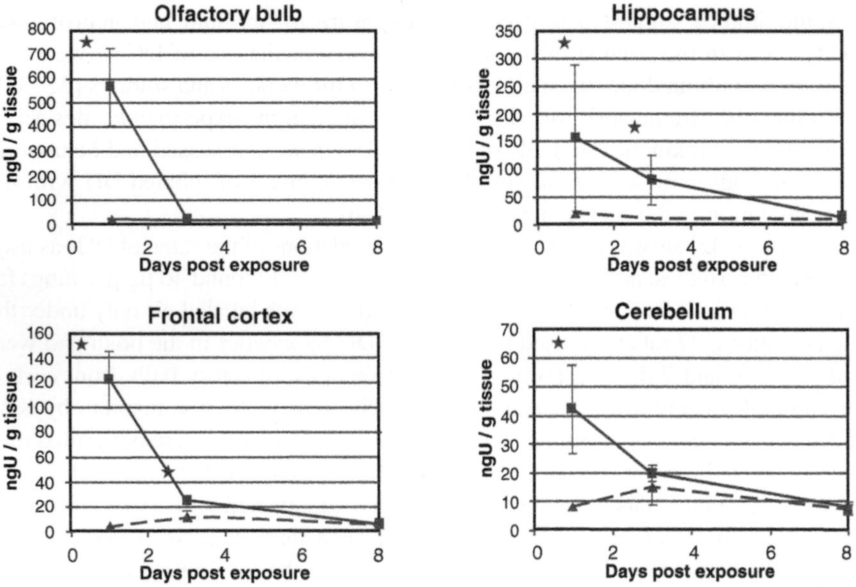

FIGURE 2.12[1] Uranium concentration in rat brain areas as a function of days post exposure. Data are represented as mean ± S.D. n = 6 for exposed rats, group U; n = 3 for control rats, group C. p values below 0.05 between the U and C groups.

[1] Reprinted from Monleau et al. Bioaccumulation and behavioral effects of depleted uranium in rats exposed to repeated inhalations, *Neuro. Lett.* 390, 31. Copyright 2005, with permission from Elsiever.

(4 ng·g^{-1} for the controls) and 42 ng·g^{-1} in the cerebellum (8 ng·g^{-1} for the controls). The DU quantities then dropped rapidly to the level of the controls. Three days after repeated exposure, only the hippocampus and to a lesser extent the cerebral cortex still showed DU concentrations far higher than for the controls.

This heterogeneous bioaccumulation could be explained by the three possible paths taken by the DU to penetrate the brain after inhalation [37]. The DU can enter the systemic circulation after dissolution of the particles in the lungs and absorption into the blood or via the GIT after mucociliary movement of the particles from the lungs. The DU then migrates into the central nervous system via the cerebral capillaries (blood–brain barrier) or via the cerebrospinal fluids (blood–spine barrier). The second possible path is via the olfactory epithelium. Last, the last path is direct transport along the olfactory nerve connecting the nasal cavity to the olfactory bulb. Although described as the favored path for the passage of Mn, Cd, Ni, Hg, Co, and Zn, this passage is not widespread for all metals [37–42]. DU in the form of UO_2 may be transported by several paths, (1) via the blood–brain barrier after dissolution and/or passage into the blood via the GIT and (2) for the soluble part of UO_2, by the direct path of the olfactory nerve via the olfactory neurones, as the highest concentrations are found in the olfactory bulb. The compound solubility was described for manganese in particular, as a parameter influencing the passage to the brain, as a soluble compound penetrates more easily than an insoluble one [43].

The passage of DU into the brain seems to modify the behavior of some animals compared with the controls. At 1 d post exposure, an increase in spontaneous locomotion and rearing was revealed in the contaminated rats in experiment A (performed 1 d post exposure) but no difference was noted for the rats in experiment B. The results are presented in Figure 2.13.

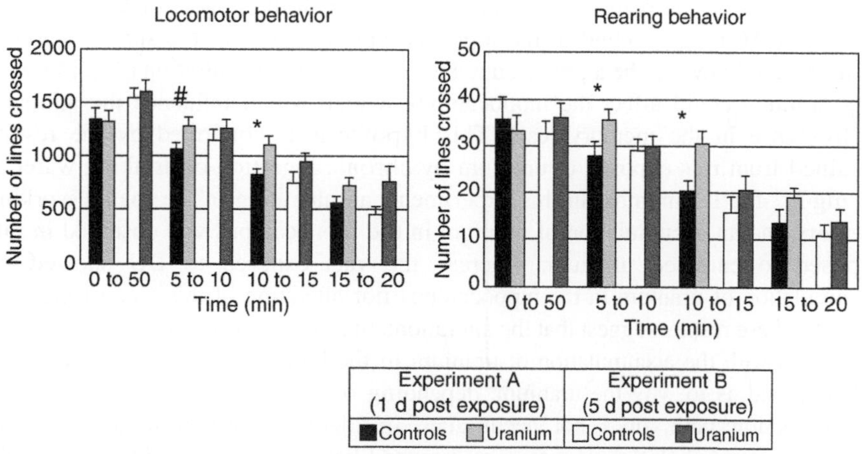

FIGURE 2.13[2] Effects of uranium inhalation on spontaneous locomotion in rats. The locomotor and rearing behaviors are presented for 5-min periods throughout the entire test (20 min). Bars represent mean ± SEM, n = 12. The exposed U group, in experiment A, is significantly different from control at $p < 0.04$ (*) and reached a significance of 0.07 (#).

[2] Reprinted from Monleau et al., Bioaccumulation and behavioral effects of depleted uranium in rats exposed to repeated inhalations, *Neuro. Lett.*, 390, 31. Copyright 2005. With permission from Elsiever.

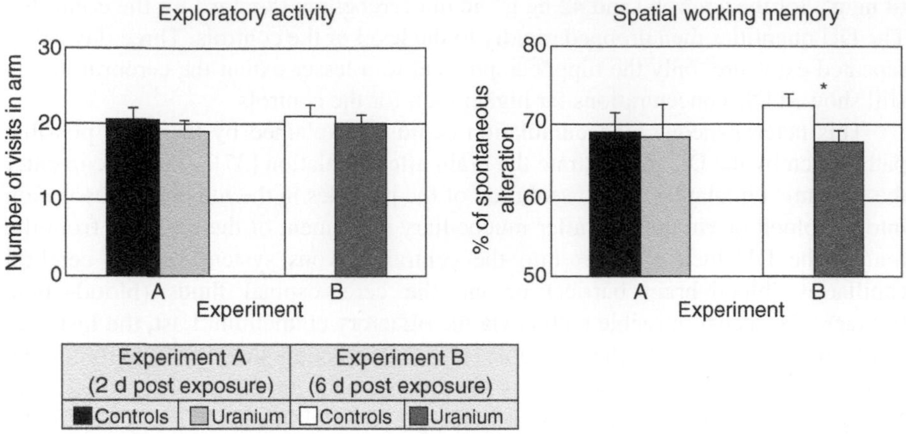

FIGURE 2.14[3] Effects of uranium inhalation on spatial working memory in rats. The exploratory activity and the spatial working memory are shown after 10 min (total length of the Y-maze test) for experiments A and B. Bars represent mean ± SEM, n = 12. The exposed U group, in experiment B, is significantly different from control at $p = 0.03$.

In neither experiment performed in the y-maze was a difference found with the controls for the exploratory activity, which is not contradictory with the previous, more sensitive test.

On the other hand, the spatial working memory 6 d post exposure showed signs of reduction for the rats in experiment B compared with the controls, as shown in Figure 2.14.

These results on behavior may be correlated with the larger concentrations of uranium (156 $ng.g^{-1}$) found in the hippocampus, involved in the spatial working memory and known to be a preferred zone for uranium concentration [28], [44–46]. The uranium could affect the hippocampus, which would influence the animals' performance in the memory tests. This hypothesis is confirmed by the results obtained from rats exposed to uranium by chronic ingestion of drinking water at 40 $mg \cdot L^{-1}$ for 1.5 months. In this experiment, an alteration in the spatial working memory and an accumulation of uranium in the hippocampus was observed in rats exposed to enriched uranium, whereas the rats exposed to DU showed no accumulation of uranium in the hippocampus nor alteration of this type of memory [47]. All these results suggest that the alterations in spatial working memory in rats are correlated with the accumulation of uranium in the hippocampus. An explanation is still required as to why the uranium, depending on its degree of enrichment or its contamination path, does not accumulate in identical fashion in this cerebral structure. After inhalation, it seems even more likely that the DU will pass directly from the nasal cavity to the olfactory bulbs via the olfactory nerve.

---

[3] Reprinted from Monleau et al. Bioaccumulation and behavioral effects of depleted uranium in rats exposed to repeated inhalations, *Neuro. Lett.*, 390, 31. Copyright 2005, with permission from Elsiever.

Genotoxicity and Inflammatory Effects after Inhaling DU

Uranium inhalation can therefore affect behavior, but other more described effects may also be the result. In particular, DU inhalation may cause pulmonary fibrosis and neoplasia. However, little is known about the molecular mechanisms producing these frequently described pathologies [48]. Recent studies do, however, exist on blood samples from soldiers wounded by DU fragments as well as humans exposed to uranyl nitrate and uranium miners, which have shown chromosomal aberrations and genetic damage. *In vitro*, DU produces genomic instability which shows itself by forming micronuclei [49,50].

Experiments on contamination by DU inhalation have therefore been performed to study the genotoxicity and inflammatory response as a function of the solubility of the uranium-bearing compound inhaled, the dose administered and whether inhalation is acute or repeated [51].

The uranium oxides used were identical to the studies described previously. Several series of inhalation were performed. Three acute inhalations of UO_2 at increasing concentrations (AcUO_2-1, AcUO_2-2, AcUO_2-3), repeated inhalation of UO_2 over 3 weeks with a total lung intake similar to the strongest acute inhalation (RepUO_2) and acute inhalation of UO4 similar to the weakest inhalation of UO_2 (AcUO_4). A group of rats inhaling air only was also controlled. These protocols are used to assess the effect on the genotoxicity and pulmonary inflammation of the dose administered by comparing the three acute experiments, the effect of the solubility by compared the UO_2 and UO_4 acute inhalations with equal lung intake and the effect of the type of inhalation by comparing the UO_2 acute and repeated experiments. The concentrations of biochemical parameters like ALT, AST, creatinine, and urea were measured on blood serum samples from the groups of rats contaminated with UO_4 and UO_2 by repeated inhalation only.

The nasal epithelial cells, broncho alveolar lavage (BAL) cells and kidney cells were isolated for each series of experiments. The comet test under neutral and alkaline conditions was used on each cell type to detect the double-strand (revealing radio-induced damage) and single-strand breaks in the DNA and thus assess the genotoxicity of the inhaled DU. The pulmonary inflammation was quantified by the RT-PCR dose and the following pro- and antiinflammatory cytokines: TNF-α, IL-8, MIP-2, IFN-γ, IL-10, and HPRT gene. Last, the hydrogen peroxide rate was determined from samples of rat lungs contaminated by repeated inhalation of UO_2 and by acute inhalation of UO_2 at the higher concentration.

The results (Table 2.4) showed that the animals' general health parameters depended on the dose, the solubility and whether inhalation was acute or repeated.

The biochemical dosages (increase in creatinine and urea) revealed a kidney disorder only for the rats contaminated with UO_4, and no effect on the liver (stable ALT and AST) was detected regardless of the group analyzed.

Solubility produced no effect on the damage to the DNA at the dose tested. Similarly, no DNA damage was noted in the nasal epithelial cells regardless of the experiments or test conditions.

DNA damage was noted under alkaline conditions only in the BAL cells for the strongest acute inhalation of UO_2, 1 d and 8 d post exposure (Figure 2.15). This could

TABLE 2.4
General Health Parameters of Different Rat Groups

	Days Post Exposure	Control	AcUO4	AcUO2-1	AcUO2-2	AcUO2-3	RepUO2
Rat weight (g)	0	563 ± 25	554 ± 35	577 ± 26	552 ± 50	513 ± 57	565 ± 32
	10	578 ± 29	431# ± 36	590 ± 37	528# ± 36	469# ± 60	575 ± 36
	14	592 ± 31	483# ± 27	600 ± 29	541# ± 38	520# ± 81	610 ± 29
Food (g)	0	21 ± 2	20 ± 3	20 ± 3	21 ± 6	21 ± 6	18 ± 4
	4	24 ± 3	1# ± 1	23 ± 5	15# ± 4	5# ± 6	20 ± 3
	10	25 ± 3	30 ± 2	26 ± 5	21.5 ± 2	17 ± 5	25 ± 2
Water (g)	0	35 ± 7	31 ± 6	30 ± 7	32 ± 4	37 ± 9	38 ± 9
	4	37 ± 6	14# ± 5	38 ± 4	35 ± 7	18# ± 10	35 ± 6
	10	38 ± 5	36.5 ± 5	37 ± 4	35 ± 9	32 ± 6	36 ± 12

Note: The values are given as mean ± SEM, n > 3. # p<0.05 between the exposed groups and the control group.

suggest that the genotoxic effect only appears for a threshold dose. Similarly, damage was also noted for repeated inhalation for extended periods.

The damage appears equal for both these same groups in neutral conditions. These UO_2 inhalations therefore produce simple- and double-strand DNA breaks in the BAL cells. Miller et al. have already described that radiation could play a part in inducing the biological effects of DU [49]. In the cases studied here, the chemical effects and radiation could therefore be synergic.

However, the alkaline conditions alone could detect the DNA breaks in the kidney cells of the group of rats contaminated repeatedly by UO_2, 3 and 8 d post exposure, as shown in Figure 2.16. Repeated inhalation could therefore induce a potentiation effect of the toxicity of the uranium in the kidney and BAL cells.

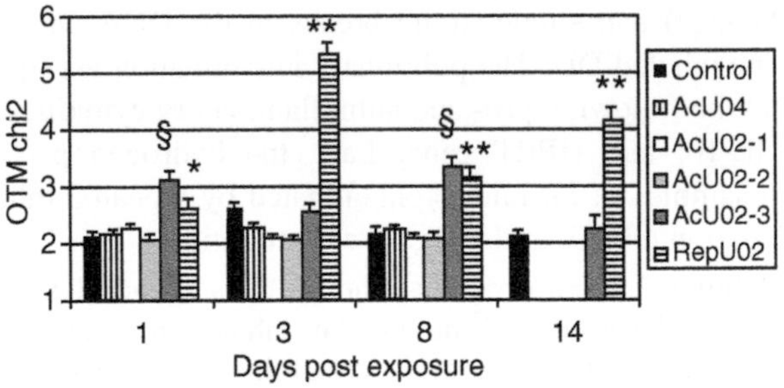

FIGURE 2.15[4] Comet assay under alkaline conditions in BAL cells. Mean of OTMchi2 values are presented in relation to days postexposure; * p < 0.01 and ** p < 0.001 between control and RepUO2. § p < 0.001 between control and AcUO2-3.

[4] Reprinted from Monleau et al., Genetic and inflammatory effects of depleted uranium particles inhaled by rats, *Toxicological Sciences*, 2006, 89, 287, by permission of Oxford University Press.

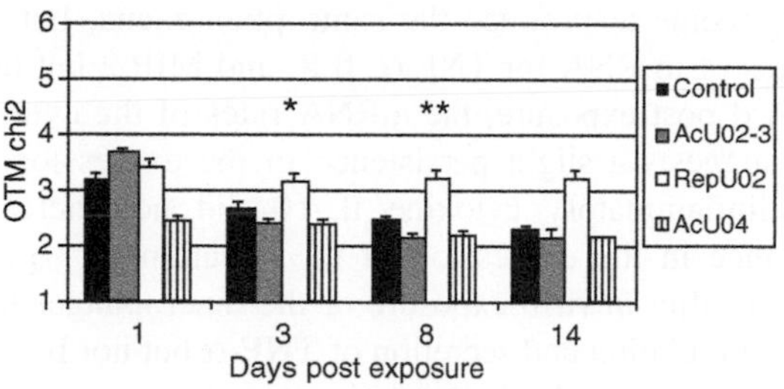

FIGURE 2.16[5] Comet assay under alkaline conditions in kidney cells. Mean of OTMchi2 values are presented in relation to days post exposure; * $p < 0.05$ and ** $p < 0.001$ between control and RepUO2.

The pulmonary inflammation results are presented in Figure 2.17. The acute inhalation of a stronger dose of UO_2 caused a rapid increase in the expression of the inflammatory (increase of cytokines TNF-α, IL8 but not IFN-γ nor MIP-10) and antiinflammatory (IL-10) genes, only 1 day post exposure. Similarly, repeated

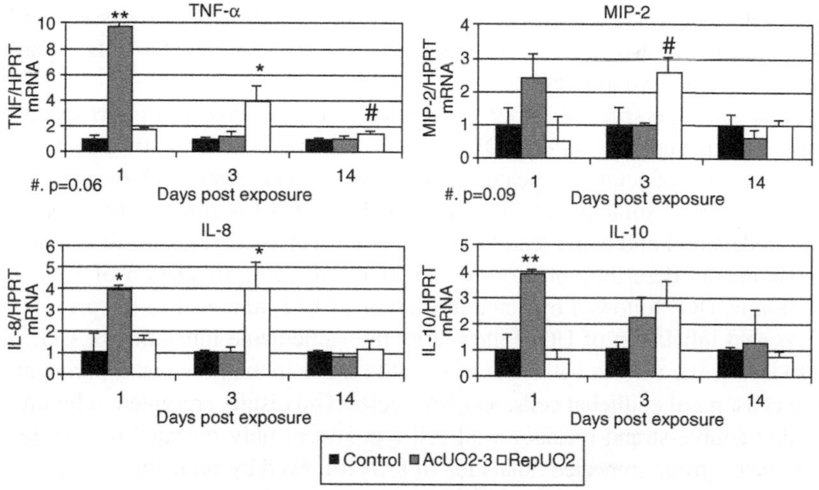

FIGURE 2.17[6] Relative mRNA expression of the cytokines TNF-α, MIP-2, IL-8, and IL-10 in lung tissues at 1, 3, and 14 d post exposure for the group control (n = 5), AcUO2-3 (n = 3) and RepUO2 (n = 3); ** $p < 0.01$, * $p < 0.05$, and # reached a significance indicated on the graph between the exposed and control groups. The results are expressed as a ratio to the mRNA levels of the reference gene hypoxanthine-guanine phosphoribosyltransferase (HPRT). Data are the means ± SEM.

[5,6] Reprinted from Monleau et al., Genotonic and inflammatory effects of depleted uranium particles inhaled by rats. *Toxicological Sciences*, 2006, 89, 287, by permission of Oxford University Press.

inhalation of UO_2 seems to produce the same phenomena, but later on, with an increase in the rates of m-RNA for TNF-α, IL8, and MIP-2 but no IFN-γ, 3 d post exposure. Last, 14 d post exposure, the mRNA rates of the cytokine TNF-α were still high, which suggests a slight persistence in the expression of this cytokine. However, the antiinflammatory cytokine IL-10 did not increase, which could suggest an imbalance in the expression of the inflammatory genes. Gazin et al. found similar results after *in vitro* exposure of the macrophages to uranium, where induction of the accumulation and secretion of TNF-α but not IL-10 was noted [52].

Last, the hydrogen peroxide dose rate increased for several days in the lung tissue after repeated exposure, unlike the acute exposure. This could imply that repeated inhalation does not lead to an adaptive response. All these results therefore suggest that DU could induce oxidative stress.

To summarize, DU inhalations could cause damage to the DNA in various types of cell in rats. In BAL cells, the lesions relate to the dose, independent of the solubility, are linked to the type of inhalation and composed partially of double-strand breaks suggesting that radiation could contribute to the genotoxic effects of DU *in vivo.*

In kidney cells, only repeated exposure could induce DNA strand breaks. Secondly, insoluble DU particles induced time-dependent increases in mRNA levels of different cytokines and in hydroperoxide production in rat lungs; the pattern differs between acute and repeated inhalation exposure. These results suggest that the DNA damage was partly a consequence of the inflammatory processes and ROS production. They also suggest that repeated exposure of insoluble DU particles could induce a potentiation effect.

In the same perspective, other genotoxicity studies have been performed after successive exposure to various DU oxides. The purpose was to determine the genotoxic effects of acute or repeated prior exposure to insoluble UO_2 on a subsequent inhalation of more soluble UO_4. The combined effects of multiple exposures are not clearly understood and could lead to synergic or antagonistic effects. To achieve this, the experiments described previously [29] (namely acute inhalation of UO_4, acute inhalation of UO_2 followed by acute inhalation of UO_4 identical to the previous one and repeated inhalation of UO_2 followed by the same acute inhalation of UO_4) were used to isolate the various cell types identical to those in the previous experiment, i.e., kidney cells, nasal epithelial cells, and BAL cells. The results presented in Figure 2.18 show that double-strand breaks for all cell types were only revealed for the doubly-contaminated group (repeated inhalation of UO_2 followed by acute inhalation of UO_4). The same group also showed DNA damage in neutral comet test conditions, i.e., single- and double-strand breaks (Figure 2.19). This could support the hypothesis put forward earlier on the genotoxic effects of DU radiation.

In addition, the number of BAL cells increases significantly by a factor of 2.5, compared with the control group and the group inhaling UO_4 only. This could reveal a recruitment of inflammatory cells in the lung tissue, suggesting an inflammatory response partially responsible for the DNA damage.

The genotoxic response obtained only from the doubly-contaminated group after repeated inhalation of UO_2 but not acute inhalation of UO_2 (with comparable lung intake) could suggest that the repetition of the inhalations is the basic cause for these effects. This correlates with the results described previously and establishes

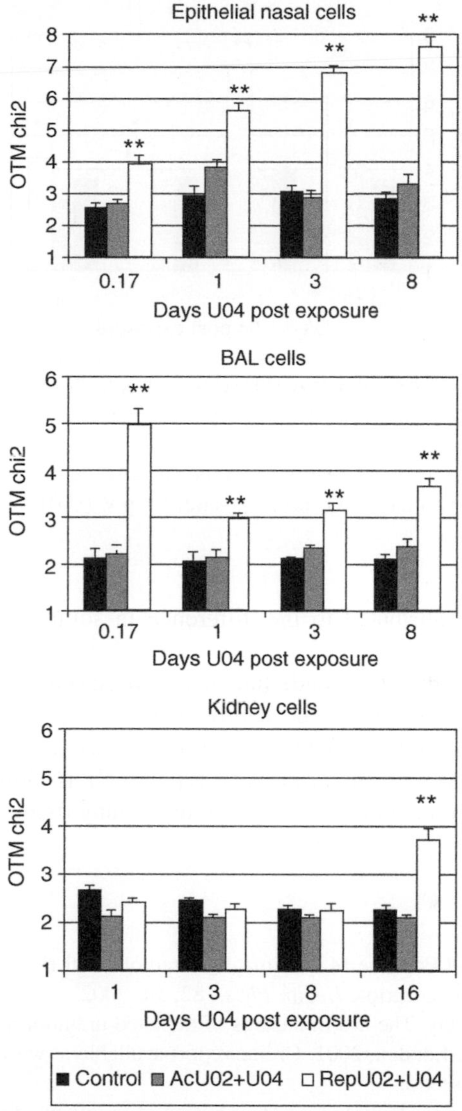

FIGURE 2.18 Comet assay under alkaline conditions in different cell types. Mean of OTMchi2 values are presented in relation to days UO4 post exposure; ** $p < 0.001$ between the control and exposed groups.

the hypothesis of a potentiation of damages noted in the kidney and BAL cells after repeated inhalation.

Unlike the previous study that showed no DNA damage after repeated inhalation of UO_2 or acute inhalation of UO_4, damage was noted after double repeated contamination of UO_2 followed by acute inhalation of UO_4. A synergic effect could therefore appear between UO_2 and UO_4 like the one described [53] for welding

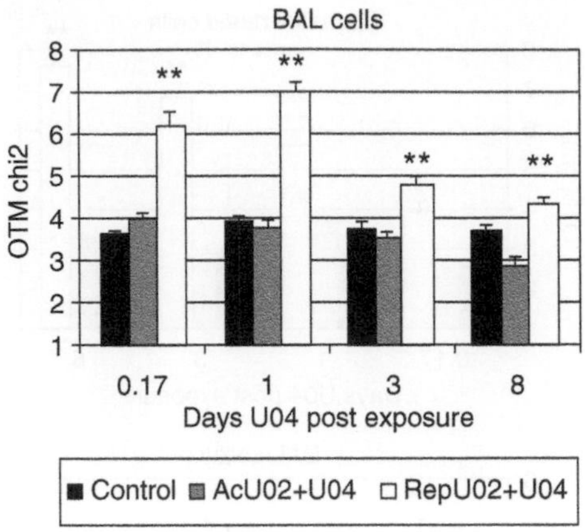

FIGURE 2.19 Comet assay under neutral conditions in BAL cells. Mean of OTMchi2 values are presented in relation to days UO_4 postexposure; ** $p < 0.001$ between the control and exposed groups.

fumes and would be attributed to the difference in solubility of the two oxides inhaled successively.

All these studies therefore underline the importance of taking the type of inhalation, the amount taken in, the nature of the compound, and the exposure history into account when assessing the risk of internal contamination from DU. DU can cause genotoxic effects on various cell types such as kidney, nasal epithelial, and BAL cells, and an inflammatory response in the lung tissue.

REFERENCES

1. Durante, M. and Pugliese M., Estimates of radiological risk from depleted uranium weapons in war scenarios, *Health Phys.*, 82, 14, 2002.
2. The Royal Society, The health hazards of depleted uranium munitions part I. Policy document 6/01, London, 2001. Online report available at www.royalsoc.ac.uk in the science policy section.
3. Hanson, W.C. et al., Particle Size Distribution of Fragments from Depleted Uranium Penetrators Fired against Armor Plate Targets, Report LA-5654, Los Alamos Scientific Laboratory, Los Alamos, 1974.
4. Patrick, M.A. and Cornette, J.C., Morphological Characteristics of Particulate Material Formed from High Velocity Impact of Depleted Uranium Projectiles with Armor Targets, Report AFATL-TR-78-117, Air Force Armament Laboratory, November 1978.
5. Parkhurst, M.A. et al., Depleted Uranium Aerosol Doses and Risks: Summary of US Assessments, Report PNWD-3476, prepared for the U.S. Army by Battelle, 2004.
6. Glissmeyer, J.A. and Mishima, J., Characterization of Airborne Uranium from Test Firings of XM774 Ammunition, Report PNL-2944, Pacific Northwest Laboratory, Richland, WA, 1979.

7. Chambers, D.R. et al., Aerosolization Characteristics of Hard Impact Testing of Depleted Uranium Penetrators, Technical Report ARBRL-TR-023435, Aberdeen Proving Ground, MD: Ballistic Research Laboratory, October 1982.
8. Brown, R., Depleted Uranium Munitions and Assessment of the Potential Hazards. Notes of a presentation to the Royal Society Working Group on January 19, 2000.
9. Scripsick, R.C. et al., Preliminary study of uranium oxide dissolution in simulated lung fluid, *Report LA-10268-MS*, Los Alamos National Laboratory, 1985. http://lib-www.lanl.gov/la-pubs/00318819.pdf.
10. Scripsick, R.C. et al., Differences in in vitro dissolution properties of settled and airborne uranium material, in *Occupational Radiation Safety in Mining*, Stocker, H., Ed., Canadian Nuclear Association 255, 1985. http://lib-www.lanl.gov/la-pubs/00374828.pdf.
11. Depleted Uranium in Kosovo, post conflict environmental assessment, *UNEP*, 157–162, 2001.
12. Larousserie, D., Enquête sur l'U.A, *Sciences et Avenir*, p.105, décembre 2000.
13. Chazel, V. et al., Characterisation and dissolution of depleted uranium aerosols produced during impacts of kinetic energy penetrators against a tank. *Radiat. Prot. Dosimetry*, 105, n° 1-4, 2003.
14. Jarvis, N.S. and Birchall, A., Ludep 2.0. Personal computer program for calculating internal doses using ICRP publication 66 respiratory tract model, *NRPB-SR287*, 1996.
15. Miller, A., Neoplastic transformation of human osteoblast cells to the tumorigenic phenotype by heavy metal-tungsten alloy particles: induction of genotoxic effects, *Carcinogenesis*, 22(1), 115, 2001.
16. Chazel, V. et al., Effect of U_3O_8 specific surface area on in vitro dissolution, biokinetics and dose coefficients, *Radiat. Prot. Dosimetry*, 79, 39, 1998.
17. Chazel, V. et al. Variation of solubility, biokinetics and dose coefficient of industrial uranium oxides according to specific surface area, *Radiat. Prot. Dosimetry,* 88, 223, 2000.
18. ICRP, Dose coefficients for intakes of radionuclides by workers, ICRP Publication 68, *Annals of the ICRP* 24(4), Elsevier Science Ltd., Oxford, 1994.
19. UNSCEAR, Sources and Effects of Ionizing Radiation, Report to the General Assembly, with Annexes (New York: United Nations), 1977.
20. Ansoborlo, E. et al., Determination of the physical and chemical properties, biokinetics and dose coefficients of uranium compounds handled during nuclear fuel fabrication in France, *Health Phys.*, 82, 279, 2002.
21. Monleau, M. et al., The effect of repeated inhalation on the distribution of uranium in rats, *JTHE*, in press.
22. ICRP, Limits for intakes of radionuclides by workers, ICRP Publication 30, Part 1, *Annals of the ICRP* 2(3/4), Pergamon Press, Oxford, 1979.
23. ICRP, Age-dependent doses to members of the public from intake of radionuclides: Part 4 Inhalation dose coefficients, ICRP Publication 71, *Annals of the ICRP* 25(3–4), Elsevier Science Ltd., Oxford, 1995.
24. Birchall, A., and James, A.C., A general algorithm for solving compartmental models with constant coefficients and its implementation on a microcomputer, *NRPB-R216*, 1987.
25. Guglielmotti, M.B. et al., Effects of acute intoxication with uranyl nitrate on bone formation, *Experientia*, 40, 474, 1984.
26. Diaz Sylvester, P.L. et al., Exposure to subcutaneously implanted uranium dioxide impairs bone formation, *Arch. Environ. Health*, 57, 320, 2002.

27. Ubios, A.M. et al., Uranium inhibits bone formation in physiologic alveolar bone modeling and remodeling, *Environ. Res.*, 54, 17, 1991.
28. Paquet, F. et al., Accumulation and distribution of uranium in rats after chronic exposure by ingestion, *Health Phys.* 90, 139, 2006.
29. Monleau, M. et al., Distribution and genotoxic effects after successive exposures to different uranium oxides particles inhaled by rats, *Inhalation Toxicol., in revision,* 2006.
30. Benson, J.M. et al., Effects of Repeated Inhalation Exposure of F344 Rats and B6C3F1 Mice to Nickel Oxide and Nickel Sulphate Hexahydrate on Lung Clearance, LRRI report, Albuquerque, NM, 1992.
31. Benson, J.M. et al., Particles clearance and histopathology in lungs of F344/N rats and B6C3F1 mice inhaling nickel oxide or nickel sulfate. *Fundam. Appl. Toxicol.*, 28, 232, 1995.
32. Fulco, C.E., Liverman, C.T., and Sox, H.C., Depleted uranium, in Fulco, C.E., Liverman, C.T., and Sox, H.C., Eds., *Gulf War and Health, Vol. 1, Depleted uranium, Pyridostigmine bromide, Sarin and Vaccines*, National Academies Press, Washington, DC, 2000, pp. 89–168.
33. Durakovic, A., Medical effects of internal contamination with uranium, *Croat. Med. J.*, 42, 49, 1999.
34. Durakovic, A. et al., Estimate of time zero lung burden of depleted uranium in Persian Gulf War veterans by the 24-hours urinary excretion and exponential decay analysis, *Mil. Med.*, 168, 600, 2003.
35. Salbu, B. et al., Oxidation states of uranium in depleted uranium particles from Kuwait, *J. Environ. Radioact.*, 78, 125, 2005.
36. Monleau, M. et al., Bioaccumulation and behavioural effects of depleted uranium in rats exposed to repeated inhalations, *Neuro. Lett.*, 390, 31, 2005.
37. Tjalve, H. et al., Uptake of manganese and cadmium from the nasal mucosa into the central nervous system via olfactory pathways in rats, *Pharmacol. Toxicol.*, 79, 347, 1996.
38. Henriksson, J., Tallkvist, H., and Tjalve, H., Uptake of nickel into the brain via olfactory neurons in rats, *Toxicol. Lett.*, 91, 153, 1997.
39. Henriksson, J., and Tjalve, H., Uptake of inorganic mercury in the olfactory bulbs via olfactory pathways in rats, *Environ. Res.*, 77, 130, 1998.
40. Persson, E. et al., Transport and subcellular distribution of intranasally administered zinc in the olfactory system of rats and pikes, *Toxicology*, 191, 97, 2003.
41. Persson, E., Henriksson, J., and Tjalve, H., Uptake of cobalt from the nasal mucosa into the brain via olfactory pathways in rats, *Toxicol. Lett.*, 145, 19, 2003.
42. Rao, D.B. et al., Inhaled iron, unlike manganese is not transported to the rat brain via the olfactory pathway, *Toxicol. Appl. Pharmacol.*, 193, 116, 2003.
43. Dorman, D.C. et al., Influence of particle solubility on the delivery of inhaled manganese to the rat brain: manganese sulfate and manganese tetroxide pharmacokinetics following repeated (14-d) exposure, *Toxicol. Appl. Pharmacol.*, 170, 79, 2001.
44. Pellmar, T.C. et al., Distribution of uranium in rats implanted with depleted uranium pellets, *Toxicol. Sci.*, 49, 29, 1999.
45. Pellmar, T.C. et al., Electrophysiological changes in hippocampal slices isolated from rats embedded with depleted uranium fragments, *Neurotoxicology*, 20, 785, 1999.
46. Barber, D.S., Ehrich, M.F., and Jortner, B.S., The effect of stress on the temporal and regional distribution of uranium in rat brain after acute uranyl acetate exposure, *J. Toxicol. Environ. Health*, 68, 99, 2005.

47. Houpert, P. et al., Enriched uranium but not depleted uranium affects central nervous system in long-term exposed rat, *Neurotoxicology*, 26, 1045, 2006.
48. ATSDR, Toxicological Profile for Uranium, Public health service, agency for toxic substances and disease registry, U.S. Department of health and human services, Atlanta, GA, 1999.
49. Miller, A., Potential late health effects of depleted uranium and tungsten used in armor-piercing munitions: comparison of neoplastic transformation and genotoxicity with the known carcinogenic nickel, *Mil. Med.*, 167, 120, 2002.
50. Miller, A. et al., Genomic instability in human osteoblast cells after exposure to depleted uranium: delayed lethality and micronuclei formation, *J. Environ. Radioact.*, 64, 247, 2003.
51. Monleau, et al., Genotoxic and inflammatory effects of depleted uranium particles inhaled by rats, *Toxicol. Sci.*, 89, 287, 2006.
52. Gazin, V. et al., Uranium induces TNF alpha secretion and MAPK activation in a rat alveolar macrophage cell line, *Toxicol. Appl. Pharmacol.* 194, 49, 2004.
53. Taylor, M.D. et al., Effects of welding fumes of differing composition and solubility on free radical production and acute lung injury and inflammation in rats, *Toxicol. Sci.,* 75, 181, 2003.

3 Carcinogenesis of Depleted Uranium: Studies in Animals

Fletcher F. Hahn

CONTENTS

INTRODUCTION

Interest in Depleted Uranium

A number of U.S. veterans of the Operation Desert Storm (ODS) have fragments of depleted uranium (DU) metal containing 0.75% titanium (Ti) embedded in their tissues. Several dozen of these veterans are in a medical surveillance program of the U.S. Department of Veterans Affairs to detect any untoward health effects (1). Some had elevated concentrations of uranium in the urine 12 years after wounding. The persistence of these elevated urine concentrations suggested an ongoing mobilization of uranium from the embedded fragments, which resulted in a chronic systemic exposure. This observation indicates that DU fragments in the tissues are not inert foreign bodies and may react differently in the body compared with other embedded shrapnel.

Concern has also been expressed for military personnel exposed to aerosols of DU. The Capstone Program of the U.S. Army was developed to address such concerns (2). The program required the characterization of the DU aerosols collected in Abrams tanks and Bradley fighting vehicles after perforation by DU munitions. In a second phase of the program, the radiation and chemical doses of DU to critical organs of the body of individuals in various exposure categories were calculated and estimates of health risk determined. These risk estimates indicate that the only effect of high concentrations of DU after a perforation is possible renal damage. Cancer from this type exposure is not predicted.

Early Studies of Natural Uranium

The chemical toxicity of uranium compounds is well known when compared to the toxicity of most other compounds. In 1824, a treatise described uranium salts as "feeble poisons" when given by mouth to animals. In the late 1800s, uranium salts were used as homeopathic therapeutic agents in humans, primarily for treatment of diabetes. In the early 1900s, the renal toxicity of uranium became apparent in humans, and the use as a therapeutic agent ceased (3).

Toxicity studies of uranium compounds were initiated during World War II as nuclear weapons were developed. From studies in animals it became clear that the amount of soluble uranium salts depositing and remaining in the lung or bone would never constitute a sufficient radiation hazard to override the chemical toxicity to the kidney. Based on these studies in animals a threshold concentration of 3 μgU/g kidney and a limiting air concentration of 50 μg/m^3 were recommended for soluble uranium (4).

ACCUMULATION OF DEPLETED URANIUM IN THE BODY

Exposure to Uranium Compounds

For the general population, inhalation and ingestion are the primary ways for uranium to enter the body. Uranium is ubiquitous in the environment and is present in all soils and rocks in the form of a variety of minerals. Trace amounts of uranium are found in all foods. Drinking water is also a source of uranium and may contribute more than food to the human intake. The quantity of uranium in drinking water varies widely among locales (5).

Industrial processes, such as mining or milling of uranium ore and nuclear manufacturing facilities, can increase the uranium concentrations present in the air, resulting in potential occupational exposures to uranium (6).

For military personnel, wounding with DU fragments is another potential route of exposure to DU (7). The DU fragments, because of their size and mass, can serve as a source of U in the body for many years. For example, some of the ODS veterans wounded with DU fragments were still excreting depleted uranium 8 years after wounding. In addition, the estimated renal concentration of U is still increasing in some (8).

Distribution of Depleted Uranium Among Organs

The three isotopes of uranium, all radioactive, occur together in both natural and depleted uranium: U-238, U-234, and U-235. The chemistry of these isotopes, which is identical, determines the reactions of the isotopes within the environment as well their transport and reactions within the body.

After deposition in the lung (9) or in wound sites (10), the movement or translocation of uranium depends primarily on the solubility of the uranium compound. Relatively insoluble compounds may be retained in the lung or a wound with a half-time of years. After absorption to blood from the lungs, a wound site or intestines, uranium is deposited systemically or is excreted by the kidneys. A substantial fraction of the metal ions filtered by the kidneys is retained in the renal tubules before it is passed into the urinary bladder. Over 90% of the uranium remaining in systemic tissues at 1 d is excreted with half-time ranging from 2 to 6 d and the remainder with half-times ranging from 30 to 340 d. After a few days, most of the remaining uranium in the body is found in the kidneys, skeleton and, in the case of insoluble compounds, the site of entry (lung or wound).

STUDIES OF DEPLETED URANIUM CARCINOGENESIS IN ANIMALS

Short-Term Studies of Embedded Depleted Uranium Fragments

Short-term studies in rats have provided evidence that embedded DU fragments may be carcinogenic. Rats implanted in the muscles with 20 DU pellets (2.0 × 1.0 mm diameter) excreted uranium in the urine for at least 18 months (11). Urine from these rats had enhanced mutagenic activity in *Salmonella typhimurium* strain TA98 and the Ames II™ mixed strains (TA7001-7006) (12). The mutagenicity increased in a dose- and time-dependant manner with a strong positive correlation with urinary uranium concentration.

Life-Span Studies of Embedded Depleted Uranium Fragments

A lifetime study in laboratory rats has confirmed that embedded DU fragments are carcinogenic in these animals (13). The carcinogenicity of DU metal, containing 0.75% titanium, implanted in muscle tissues was determined using a lifespan bioassay in rats. Tissue reactions were compared to a nonradioactive metal tantalum (Ta), as a foreign body control, and to a colloidal suspension of radioactive thorium dioxide, Thorotrast®, as a positive control. DU was surgically implanted in the thigh muscles of male Wistar rats, as four squares (2.5 × 2.5 × 1.5 mm or 5.0 × 5.0 × 1.5 mm) or four pellets (2.0 × 1.0 mm diameter) per rat. Ta was similarly implanted, as four squares (5.0 × 5.0 × 1.1 mm) per rat. Thorotrast® was injected at two muscle sites of each rat. Control rats had only a surgical implantation procedure. Each of the six groups included 50 rats. The rats were observed for their lifespan and given a complete necropsy at death. Details of the methods have been reported (13).

At the time of implantation, the fragments were smooth squares with regular, sharp, well-defined edges as indicated by radiography (Figure 3.1A). At 21 d after implantation, small, dense blebs extended from the edges of the fragments, making them appear larger than at time of implantation. The jagged appearance disrupted the sharp profile of the fragments (Figure 3.1B). At 1 year after implantation, the radiographic profiles of the fragments were rounded, with no corners and a fine, jagged edge (Figure 3.1C). At the time of death, many of the profiles were enlarged

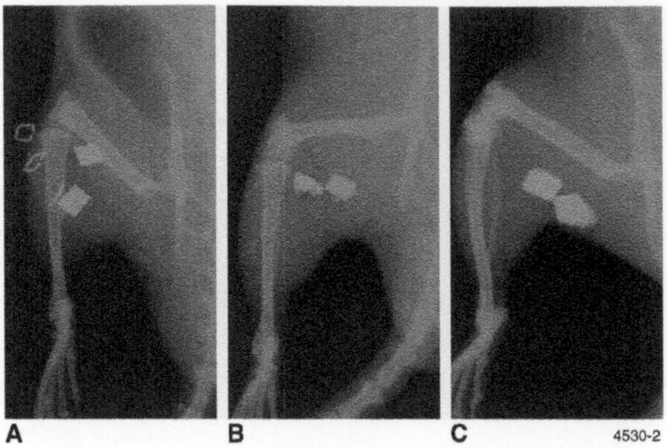

FIGURE 3.1 Radiograph of 5.0 × 5.0 × 1.5 mm DU fragments in a rats (A) on day of implantation with radio-opaque wound clips visible on the skin; (B) 3 weeks after implantation; (C) 1 year after implantation.

up to 1.5 times in linear dimensions. These radiographic changes were related to surface corrosion of the DU *in vivo* and tissue capsule formation around the implant.

In contrast, the radiographic profiles of the Ta fragments were smooth at all times with sharp, well-defined edges and did not increase in size with time. Initially, the Thorotrast® injections gave a spherical radiographic outline. At 4 weeks the profiles were irregular and diffuse, with no distinct boundary and were similar in appearance at 1.5 years after injection.

The histologic reaction to the DU implants were shown in a separate set of rats implanted with 2.5 × 2.5mm squares and sacrificed 1 week and 1, 2, 4, 6, 9, 12, and 18 months after implant. Corrosion and inflammation occurred as early as 7 d after implant. Inflammation was moderate to marked through 1 month and then waned but was still present at death. Corrosion was minimal at 1 week but increased to moderate or marked by 18 months and at death. Fibrosis forming a capsule was minimal at 1 week but increased in severity with time. The increased severity of fibrosis was related to the size and degree of corrosion of the implant.

At death, the DU implants were encapsulated by a dense fibrotic tissue with inflammation, degeneration, and mineralization (Figure 3.2A). The capsules were 0.1 to 0.2 mm thick. Shards of black, presumably oxidized DU were embedded in the fibrous tissue capsules around the larger fragments. The amount of chronic inflammation in and around the capsules was generally similar with DU implants of all sizes. Degeneration of the fibrous tissue in the capsule wall was frequent at the interface with the implant. With this reaction, the tissue on the inner surface of the capsule was devitalized, necrotic, and occasionally mineralized. The lumen of the capsules contained necrotic and proteinaceous debris, scattered acute inflammatory cells, and varying amounts of black shards or particles.

At death, the Ta implants were encapsulated by a thin fibrous capsule with scant inflammation and no degeneration or mineralization (Figure 3.2B). The capsule walls were less than 0.1 mm thick with a smooth inner surface. No shards or particles of corroded metal were present.

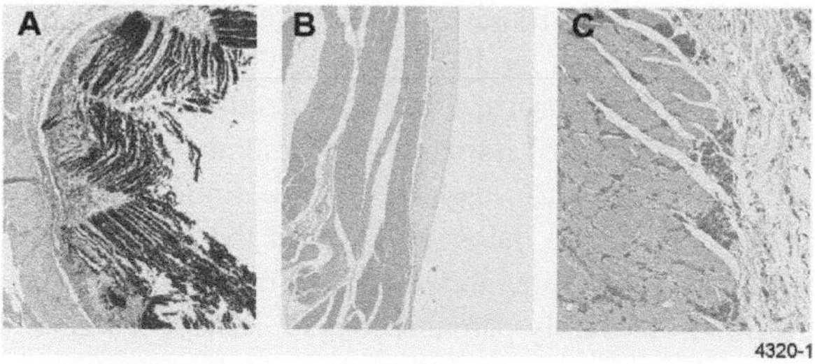

FIGURE 3.2 (See color insert following page 80) (A) DU fragment: Black shards of corroded DU lined the thick cellular fibrotic capsule surrounding the fragment 520 d after implantation; (B) Ta fragment: A thin acellular fibrotic capsule with no metal pieces present 603 days after implantation; (C) Thorotrast® injection: no capsule or inflammation; Thorotrast®-laden macrophages infiltrated skeletal muscle 792 d after injection.

The Thorotrast® lesion was an accumulation of macrophages between muscle fibers and adjacent to muscles (Figure 3.2C). The macrophages were filled with a tan, coarsely granular material, presumably Thorotrast®. These macrophages were not associated with inflammation or fibrosis, and there was no capsule formation.

A total of 40 soft tissue tumors of various types were associated with many of the implants (Table 3.1). The most commonly found tumor types were malignant fibrous histiocytoma and fibrosarcoma. Three osteosarcomas were identified but were not associated with the skeleton. Although there was a higher number of fibrosarcomas in the Thorotrast®-treated rats and a broader range of tumors in the DU-treated rats, a specific tumor type could not be correlated with a specific treatment. Based on their morphology, all of the tumors were most likely derived from primitive mesenchymal stem cells (14).

TABLE 3.1
Implant-Associated Soft Tissue Tumor Types

	Thorotrast®	Depleted Uranium	Tantalum	Surgical Control
		Benign Tumors		
Benign fibrous histiocytoma	0	1	0	0
Fibroma	0	1[a]	0	0
Granular cell myoblastoma	1	0	0	0
		Malignant Tumors		
Malignant fibrous histiocytoma	13	7[a]	2	0
Fibrosarcoma	10	2	0	0
Osteosarcoma	1	2	0	0
Total	25	13	2	0

[a]One rat had two tumors associated with separate implants.

All tumors were in the soft tissues of the hind legs, directly associated with the implanted DU or Ta fragments or the injected Thorotrast®. Three tumors were localized to the wall of the fibrous capsules surrounding the implants. In other tumors, black shards, particles of implanted fragments, or Thorotrast®-filled macrophages could be seen scattered through most of the tumor tissues. These histologic findings lend further credence to the association of the implants with the tumors.

Biologically, these tumors were moderately aggressive. Many were rapidly growing, expanding in size from barely palpable to 3 or 4 cm in 2 weeks. However, none invaded bone and only one ulcerated the skin. Only two tumors metastasized. Twenty-nine of the 40 were large enough to result in euthanasia. The other 11 were discovered at necropsy or in tissue sections of the capsules surrounding the implants.

The incidence of the tumors was increased in the rats with the largest DU implants when compared with the sham or foreign-body (Ta) controls (Figure 3.3). The difference was significant ($p \leq 0.03$) using a Fisher's exact test. The radioactive-material control animals, injected with Thorotrast®, had a significant increase in number of tumors compared with the DU-implanted rats ($p < 0.001$).

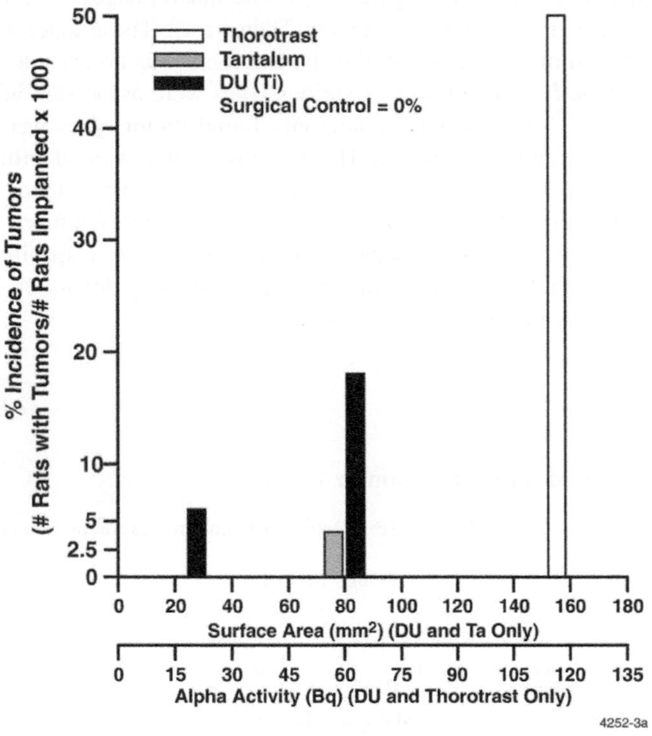

FIGURE 3.3 Incidence of soft tissue tumors in rats implanted with DU, Ta, or Thorotrast®. The incidence of tumors in rats with 5.0 × 5.0 mm DU fragments was significantly increased over the incidence in rats with Ta fragments of the same size. The positive control, Thorotrast®, produced the highest incidence of tumors. Thorotrast® > DU 5.0 × 5.0 mm, $p = 0.0014$; DU 5.0 × 5.0 mm > sham control, $p = 0.0012$; DU 5.0 × 5.0 mm > Ta, $p = 0.028$; Fischer's exact test.

In addition, increased tumor incidence was related to increased fragment size in the DU-treated rats. The response could not be explained by physical surface area alone because the tumor incidence was much lower with the Ta implants of similar size. There was a correlation with the initial surface alpha radioactivity (Figure 3.3). The initial activity was calculated from the physical characteristics of the DU fragments and the Thorotrast® colloid. The injected colloid was considered a sphere for calculation purposes. Radiographs showed that the physical shape of the fragments and the colloid changed within 4 weeks after implantation. Thus, the surface alpha radioactivity changed with time as the shape of the implants changed.

Comparison of the radiographs of the DU-associated lesions with the histologic appearance of the lesions showed a correlation. After being implanted for a year or more, all of the DU fragments were enlarged and rounded in radiographic profile (Figure 3.4A). These radiographic features generally correlated with a dense,

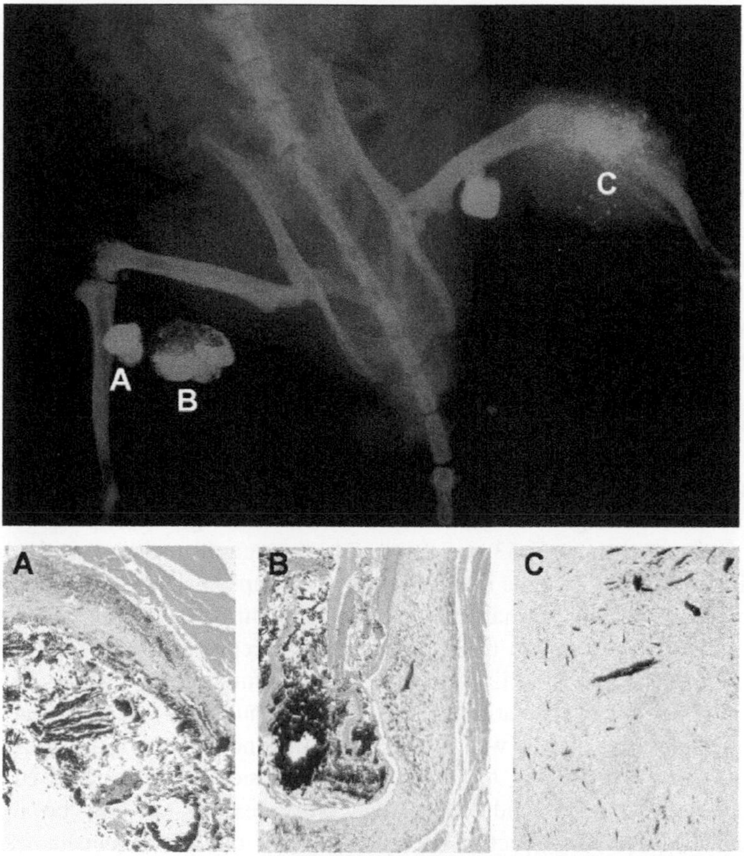

FIGURE 3.4 Correlation of radiographic appearance with histologic appearance: (A) thick fibrotic capsule with shards of corroded DU in lumen; (B) thick cellular capsule lined by squamous metaplasia, particles, and shards of corroded DU in wall and lumen; (C) particles and shards of disintegrated DU fragment scattered throughout a soft tissue sarcoma.

connective tissue capsule. However, disruption of the smooth-edge profile and focal loss of density in the DU fragment were associated with proliferative lesions or small tumors in the capsules (Figure 3.4B). Disintegration and breakup of the DU fragment were apparent on radiographs when frank tumors were present. On the histologic sections, black shards of DU could be seen throughout the tumor tissue (Figure 3.4C). These radiographic changes associated with the DU fragments may be important indicators of accompanying proliferative lesions and may have prognostic value in clinical evaluations.

It can be concluded from this study that DU fragments of sufficient size cause local tissue reactions that can be detected with radiography of the muscles of rats. The tissue reactions can also lead to localized soft tissue sarcomas in these rats. Whether or not they are carcinogenic for humans is not clear.

Mechanistic Studies of DU Fragment Carcinogenesis

Materials carcinogenic in rats are frequently, but not always, carcinogenic in humans. Exceptions may occur, particularly when the mechanism of carcinogenesis in rats is not operative in humans (15). Thus, the mechanism by which DU induces tumors takes on added importance. If DU is considered to induce tumors by a foreign-body mechanism, then soft tissue tumor incidence for foreign bodies should be used in extrapolations. If a radiation mechanism predominates, then soft tissue tumor incidence for radioactive compounds, such as Thorotrast®, should be used for extrapolations. If yet another mechanism is involved, other tumor databases may be required in extrapolation.

Gene changes in the tumors induced by DU and Thorotrast®, a radioactive compound, were compared looking for a similarity that might indicate a radiation mechanism for DU carcinogenesis. The specific gene changes sought were the expression of p53, MDM2, p21 or c-myc, and the mutation of K-ras. P53 was chosen because of its relatively high frequency of mutation in human soft tissue sarcomas and in rodent soft tissue sarcomas induced with materials placed subcutaneously. The regulation of cell growth by p53 depends on activating transcription of MDM2 by p53. Overexpression of p53 was detected by immunohistochemistry (16). Overexpression of the MDM2 gene interferes with transcriptional activation of wild-type p53, thereby abrogating normal p53 function. In addition to this close correlation, several studies of human soft tissue sarcomas have reported the overexpression of MDM2. Expression of MDM2 was detected by immunohistochemistry (17). P21 protein is important in regulation of the cell cycle and falls in the p53 activation pathway. Expression of p21 was detected by immunohistochemistry (Santa Cruz Biotechnology, Santa Cruz, CA). The c-myc oncogene was selected based on evidence from both humans and rodents that c-myc expression may be an aid in separating the radiation-induction mechanism from other mechanisms of cancer induction C-myc expression was detected by immunohistochemistry (Santa Cruz Biotechnology). *In vitro* studies with DU suggest that K-ras may be important in carcinogenesis. For example, DU causes transformation of human osteoblasts and expression of high levels of K-ras by these cells (18). In rats, pellets of DU embedded in the muscles caused increased levels of K-ras in the muscles (19). K-ras codon 12

TABLE 3.2
Frequency[a] of Gene Changes in Soft Tissue Tumors Induced by DU or Thorotrast®

	p53 Expression	MDM2 Expression	c-myc Expression	p21 Expression	K-ras Mutation
DU	3/9[b]	1/8	0/8	0/8	0/6
Thorotrast®	0/18	2/17	2/20	0/21	0/20
Tantalum	0/1	1/1	1/1	1/1	0/1

[a] Number with positive expression/number examined.
[b] Significantly different from Thorotrast® ($p < 0.03$ Fisher's exact test).

mutations were detected by a restriction fragment length polymorphism analysis using the B*st*N1 test. (16).

The results of the IHC determinations of gene expression of p53, MDM2, and c-myc, as well as the B*st*N1assay for K-ras codon 12 mutations are shown in Table 3.2. Overexpression of p53, indicative of p53 gene mutation, occurred in 3/11 (27%) of the DU-induced malignant soft tissue tumors and was significantly increased over the incidence in the Thorotrast®-induced soft tissue tumors ($p < 0.03$ two-tailed Fisher's exact test).

The expression of MDM2, p21, and c-myc were 10% or less of the malignant tumors tested in any of the groups. No mutation in K-ras codon 12 was noted in any of the tumors tested. This lack was not expected as K-ras mutations are the most frequent mutations in rodent tumors and mutations of codon 12, the most frequent of the k-ras mutations. To verify the BstN1 result, codons 12 and 13 were sequenced but mutations were found.

The tissue reactions to DU implants in rats generally follow the time course of inflammation, wound healing, and foreign-body response as classically outlined (20). With DU the chronic inflammation and fibrosis follows the acute inflammation relatively rapidly and continues. Although persistent, the inflammation is not particularly severe. The feature of degeneration with subsequent mineralization is not one included in the classical description of foreign body responses. An important phenomenon following implantation of DU in the tissues is the marked corrosion and fragmentation of the DU that occurs within days and continues for the lifespan of the rats. Corrosion may well cause a redox reaction as the interface of the implant leading to degeneration of the adjacent tissue (21). This phenomenon may also play a role in the persistent low-grade inflammation and the production of the fibrous capsule.

The tissue reactions to DU implants differ from the reactions to the Ta implants and the injections of Thorotrast®. The DU reactions have markedly more inflammation and fibrosis than do the reactions to the radioactive material, Thorotrast®, or the foreign-body, Ta implants. The differences in tissue reactions may well be related to different mechanism for carcinogenesis. The finding of a significant increase in the frequency of p53 overexpression in the tumors induced by the DU compared with Thorotrast® is evidence that the mechanism for carcinogenesis for these two compounds is not the same.

CONCLUSIONS

The results of these studies demonstrate that DU embedded in the muscles of rats causes local soft tissue tumors. In addition, evidence is provided that the mechanism for the tumor induction is not related to the radioactivity of the DU but to its corrosion in the tissues that results in prolonged inflammation and fibrosis.

REFERENCES

1. McDiarmid, M.A., Engelhardt, S.M., Oliver, M., Gucer, P., Wilson, P.D., Kane, R., Kabat, M., Kaup, B., Anderson, L., Hoover, D., Brown, L., Albertini, R.J., Gudi, R., Jacobson-Kram, D., Thorne, C.D., and Squibb, K.S. Biological monitoring and surveillance results of Gulf War I veterans exposed to depleted uranium. *Int. Arch. Occup. Environ. Health*, 2, 1–11, 2005.
2. Parkhurst, M.A., Daxon, E.G., Lodde, G.M., Szrom, F., Guilmette, R.A., Roszell, L.E., Falo, G.A, and McKee, C.B. *Depleted Uranium Aerosol Doses and Risks: Summary of U.S Assessments.* Battelle Press, Columbus, OH, 2005.
3. Hodge, H.C. A history of uranium poisoning (1824–1942). In Hodge, H.C., Stannard, J.N., and Hursh, J.B., Eds., *Uranium, Plutonium, Transplutonium Elements: Handbook of Experimental Pharmacology.* Vol. 36, Springer-Verlag, New York, 1973, chap. 1.
4. Spoor, N.L. and Hursh, J.B. Protection criteria. In Hodge, H.C., Stannard, J.N., Hursh, J.B. Eds., *Uranium, Plutonium, Transplutonium Elements: Handbook of Experimental Pharmacology.* Vol. 36, Springer-Verlag, New York, 1973, chap. 3.
5. Hahn, F.F. and Guilmette R.A. Uranium. In *Encyclopedia of Toxicology*, Vol. 4, P. Wexler (Ed. in chief), 2nd ed., Elsevier Science, Oxford, U.K., 2005, pp. 406–409.
6. West, C.M., Scott, L.M., and Schultz, N.B. 16 years of uranium personnel monitoring experience — in retrospect. *Health Phys.* 17, 781–791, 1979.
7. OSAGWI (Office of the Special Assistant for Gulf War Illness) Depleted Uranium in the Gulf (II). U.S. Department of Defense, OSAGWI, Washington, DC, 2000.
8. Gwiazda, R.H., Squibb, K., McDiarmid, M.A., and Smith, D. Detection of depleted uranium in urine of veterans from the 1991 Gulf War. *Health Phys.* 86, 12–18, 2004.
9. Morris, K.J., Khanna, P., and Batchelor, A.L. Long-term clearance of inhaled UO2 particles from the pulmonary region of the rat. *Health Phys.* 58, 477–485, 1990.
10. Leggett, R.W. and Pellmar, T.C. The biokinetics of uranium migrating from embedded DU fragments. *J. Environ. Radioact.* 64, 205–225, 2003.
11. Pellmar, T.C., Fuciarelli, A.F., Ejnik, J.W., Hamilton, M., Hogan, J., Strocko, S., Emond, C., Mottaz, H.M., and Landauer, M.R. Distribution of uranium in rats implanted with depleted uranium pellets. *Toxicol. Sci.* 49, 29–39, 1999.
12. Miller, A.C., Fuciarelli, A.F., Jackson, W.E., Ejnik, E.J., Emond, C., Strocko, S., Hogan, J., Page, N., and Pellmar, T. Urinary and serum mutagenicity studies with rats implanted with depleted uranium or tantalum pellets. *Mutagenesis* 13, 643–648, 1998.
13. Hahn, F.F., Guilmette, R.A., and Hoover, M.D. Implanted depleted uranium fragments cause soft tissue sarcomas in the muscles of rats. *Environ. Health Perspect.* 110, 51–59, 2002.
14. Brooks, J.J. The significance of double phenotypic patterns and markers in human sarcomas: a new model of mesenchymal differentiation. *Am. J. Pathol.* 125, 113–123, 1986.

15. MacDonald, J.S., Scribner, H.H. The maximum tolerated dose and secondary mechanisms of carcinogenesis. In K.T. Kitchin, Ed. *Carcinogenicity: Testing, Predicting and Interpreting Chemical Effects*. Marcel Dekker, Inc., New York, 1999, pp. 125–144.
16. Nickell-Brady, C., Hahn, F.F., Finch, G.L., and Belinsky, S.A. Analysis of K-ras, p53, and c-raf-1 mutations in beryllium-induced rat lung tumors. *Carcinogenesis* 15: 257–262, 1994.
17. Swafford, D.S., Nikula, K.J., Mitchell, C.E., and Belinsky, S.A. Low frequency of alterations in p53, K-ras, and mdm2 in rat lung neoplasms induced by diesel exhaust or carbon black. *Carcinogenesis* 16: 1215–1221, 1995.
18. Miller, A.C., Blakely, W.F., Livengood, D., Whittaker, T., Xu, J., Ejnik, J.W., Hamilton, M.M., Parlette, E., St. John, T., Gerstenberg, H.M., and Hsu, H., Transformation of human osteoblast cells to the tumorigenic phenotype by depleted uranium-uranyl chloride. *Environ. Health Perspect.* 106: 465–471, 1998.
19. Miller, A.C. Depleted uranium health effects: transformation, mutagenicity and carcinogenicity. In *Health Effects of Embedded Depleted Uranium Fragments,* Livengood, E.R., Ed. AFRRI Special Publication 93-3, Bethesda, MD, 1988, pp. 11–13.
20. Anderson, J.M., Gristina, A.G., Hanson, S.R., Harker, L.A., Johnson, R.J., Merritt, K., Naylor, P.T., and Schoen, F.J. Host reactions to biomaterials and their evaluation. In Ratner, B.D., Hoffman, A.S., Schoen, F.J., and Lemons, E., Eds. *Biomaterials Science: An Introduction to Materials in Medicine.* Academic Press, San Diego, CA, 1996, pp. 165–214, chap. 4.
21. Steinemann, S.G. Metal implants and surface reactions. *Injury* 27(Suppl. 3): S-C16-22, 1966.

4 Neurotoxicology of Depleted Uranium in Adult and Developing Rodents

Wayne Briner

CONTENTS

GENERAL PHYSIOLOGIC EFFECTS OF DU IN THE CNS

Uranium (U) and depleted uranium (DU) are only now being investigated in detail for their potential neurotoxicity. It is reasonable to assume that detailed studies of the clinical neurotoxicity of uranium are lacking because the kidney is historically considered to be the target organ of clinical uranium toxicity. The resultant renal failure and tumor formation from uranium poisoning were more important in uranium workers than potential CNS effects. The entry of DU into the environment, and the potential for chronic low-dose exposure to a slightly radioactive form of uranium, is shifting research to examine subclinical effects, with the brain a likely target. DU and native U are chemically identical. It is suspected that the neurotoxicity of DU is largely from its chemical activity.

An early study by Domingo et al. [1] produced tremors in rats give high doses of uranium. Work by Lin et al. [2] at the neuromuscular end plate suggests that U may compete with calcium at the synapse. Evidence that DU crosses the blood brain barrier and accumulates in the brain comes from work done by Pellmar et al. [3] Recent work by Barber et al. [4] demonstrated that DU accumulates preferentially in particular brain regions. Specifically, the hippocampus and striatum accumulate DU more readily than cerebellum and cortex. Recently, work by Monleau et al. [5]

using a DU dust exposure protocol, demonstrated the accumulation of DU in the CNS with olfactory bulb, hippocampus, cortex, and cerebellum demonstrating the accumulation of DU, in that order.

DU not only accumulates in the CNS but also has physiologic activity there. In an important study Pellmar et al. [6] demonstrated that implanted DU pellets inhibited spike formation in the hippocampus of rats with no evidence of renal damage. Lestaevel et al. [7,8] have demonstrated that exposure to DU in drinking water alters the electroencephalographic architecture of the EEG in free moving rats with accompanying changes in the sleep wake cycle and REM sleep. Weight gain was also affected. In addition to documenting the accumulation of DU in the brain via inhalation. Monleauís group [5] also demonstrated changes in the behavior of DU-exposed rats in the open-field and the Y maze, implying that DU has neurophysiologic effects. Another study examined the exposure of rats to either DU or enriched uranium did not find altered sleep wake cycles or changes in spatial activity associated with DU exposure [9].

ADULT RAT AND MOUSE STUDIES, BEHAVIOR, AND LIPID OXIDATION

Work in this laboratory has focused on behavioral effects of exposure to DU in drinking water and the manner in which oxidation of CNS lipids may mediate those effects. Our studies indicate that DU exposure produces behavioral changes that are dependent both on the exposure dosage and the length of exposure. There also appears to be important gender differences in response to DU exposure.

Behavioral differences between DU exposed and control animals imply that DU enters the CNS and exerts some as of yet unknown physiologic effect or effects on neuronal functioning. One possible cause for this effect is lipid oxidation. The oxidation of lipids is known to be one mechanism of action for a variety of metals [10]. In our studies we examined the degree of lipid oxidation using the thiobarbituric acid (TBA) assay [11].

Dosage is, as would be expected, important in the organisms response to DU. Mice exposed to DU at doses of 0, 19, 37, or 75 mg/l for 2 weeks demonstrated greater activity in the open-field maze following an inverted U-shaped dose-response curve. Assessment of the animals' reactivity to novel stimuli and visual placing ability were also dose dependent, with higher dosages demonstrating greater differences from control. Oxidation of CNS lipids also occurred in DU-exposed animals in a dose-dependent fashion (Figure 4.1). Interestingly, the degree of lipid oxidation was correlated with the reactivity and visual placing behavior seen for these animals [12]. Another study using rats exposed at doses of 0, 75, or 150 mg/l of DU demonstrated dose dependent alterations in open-field behavior, weight gain, and lipid oxidation. In a manner similar to the study with mice, lipid oxidation was correlated with open-field behaviors (Figure 4.2) [13].

Length of exposure to DU is another variation of dosage, except that homeostatic mechanisms may be allowed to compensate for some of the effects of exposure. In a simple time-exposure model, rats exposed to DU at 0 or 75 mg/l for either 2 weeks

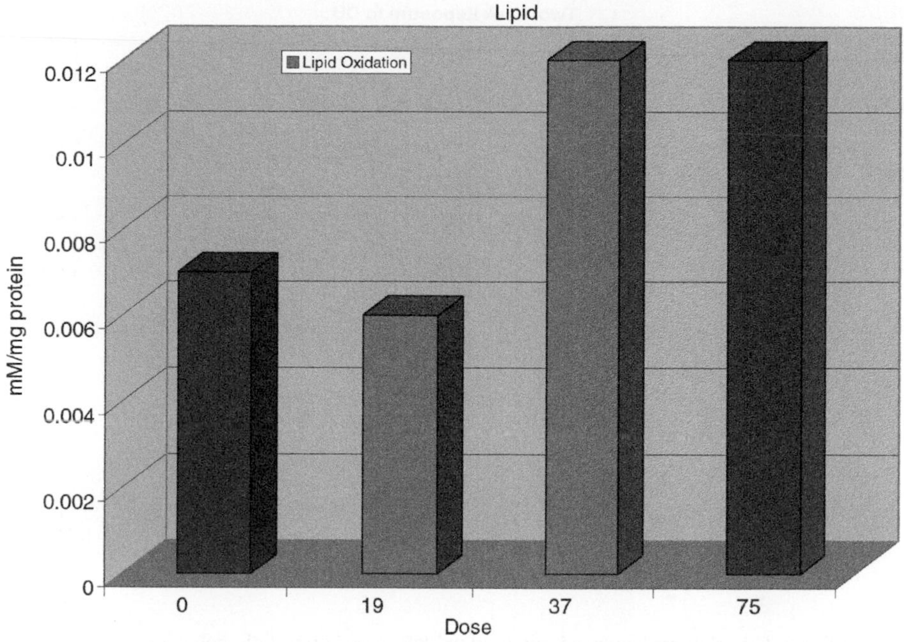

FIGURE 4.1 Oxidation of brain lipids in mice exposed to DU in drinking water for 2 weeks. The horizontal axis indicates dosage of DU. The vertical axis indicates amount of oxidation compared to the standard MDA. (From Briner, W. and Davis, D. Lipid oxidation and behavior are correlated in depleted uranium exposed mice. In *Metal Ions in Biology and Medicine*, Khassanova, L. et al., Eds. John Libby Eurotext, Paris, 2002, pp. 59–63. With permission of authors.)

or 6 months demonstrated increases in open-field behavior that was directly related to the length of exposure; animals exposed for 6 months demonstrated more activity in the open-field than animals exposed for 2 weeks or controls. Both exposed groups demonstrated greater activity than the control animals [14]. Length of exposure was also an important factor that interacted with dose in a study where rats were exposed to DU at 0, 75 or 150 mg/l for either 2 weeks or 6 months. In general, the longer the length of exposure the greater an effect there was on open-field behavior. However, rats exposed to DU for 6 months demonstrated less lipid oxidation than those exposed for 2 weeks. In addition, the correlations between open-field behavior and lipid oxidation seen at 2 weeks of exposure were no longer apparent after 6 months of exposure [13]. This suggests that a homeostatic mechanism has been brought into play with prolonged exposure to DU, ameliorating the oxidative effects of DU. This data also suggests that the behavioral effects of DU may be independent of its oxidative activity.

The effects of DU may also be mediated by the sex of the animal. DU-exposed male rats tend to exhibit less pronounced behavioral effects than females regardless of dosage or length of exposure [14]. However, in a later study these differences, while present, had no overall bearing on the effect of DU and behavior [13]. We have not found any gender effects concerning DU exposure and CNS lipid oxidation.

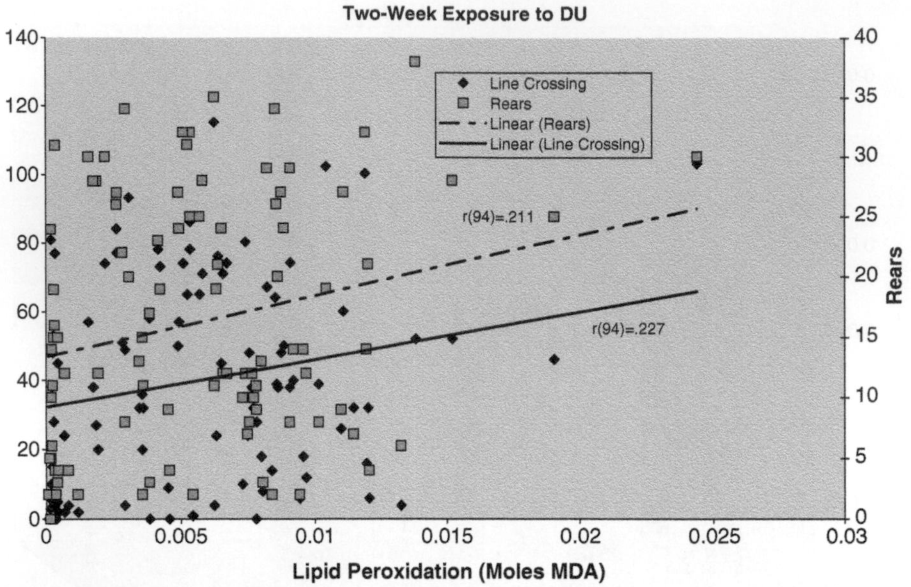

FIGURE 4.2 Correlations between brain lipid oxidation and open-field behavior after 2 weeks' exposure to DU in drinking water. Both correlations are significant at $p < .05$. (From Briner, W. and Murray, J. Effects of short-term and long-term depleted uranium exposure on open-field behavior and brain lipid oxidation in rats. *Neuotoxicol. Teratol.*, *27*, 135–144, 2005. With permission of authors.)

GENERAL EFFECTS OF DU ON REPRODUCTION AND DEVELOPMENT

Developing animals are often the most sensitive to the effects of metals. The indication that DU may produce adverse effects on developing organisms includes evidence that DU reduces litter size, viability, and lactation in rodents with orally administered uranium acetate in doses of 25–50 mg/kg (native U) [15], and may produce maternal toxicity, reduced fetal weight, and skeletal malformations (0.5 mg/kg injection; native U) [16]. Similarly, embryolethality and delayed growth was seen in rodents exposed to native U (25–50 mg/kg orally) [17]. Other adverse effects include lower maternal weight gain with increased maternal liver weight, reduced fetal weight and body length, renal hypoplasia, and up to a 15% occurrence of skeletal malformations (native U) [18]. Gestational day 10 (neural tube formation) appears to be the most vulnerable time for DU exposure [19].

RAT AND MOUSE STUDIES ON DEVELOPMENT

We have examined the effects of orally administered DU on the development of mice [20]. DU in drinking water was administered to female mice for 2 weeks in dosages of 0, 19, 37, or 75 mg/kg, and then the females were mated. DU administration was continued through development and testing of offspring until the offspring were 21 d of age. The development of the offspring was followed using the Fox developmental scale [21]. When the pups reached 21 d of age they were examined with a standard neurotoxicology battery [22].

The DU-exposed mouse pups experienced accelerated developed on a number of indices, when compared to controls. This included righting reflexes, forelimb placing and grasping, and swimming development. Hind-limb placing was at first accelerated but then later delayed. Weight gain was also accelerated for DU-exposed animals [20].

At 21 d of age postnatal the DU-exposed mouse pups differed from control on the standard neurotoxicology battery with fewer spontaneous vocalizations, and demonstrated more freezing and jerking behavior for the touch response test. There was also a trend for the DU-exposed animals to differ in their responses to the tail pinch test, to display more vigorous responses to the arousal test, and to display either more quiet responses or more aggressive behavior in the reactivity test. Uranium-exposed mouse pups had significantly higher body and brain weights at the time of sacrifice. However, the brain as a percentage of body weight was smaller in the uranium exposed groups [20].

In a similar 2002 study [23], we repeated the findings of our earlier study, again demonstrating accelerated development in mouse pups and similar differences in brain and body weight. The development of swimming behavior, a general indicator of neuromuscular development, was also accelerated for DU- exposed mice (Figure 4.3).

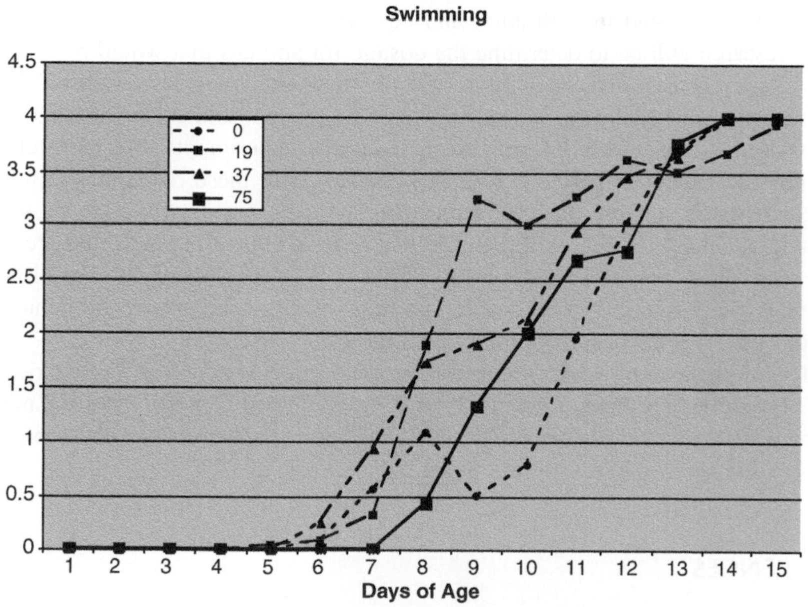

FIGURE 4.3 Development of swimming behavior in young mice exposed to DU in drinking water. Legend indicates concentration of DU in drinking water. (From Briner, W. and Abboud, B. Behavior of juvenile mice chronically exposed to depleted uranium, In *Metal Ions in Biology and Medicine*, Khassanova, L. et al., Eds. John Libby Eurotext, Paris, 2002, p. 353–356. With permission of authors.)

Behavior at 21 d of age also differed for DU-exposed animals. We found that general measures of arousal were significantly different from controls in what appears to be an inverted U-shaped response. The approach response was altered in a dose-dependent fashion, as was the tail-pinch response. Swim maze learning, a task of working memory, was also affected by DU, with DU-exposed animals taking longer to solve the task and showing less improvement on a second trial. In this study we also examined brain lipid oxidation and found DU-exposed pups had greater lipid oxidation levels.

On the surface it would seem that accelerated development of DU-exposed mouse pups would not be problematic. However, accelerated development may interfere with the overall development of other neural systems, perhaps sacrificing the development of some systems in preference for others. This is supported by our findings that developing mice exposed to DU perform abnormally on a standard neurotoxicology battery [20, 23]. We also demonstrated that these mice differ in their open-field behavior with fewer line crossings than control animals [23]. This suggests that animals exposed to DU during development may later have difficulty modulating their responses to environmental stimuli.

CONCLUSION

The evidence presented above indicates that DU may fit the criteria as a potential neurotoxic compound in both adult and developing organisms. The challenge for future research will be to determine the dosage for animals that would reflect what humans may be exposed to. In addition to knowledge about dose we would need information about the length and manner of exposure that soldiers and civilians may be exposed to. Once this information is available experiments that more closely resemble the human experience will be possible. With these caveats in mind, the current research suggests that the effects of DU exposure are likely to be subtle. Behavioral changes are often difficult to detect in individual animals and are often only seen when looking at large groups. This will be even more difficult in determining if DU has an effect on developing humans where cognitive and behavioral abilities have a great deal of normal variance.

The mechanism by which DU exerts its toxic effects is unclear. Lipid oxidation may play a role, but there are likely to be a variety of mechanisms that have not yet be elucidated. It would be useful in the design of future behavioral and developmental research to have some insight into the cellular effects of DU.

REFERENCES

1. Domingo, J.L. et al. Acute toxicity of uranium in rats and mice. *Bull. Environ. Contam. Toxicol.*, *39*, 168, 1987.
2. Lin, R.H., Fu, W.M., and Lin-Shiau, S.Y. Presynaptic action of uranyl nitrate on the phrenic nerve-diaphragm preparation of the mouse. *Neuropharmacology*, *27*, 857, 1988.
3. Pellmar, T.C. et al. Distribution of uranium in rats implanted with depleted uranium pellets. *Toxicol. Sci.*, *49*, 29, 1999.

4. Barber, D.S., Ehrich, M.F., and Jortner, B.S. The effect of stress on the temporal and regional distribution of uranium in rat brain after acute uranyl acetate exposure. *J. Toxicol. Environ. Health A*, *68*, 99, 2005.
5. Monleau, M. et al. Bioaccumulation and behavioral effects of depleted uranium in rats exposed to repeated inhalations. *Neurosci. Lett.*, 390, 31, 2005.
6. Pellmar, T.C. et al. Electrophysiological changes in hippocampal slices isolated from rats embedded with depleted uranium fragments. *Neurotoxicology*, *20*, 785, 1999.
7. Lestaevel, P. et al. Changes in sleep-wake cycle after chronic exposure to uranium in rats. *Neurotoxol. Teratol.*, 27, 835, 2005.
8. Lestaevel, P. et al. The brain is a target organ after acute exposure to depleted uranium. *Toxicology*, *1*, 219, 2005.
9. Houpert, P. et al. Enriched but not depleted uranium affects central nervous system in long-term exposed rat. *Neurotoxicology*, in press.
10. Schaich, K.M. Metals and lipid oxidation: contemporary issues. *Lipids*, *27*, 209, 1992.
11. Ohkawa, H., Ohishi, N., and Tagi, K. Assay for lipid peroxides in animal tissues by thiobarbituric acid reaction. *Anal. Biochem.*, *95*, 351, 1979.
12. Briner, W. and Davis, D. Lipid oxidation and behavior are correlated in depleted uranium exposed mice. In *Metal Ions in Biology and Medicine*, Khassanova, L. et al., Eds. John Libby Eurotext, Paris, 2002, p. 59.
13. Briner, W. and Murray, J. Effects of short-term and long-term depleted uranium exposure on open-field behavior and brain lipid oxidation in rats. *Neuotoxicol. Teratol.*, *27*, 135, 2005.
14. Briner, W. Altered open-field performance in depleted uranium exposed rats. In *Metal Ions in Biology and Medicine*, Khassanova, L. et al., Eds. John Libby Eurotext, Paris, 2002, p. 342.
15. Domingo, J.L. et al. Evaluation of the perinatal and postnatal effects of uranium in mice upon oral administration. *Arch. Environ. Health*, *44*, 395, 1989.
16. Bosque, M.A. et al. Embryotoxicity and teratogenicity of uranium in mice following subcutaneous administration of uranyl acetate. *Biol. Trace Elem. Res.*, *36*, 109, 1993.
17. Paternain, J.L. et al. The effects of uranium on reproduction, gestation, and postnatal survival in mice. *Ecotoxicol. Environ. Saf.*, *17*, 291, 1989.
18. Domingo, J.L. et al. Evaluation of the perinatal and postnatal effects of uranium in mice upon oral administration. *Arch. Environ. Health*, *44*, 395, 1989.
19. Domingo, J.L. Reproductive and developmental toxicity of natural and depleted uranium: a review. *Reprod. Toxicol.*, *15*, 603, 2001.
20. Briner, W. and Byrd, K. Effects of depleted uranium on development of the mouse. In *Metal Ions in Biology and Medicine*, Centeno, J. et al. Eds. John Libby Eurotext, Paris, 2000, p. 459.
21. Fox, W.M. Reflex-ontogeny and behavioral development of the mouse. *Anim. Behav.*, 13, 234, 1965.
22. O'Donoghue, J.L. Clinical neurologic indicies of toxicity in animals. *Environ. Health Perspect.*, *104(Suppl. 2)*, 323, 1996.
23. Briner, W. and Abboud, B. Behavior of juvenile mice chronically exposed to depleted uranium, In *Metal Ions in Biology and Medicine*, Khassanova, L. et al., Eds. John Libby Eurotext, Paris, 2002, p. 353.

5 Colorimetric Determination of Uranium

John F. Kalinich and David E. McClain

CONTENTS

INTRODUCTION

There are a variety of techniques available to measure uranium levels in fluids, including neutron activation analysis [1], kinetic phosphorescence analysis (KPA) [2,3], alpha spectrometry [4], liquid scintillation spectrometry [5], and inductively coupled-plasma mass spectrometry (ICP-MS) [6]. The unifying feature of these techniques is that they either require extensive sample preparation (sometimes requiring days to complete) or utilize expensive instrumentation with which to conduct the analysis. Although the ICP-MS remains the method of choice for differentiating natural vs. depleted uranium, there are instances when a procedure to rapidly screen samples for the presence of uranium would be beneficial.

The ideal procedure would be one that could (1) be conducted rapidly and accurately, (2) would use readily available chemicals, (3) would not require extensive sample preparation, (4) would not generate excessive amounts of chemical waste, (5) would not require expensive or complicated instrumentation, (6) would require little or no technical training to conduct, and (7) could be used in a field situation if needed. In addition, the assay procedure should be applicable to a variety of samples, including water, biological fluids such as urine, extracts of metallic shrapnel, and residual dust samples from building and equipment surfaces. For the rapid screening of samples, especially in the field, colorimetric or spectrophotometric detection methods are among the most practical.

COLORIMETRIC REAGENTS FOR URANIUM DETERMINATION

Although not a definitive list, several of the more widely used methods are shown in Figure 5.1. The structures of the uranium detection reagents are similar in that

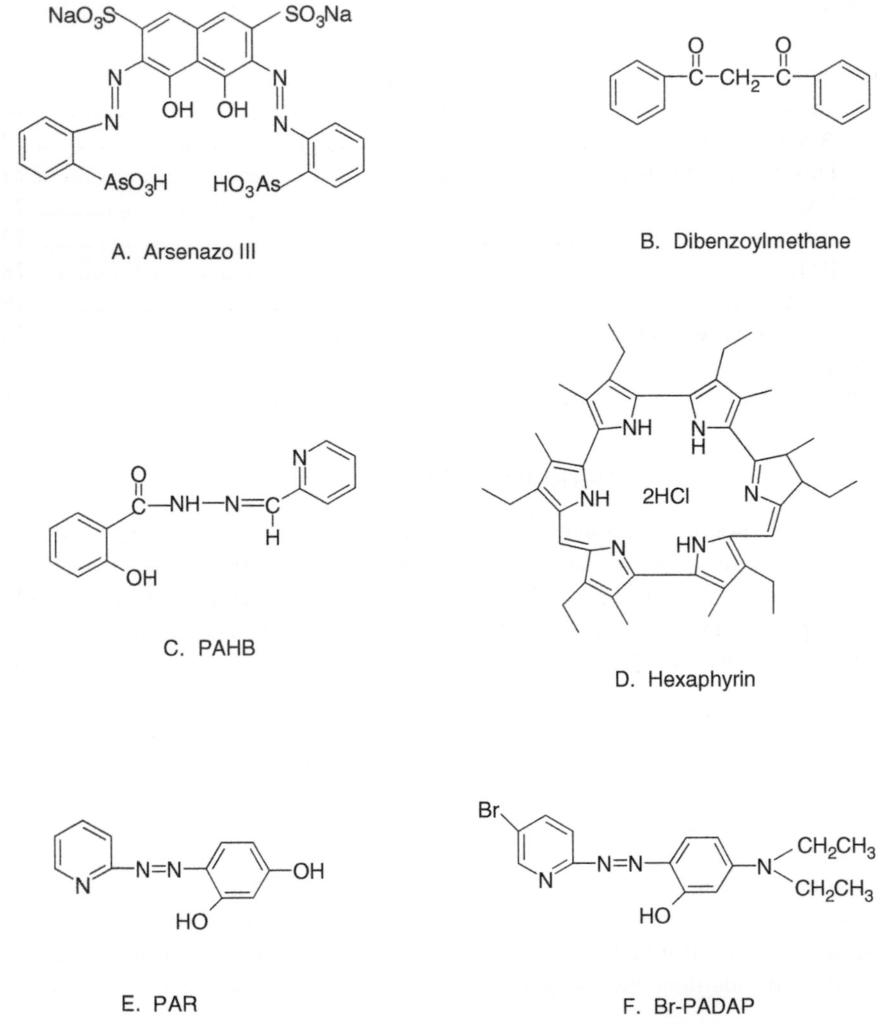

FIGURE 5.1 Structures of selected colorimetric reagents for uranium detection.

they possess a conjugated double bond system and a metal binding area. Comparison of the various detection methods published in the literature is difficult because of the different types of detection method employed, the different kinds of samples tested, and the variable use of separation/concentration techniques. However, a brief description of each of the more popular methods is given below, followed by a more detailed discussion of the method used in our laboratory.

Arsenazo III

Arsenazo III (2,7-bis[2-arsenophenylazo]-1,8-dihydroxynaphthalene-3,6-disulfonic acid) (Figure 5.1A) is a widely used reagent for the detection of uranium and has been reported by some to be one of the more sensitive reagents for uranium determination [7]. As with many of the azo compounds, it will also bind other metals, most notably calcium [8,9]. It requires a low pH (< pH 2.0) for maximum sensitivity, and the arsenazo III-uranium complex has an absorbance maximum in the 650 nm range [10]. Detection limits reported from a variety of detection procedures range from 40-400 ng/L [10–12]. The required use of strong acids to maintain the necessary low pH would be a drawback for field use.

Dibenzoylmethane

Dibenzoylmethane (Figure 5.1B) has also been used as a colorimetric reagent for the determination of uranium. After complexation with uranium, the absorbance of the dibenzoylmethane-uranium complex is determined at 405 nm. Detection limits from water samples, after a solid-phase extraction procedure, have been reported to be 100 ng/l [13]. Dibenzoylmethane is not water soluble and is generally prepared in a methanol/pyridine solution which would limit its attractiveness as a field method for uranium detection.

PAHB

Pyridine-2-carboxaldehyde 2-hydroxybenzoylhydrazone or PAHB has been reported to be useful for the extraction and detection of uranium (Figure 5.1C), with detection limits of 1–5 ppm uranium at 375 nm [14]. Currently, it must be synthesized from precursor materials [15], after which it is prepared as a stock solution in dimethylformamide. The procedure for uranium detection also requires a back-extraction using isobutyl methyl ketone. The requirement for organic solvents would limit its usefulness as a field method. In addition, PAHB has been shown to bind to other metals such as nickel, zinc, and vanadium [15,16].

Hexaphyrin

Isoamethyrin, or hexaphyrin (Figure 5.1D), is a recently synthesized expanded porphyrin shown to bind actinides effectively, including uranium. Unlike many of the reagents discussed here, hexaphyrin will not react with transition metals, but it requires a methanol/dichloromethane solution for solubility of the reagent. Reports indicate that detection limits of 28 ppm uranium at 550 nm are obtainable [17].

PAR

The pyridylazo dye 4-(2-pyridylazo)-resorcinol (PAR) (Figure 5.1E) is a colorimetric reagent for a variety of heavy metals including uranium [18,19]. PAR can be prepared in aqueous buffers, eliminating that need for organic solvents. Reports indicate a detection limit of approximately 200–250 ng/l at 530 nm with a pH maximum between 7 and 8 [20]. The nonselective nature of PAR necessitates the use of chelating or "masking" agents to prevent the binding to other metals or the inclusion of a sample cleanup step in the procedure.

Br-PADAP

The pyridylazo dye 2-(5-bromo-2-pyridylazo)-5-diethylaminophenol (Br-PADAP) (Figure 5.1F), was previously used to detect uranium in organic leach liquors from nuclear fuel reprocessing plants [21,22] as well as used histochemically to detect a variety of metals, including cobalt, nickel, zinc, and copper, in rat liver samples [23]. Its use in aqueous solutions is greatly hindered by its insolubility and nonspecificity. However, recent procedural advancements in our laboratory have enabled its use to specifically bind to uranium in such diverse aqueous environments as water and urine [24], cells [25,26], and buffered extracts of metallic shrapnel [27] (Figure 5.2).

Because Br-PADAP is capable of binding a variety of metals, a procedure was needed to eliminate binding of Br-PADAP to metals other than uranium. We accomplished that goal through the use of "masking agents," compounds that prevent the binding of the stain to metals not of interest. The addition of a mixture of EDTA and sodium citrate allows the stain to bind to uranium, but not to other metals. Table 5.1 shows the metals that do not bind Br-PADAP, those that can bind Br-PADAP but can be masked by EDTA/citrate, and those that bind Br-PADAP but are not masked.

The binding of Br-PADAP to uranium is pH dependent, with an optimal pH range of 8–12. Several common laboratory buffers, including phosphate, borate, and CAPS, are capable of maintaining the pH within the acceptable range. The limited water solubility of Br-PADAP requires stock solutions be prepared in ethanol or dimethyl sulfoxide (DMSO) prior to addition to the reaction mixture. Also, we discovered that the addition of a quaternary ammonium salt acts as a "solubilizing agent" to keep the reaction mixture components in solution. After addition of the Br-PADAP, color development is rapid if uranium is present in sufficient quantity. Binding results in a bathochromic shift of the absorption maximum from 444 nm to 578 nm (Figure 5.3). This wavelength shift can be easily detected with a visible-light

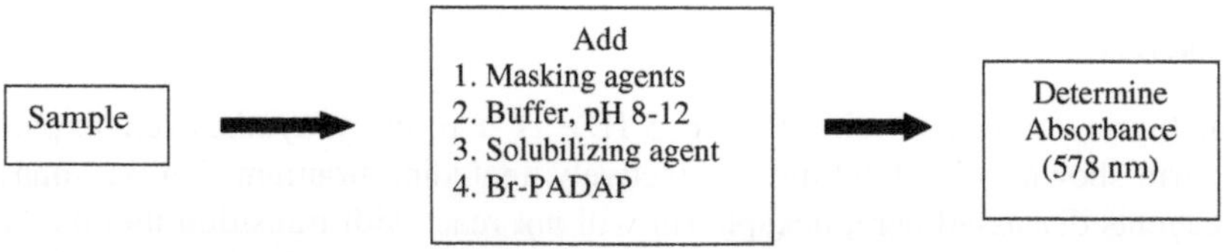

FIGURE 5.2 Procedure for the colorimetric detection of uranium using Br-PADAP.

TABLE 5.1
Metals Tested for the Ability to Bind Br-PADAP

Do Not Bind	Bind but Can Be Masked	Bind, Not Masked
Lithium	Cobalt	Uranium
Cesium	Cadmium	
Tantalum	Nickel	
Sodium	Lead	
Magnesium	Zinc	
Cerium	Iron	
Chromium	Copper	
Lanthanum		
Tungsten		
Potassium		
Calcium		
Molybdenum		
Rubidium		
Barium		
Silver		
Gadolinium		
Aluminum		

spectrophotometer or colorimeter. The procedure requires minimal technical training to conduct and can be performed in the field using commercially available, battery-powered colorimeters. Using this procedure we have been able to detect uranium levels in synthetic urine in the range of 20–30 ppm and in water to approximately 10 ppb [28]. In order to improve the procedure we are currently investigating several techniques that will hopefully increase the sensitivity of the test yet still retain its simplicity. The techniques discussed below would also be applicable to procedures using other colorimetric reagents for uranium.

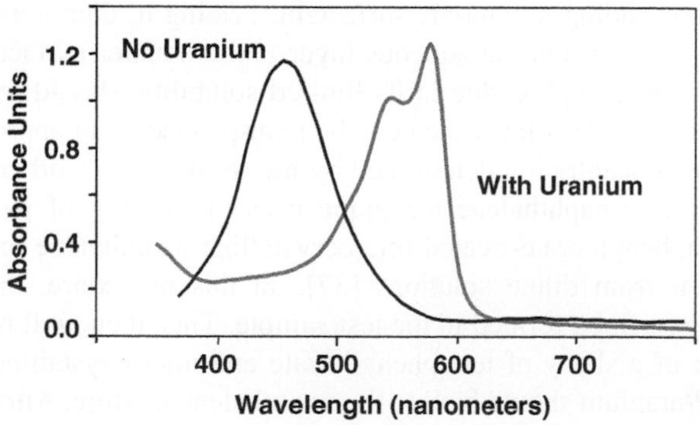

FIGURE 5.3 Spectra of Br-PADAP in the absence and presence of uranium.

CURRENT AND FUTURE STUDIES

The inclusion of a sample concentration step should greatly increase the sensitivity of the Br-PADAP technique. However, for testing samples rapidly in the field, this concentration step should be technically simple to perform, not require any complicated equipment, and not generate excessive amounts of chemical waste. There are two points in this procedure where such a step could be included. The first is prior to the formation of the Br-PADAP/uranium complex (the "pre-complexation concentration" step), whereas the second is after complexation of the stain and uranium but prior to measuring the absorbance (the "post-complexation concentration" step). The precomplexation concentration techniques that are being tested at AFRRI include solid-phase extraction procedures with commercially available ion exchange resins such as HYPHAN, UTEVA, and TRU, as well as with standard anion exchange resin. HYPHAN, UTEVA, and TRU resins have been used by others to concentrate uranium from dilute samples and may be of utility here [29,30]. The use of standard anion exchange resin involves treating the sample with concentrated HCl to convert the uranium to $UO_2Cl_4^{2-}$, a form that will be retained on the resin [31]. Both standard column procedures as well as resin-containing filter disks are being tested. Another pre-complexation concentration technique being assessed is metal chelation chromatography using C_{18} filter disks impregnated with various metal chelators. A review of the literature suggests that several compounds may prove useful in this regard, including dipicolinic acid [32,33] and tri-*n*-octylphosphine oxide [13]. The final precomplexation concentration technique under investigation in our lab is the recently developed technique of molecularly imprinted polymers. This procedure involves binding ligands to the uranium, then forming an insoluble polymer to lock the uranium recognition site in place. After removing the uranium atom via an acid wash, the polymer can act as a highly specific uranium-binding resin [34,35].

Two postcomplexation concentration techniques are being tested by us: cloud-point extraction and concentration of the Br-PADAP/uranium complex by adsorption to microcrystalline naphthalene. The technique of cloud point extraction attempts to capitalize on the solubility characteristics of Br-PADAP [36]. The solubility of many nonionic surfactants/detergents in aqueous systems is greatly reduced above a well-defined temperature called the *cloud point*. By forming the Br-PADAP/uranium complex, adding a nonionic surfactant, heating it, then letting it cool, the mixture should separate into an aqueous layer and a smaller surfactant layer. The Br-PADAP/uranium complex, due to its limited solubility, should be concentrated in the surfactant layer. This layer then can be resuspended in an appropriate solvent and the uranium concentration determined by measuring the absorbance at 578 nm. The microcrystalline naphthalene technique uses the affinity of pyridylazo compounds for tetraphenylborate-treated microcrystalline naphthalene in order to concentrate uranium from dilute solutions [37]. In this procedure, the Br-PADAP/uranium complex will be formed in the test sample. This, then, will be reacted with a small amount of a slurry of tetraphenylborate and microcrystalline naphthalene. The Br-PADAP/uranium should bind to the naphthalene mixture. After filtration, the precipitate will be dissolved in an appropriate organic solvent, and the uranium levels determined by measuring the absorbance at 578 nm.

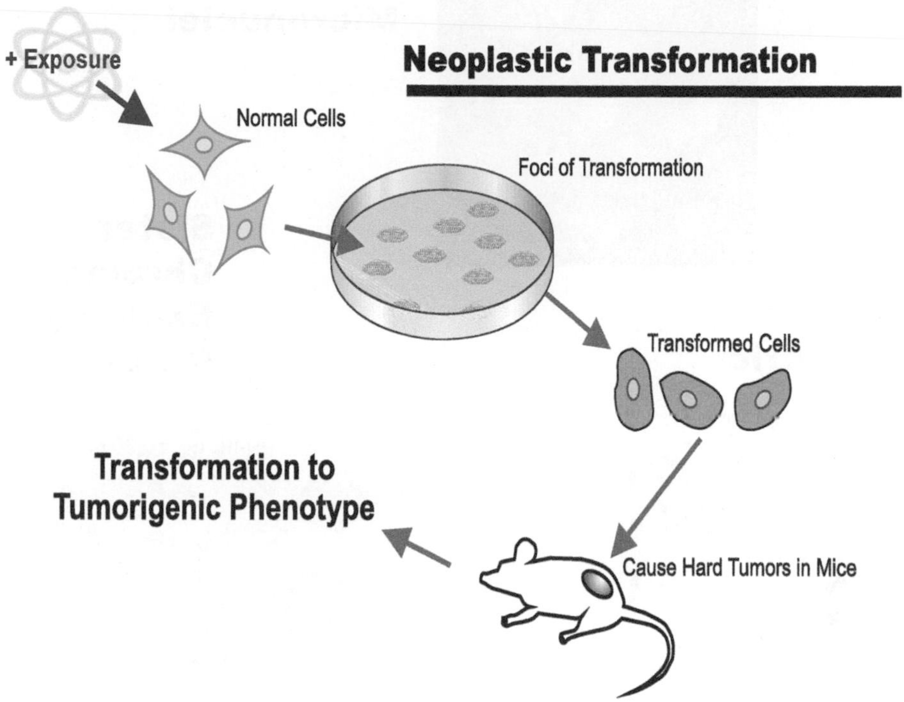

COLOR FIGURE 1.2 Carcinogenic hazard evaluation. This type of hazard evaluation is a multistep process involving *in vitro* and animal models. Human epidemiology is used as the last step in the process.

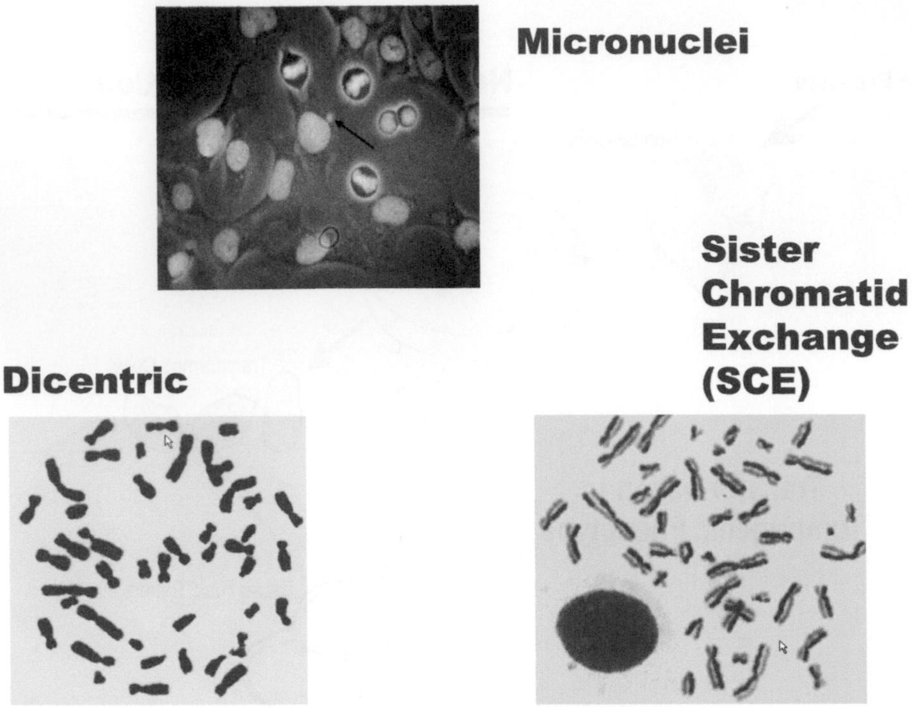

COLOR FIGURE 1.3 Examples of chromosomal damage: measures of genotoxicity.

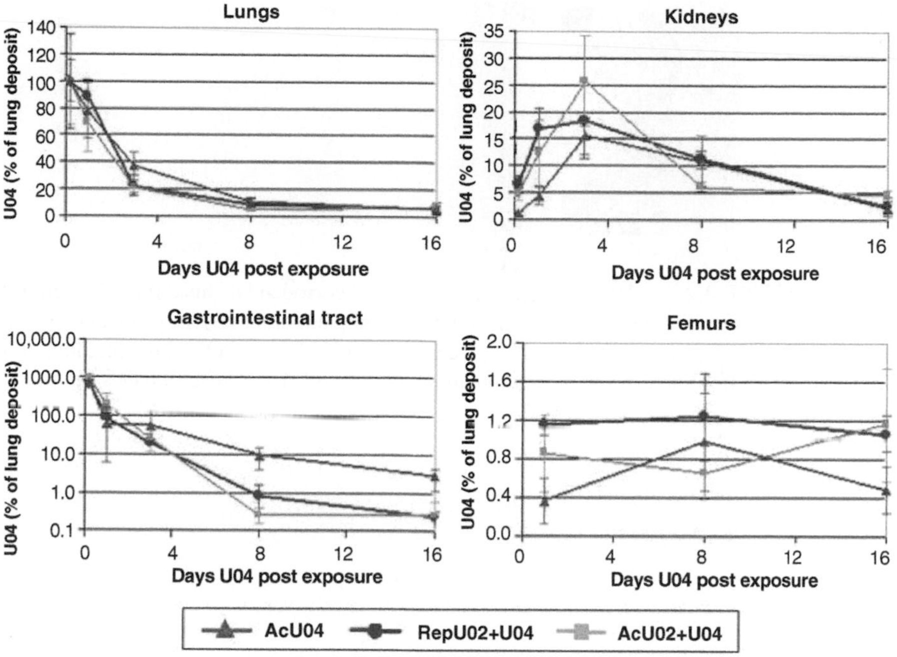

COLOR FIGURE 2.9 Biokinetics of UO_4 in the lungs, kidneys, gastrointestinal tract, and femurs as a function of days after UO_4 exposure. Percent of UO_4 lung deposit, mean ± SD. N = 5 for $RepUO_2 + UO_4$, N = 3 for $AcUO_2 + UO_4$, and N = 4 for $AcUO_4$. *: $p < 0.05$ between the $RepUO_2 + UO_4$ and $AcUO_4$ groups, #: $p < 0.05$ between the $AcUO_2 + UO_4$ and $AcUO_4$ groups.

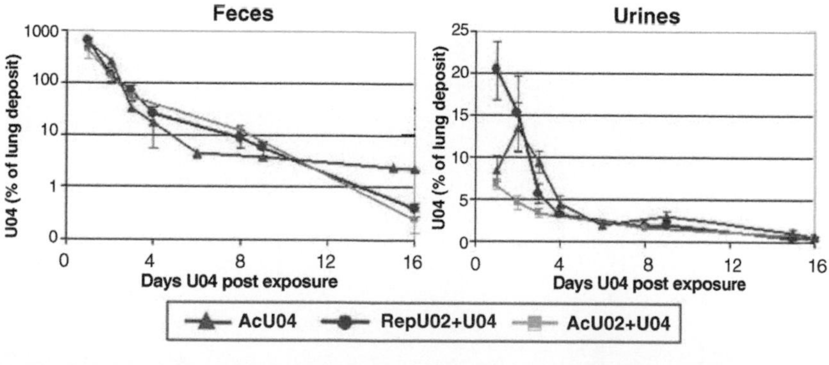

COLOR FIGURE 2.10 Daily excretion of UO_4 as a function of days after UO_4 exposure. % of UO_4 lung deposit, mean ± SD. N = 5 for $RepUO_2 + UO_4$, N = 3 for $AcUO_2$+UO_4, and N = 4 for $AcUO_4$. *: $p < 0.05$ between the $RepUO_2 + UO_4$ and $AcUO_4$ groups, #: $p < 0.05$ between the $AcUO_2 + UO_4$ and $AcUO_4$ groups.

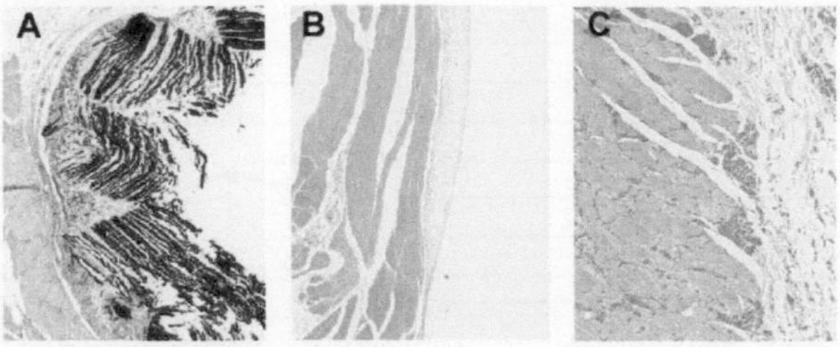

COLOR FIGURE 3.2 (A) DU fragment: Black shards of corroded DU lined the thick cellular fibrotic capsule surrounding the fragment 520 d after implantation; (B) Ta fragment: A thin acellular fibrotic capsule with no metal pieces present 603 days after implantation; (C) Thorotrast® injection: no capsule or inflammation; Thorotrast®-laden macrophages infiltrated skeletal muscle 792 d after injection.

CONCLUSION

Inductively coupled-plasma mass spectrometry remains the "gold standard" for uranium determination and differentiation of the various isotopes of uranium. However, there are situations when it would be beneficial to rapidly measure uranium levels using minimal reagents and equipment with a procedure that requires little technical expertise. Colorimetric techniques are especially suited for such a task. Whereas colorimetric procedures for the determination of uranium are not capable of differentiating natural from depleted uranium, they would be extremely useful in the screening of large number of samples prior to more extensive analysis by ICP-MS. Field testing of environmental samples in areas of suspected depleted uranium use or after detonation of a depleted uranium-containing radiological dispersion device or "dirty bomb," assessment of decontamination procedures, and rapid screening of urine samples from potentially contaminated individuals are just a few of the situations where a colorimetric procedure to rapidly measure uranium levels would be valuable.

All of the colorimetric techniques described have both positive and negative aspects to their use. Any technique chosen as a field testing procedure should use off-the-shelf reagents and instrumentation, not generate excessive waste products, and require little to no training to perform. Presently, the Br-PADAP method appears to meet most of these requirements. Continuing research in the area of sample concentration will hopefully increase the usefulness of the procedure without increasing its complexity. In addition, continuing advances in metal-complexing reagents and sensor technology may soon allow fabrication of microchips capable of detecting uranium concentrations now only attainable with the most sensitive analytical techniques [38,39].

REFERENCES

1. Zouridakis, N., Ochsenkuhn, K.M., and Savidou, A., Determination of uranium and radon in portable water samples, *J. Environ. Radioact.* 61, 225, 2002.
2. Brina, R., Uranium removal from contaminated water by enzymatic reduction with kinetic phosphorimetry detection, *Am. Lab.* May, 43, 1995.
3. Ejnik, J.W., Hamilton, M.M., Adams, P.R., and Carmichael, A.J., Optimal sample preparation conditions for the determination of uranium in biological samples by kinetic phosphorescence analysis (KPA*), J. Pharm. Biomed. Anal.* 24, 227, 2000.
4. Ethington, E.F. and Niswonger, K.R., Alpha spectrometry measurement reproducibility study for uranium, plutonium, and americium in water at Rocky Flats environmental technology site, *Health Phys.* 79, S38, 2000.
5. Salonen, L., A rapid method for the monitoring of uranium and radon in drinking water, *Sci. Total Environ.* 130/131, 23, 1993.
6. Karpas, Z., Halicz, L., Roiz, J., Marko, R., Katorza, E., Lorber, A., and Goldbart, Z., Inductively coupled plasma mass spectrometry as a simple, rapid, and inexpensive method for determination of uranium in urine and fresh water: comparison with LIF, *Health Phys.* 71, 879, 1996.
7. Khan, M.H., Warwick, P., and Evans, N., Spectrophotometric determination of uranium with arsenazo-III in perchloric acid, *Chemosphere* 63, 1165, 2006.

8. Vergara, J. and Delay, M., The use of metallochromic Ca indicators in skeletal muscle, *Cell Calcium* 6, 119, 1985.
9. Rowatt, E. and Williams, R.J.P., The interaction of cations with the dye arsenazo III, *Biochem. J.* 259, 295, 1989.
10. Rohwer, H., Rheeder, N., and Hosten, E., Interactions with uranium and thorium with arsenazo III in an aqueous medium, *Anal. Chim. Acta* 341, 263, 1997.
11. Sadeghi, S., Mohammadzadeh, D., and Yamini, Y., Solid-phase extraction-spectrophotometric determination of uranium (VI) in natural waters, *Anal. Bioanal. Chem.* 375, 698, 2003.
12. Green, P.A., Copper, C.L., Berv, D.E., Ramsey, J.D., and Collins, G.E., Colorimetric detection of uranium (VI) on building surfaces after enrichment by solid phase extraction, *Talanta* 66, 961, 2005.
13. Shamsipur, M., Ghiasvand, A.R., and Yamini, Y., Solid-phase extraction of ultratrace uranium (VI) in natural waters using octadecyl silica membrane disks modified by tri-n-octylphosphine oxide and its spectrophotometric determination with dibenzoylmethane, *Anal. Chem.* 71, 4892, 1999.
14. Bale, M.N. and Sawant, A.D., Solvent extraction and spectrophotometric determination of uranium (VI) with pyridine-2-carboxaldehyde 2-hydroxybenzoylhydrazone, *J. Radioanal. Nucl. Chem.* 247, 531, 2001.
15. Gallego, M., Garcia-Vargas, M., Pino, F., and Valcarcel, M., Analytical applications of picolinealdehyde salicylohydrazone: spectrophotometric determination of nickel and zinc, *Microchem. J.* 23, 353, 1978.
16. Gallego, M., Garcia-Vargas, M., and Valcarel, M., Analytical applications of picolinealdehyde salicylohydrazone. II. Extraction and spectrophotometric determination of vanadium (V), *Microchem. J.* 24, 143, 1979.
17. Sessler, J.L., Melfi, P.J., Seidel, D., Gorden, A.E.V., Ford, D.K., Palmer, P.D., and Tait, C.D. Hexaphyrin (1.0.1.0.0.0): a new colorimetric actinide sensor, *Tetrahedron* 60, 11089, 2004.
18. Hashem, E.Y., Spectrophotometric studies on the simultaneous determination of cadmium and mercury with 4-(2-pyridylazo)-resorinol, *Spectochim. Acta: A. Mol. Biomol. Spectrosc.* 58, 1401, 2002.
19. Mori, I., Taguchi, K., Fujita, Y., and Matsuo, T., Improved spectrophotometric determination of uranium (VI) with 4-(2-pyridylazo)-resorinol in the presence of benzyldimethylstearyltrimethylammonium chloride, *Analyt. Bioanal. Chem.* 353, 174, 1995.
20. Abbas, M.N., Homoda, A.M., and Mostafa, G.A.E., First derivative spectrophotometric determination of uranium (VI) and vanadium (V) in natural and saline waters and some synthetic matrices using PAR and cetylpyridinum chloride, *Anal. Chim. Acta* 436, 223, 2001.
21. Johnson, D.A. and Florence, T.M., Spectrophotometric determination of uranium (VI) with 2-(5-bromo-2-pyridylazo)-5-diethylaminophenol, *Anal. Chim. Acta* 53, 73, 1971.
22. Pakalns, P. and McAllister, B.R., Spectrophotometric determination in selective organic extractants with 2-(5-bromo-2-pyridylazo)-5-diethylaminophenol, *Anal. Chim. Acta* 62, 207, 1972.
23. Sumi, Y., Inoue, T., Muraki, T., and Suzuki, T., A highly sensitive chelator for metal staining, bromopyridyl-azodiethylaminophenol, *Stain Technol.* 58, 325, 1983.
24. Kalinich, J.F. Uranium-Containing/Metal Binding Complex, Process of Making and Method of Use for the Determination of Natural and Depleted Uranium in Biological Samples, U.S. Patent 6,107,098, 2000.

25. Kalinich, J.F. and McClain, D.E., Staining of intracellular deposits of uranium in cultured macrophages, *Biotechnic Histochem.* 76, 247, 2001.
26. Kalinich, J.F., Ramakrishnan, N., Villa, V., and McClain, D.E., Depleted uranium-uranyl chloride induces apoptosis in mouse J774 macrophages, *Toxicology* 179, 105, 2002.
27. Kalinich, J.F., Ramakrishnan, N., and McClain, D.E., A procedure for the rapid detection of depleted uranium in metal shrapnel fragments, *Mil. Med.* 165, 626, 2000.
28. Egan, J., personal communication, 2005.
29. Van Britson, G., Slowikowski, B., and Bickel, M., A rapid method for the detection of uranium in surface water, *Sci. Total Environ.* 173/174, 83, 1995.
30. Horowitz, E.P., Dietz, M.L., Chiarizia, R., and Diamond, H., Separation and preconcentration of uranium from acidic media by extraction chromatography, *Anal. Chim. Acta* 266, 25, 1992.
31. Millet, T.J., Development of a rapid, economical and sensitive method for the routine determination of excreted uranium in urine, *Anal. Lett.* 24, 657, 1991.
32. Shaw, M.J., Hill, S.J., and Jones, P., Chelation ion chromatography of metal ions using high performance substrates dynamically modified with heterocyclic carboxylic acids, *Anal. Chim. Acta* 401, 65, 1999.
33. Shaw, M.J., Hill, S.J., Jones, P., and Nesterenko, P.N., Determination of uranium in environmental matrices by chelation ion chromatography using a high performance substrate dynamically modified with 2,6-pyridinedicarboxylic acid, *Chromatographia* 51, 695, 2000.
34. Saunders, G.D., Faxon, S.P., Walton, P.H., Joyce, M.J., and Port, S.N., A selective uranium extraction agent prepared by polymer imprinting, *Chem. Commun.* (November/December), 273, 2000.
35. Bae, S.Y., Southard, G.L., and Murray, G.M., Molecularly imprinted ion exchange resin for purification, preconcentration and determination of UO_2^{2+} by spectrophotometry and plasma spectrometry, *Anal. Chim. Acta* 397, 173, 1999.
36. Silva, M.F., Fernandez, L., Olsina, R.A., and Stacchiola, D., Cloud point extraction, preconcentration and spectrophotometric determination of erbium (III)-2-(3,5-dichloro-2-pyridylazo)-5-dimethylaminophenol, *Anal. Chim. Acta* 342, 229, 1997.
37. Taher, M.A., Atomic absorption spectrometric determination of ultratrace amounts of zinc after preconcentration with the ion pair of 2-(5-bromo-2-pyridylazo)-5-diethylaminophenol and ammonium tetraphenylborate on microcrystalline naphthalene or by column method, *Talanta* 52, 181, 2000.
38. Collins, G.E. and Lu, Q., Radionuclide and metal ion detection on a capillary electrophoresis microchip using LED absorbance detection, *Sens. Actuators B* 76, 244, 2001.
39. Collins, G.E. and Lu, Q., Microfabricated capillary electrophoresis sensor for uranium (VI), *Anal. Chim. Acta* 436, 181, 2001.

6 Chemical and Histological Assessment of Depleted Uranium in Tissues and Biological Fluids

Todor I. Todorov, John W. Ejnik, Florabel G. Mullick, and José A. Centeno

CONTENTS

INTRODUCTION

Uranium (U) is a naturally occurring element that is found in the Earth's crust at an average concentration of 0.0004%. This heavy element consist of several isotopes, including ^{238}U (at 99.3% by mass), ^{235}U (at a 0.72% by mass), and ^{234}U (at a 0.006% by mass). In nuclear power plants, the production of electricity is based on the use of ^{235}U. During this process the concentration of ^{235}U is increased such that it can be used as a nuclear fuel. The by-product of this enrichment process is what is known as *depleted uranium* (DU), containing about 80% less ^{234}U and 70% less ^{235}U than does natural uranium [1]. DU has been used in commercial and in military applications. For example DU has been used as counterweights (flaps) in commercial aircraft, as x-ray shielding in hospitals, and in the manufacturing of a wide range of chemicals. DU exposure in civilian populations could occur in occupational settings or in unusual circumstances such as inhalation following a plane crash through burning aircraft materials [2].

In the past 10 years, concerns of occupational and environmental exposure to depleted uranium (DU) used in military and peacekeeping settings have gained intense interest [3–6]. DU used by military applications contains approximately 0.00061% ^{234}U, 0.203% ^{235}U, and 99.79% ^{238}U [1]. Because of DU's properties as a high kinetic energy penetrator, high density, availability, and low relative cost, it has been incorporated into projectiles and tank armor for military applications. Consequently, soldiers in battle may be at risk of inhaling airborne DU particles, ingesting DU particles, and/or experiencing wound contamination by DU particles. Because of the chemical toxicological properties of uranium [7,8], the long-term health effects of DU exposure are of significant concern. This concern has led to renewed studies of uranium's chemical and radiological health effects as an internal toxicant [9,10].

This chapter provides an overview of currently available techniques used on the analysis of uranium and, more specifically, DU in biological specimens. The various analytical techniques require the samples to be in liquid (biological fluids, dissolved, or digested tissues) or solid form (tissues). Particular emphasis has been placed on the use of inductively coupled plasma mass spectrometry (ICP-MS) as the preferred method for the chemical assessment of uranium and depleted uranium in biological samples. In addition, the analysis of DU metal fragments in tissues by scanning electron microscopy coupled to energy dispersive spectroscopy will be demonstrated.

AVAILABLE TECHNIQUES FOR THE DETERMINATION OF URANIUM AND DEPLETED URANIUM

Several methods have been described to measure uranium in environmental, geological, and biological samples including thermal ionization mass spectrometry (TIMS), instrumental neutron activation analysis (INAA), delayed neutron counting (DNC), inductively coupled plasma-mass spectrometry (ICP-MS), inductively coupled plasma atomic emission spectroscopy (ICP-AES), α-spectroscopy, spectrophotometry, fluorometry, and kinetic phosphorescence analysis (KPA) [11–44]. Most of the analysis of uranium has been performed using ICP-MS or α-spectroscopy, on which the focus of the chapter will be based, but the above mentioned techniques deserve a brief description.

Neutron activation analysis (NAA) is technique capable of measuring the total and isotopic composition of uranium in biological samples [45]. Neutron activation analysis uses neutron irradiation of a sample to convert the elements into radioactive isotopes. While these isotopes decay, having half-lives varying from seconds to years, they are emitting different kinds of electromagnetic radiation, among which gamma radiation. In INAA, the emitted gamma radiation is measured with semiconductor gamma-ray spectrometers. Each radionuclide emits gamma radiation of a certain wavelength or energy, and accordingly, peaks in a gamma-spectrum will correspond to individual radionuclides within a particular sample. In INAA, the evaluation of the area under each peak provides information about the concentration of each element present in the sample. In DNC, on the other hand, the uranium isotope absorbs a neutron and splits to produce fission products many of which emit a delayed neutron [46]. The emitted neutrons are counted and the concentration and isotopic composition of uranium is measured. Detection limits are typically in the range of 10–50 ng/l [47]. The main advantage of both techniques is the minimal sample preparation required; however, INAA require long counting times to measure U isotopes, whereas DNC is sensitive to ^{235}U and less sensitive to ^{238}U, making data interpretation of DU exposure difficult.

Kinetic phosphorescence analysis (KPA) is a nondestructive method for the measurement of total uranium concentration in solutions [39,48]. KPA measures the intensity of the lifetime of the phosphorescence of the uranyl cation with an excitation wavelength of 425 nm and emission at 515 nm. Main advantages of the KPA technique over other methods are: chemical separation of uranium for many samples is not necessary, the small volumes used for the analysis (1–10 ml), and the low cost of the equipment. Detection limits for the KPA method in the range 10 ng/l have been reported [39]. Major disadvantages for the KPA method are the lack of discrimination between the different uranium isotopes and the matrix effect that could quench the uranyl phosphorescence due to the presence of residual salts, organics, and chlorides [39,48–50].

Thermal ionization mass spectrometry is the method of choice for the precise analysis of isotope ratios [51]. In TIMS the sample is deposited in a filament, which is heated to a high temperature resulting by an electric current. The ions are formed by electric transfer between atom and filament or *vice versa*. The created ions are then transferred to the mass spectrometer for mass-to-charge separation and quantification. The most common type of mass analyzer used in TIMS is a magnetic sector mass separator. Detectors could either be based on a single or multicollector systems (MC-TIMS) or both. The method provides very low detection limits (less than 1ng/l), excellent precision (down to 0.002 % RSD), and accuracy for isotope ratio determinations. However, biological samples require long and tedious sample preparation [13]. Additionally, the instrumentation is expensive, requires much care during the analysis, and the technique is not suited for large-scale biological monitoring.

α-Spectrometry is a common method for uranium urinary bioassay that takes advantage of the natural radioactivity of the uranium isotopes [21,52]. It is suitable for the analysis of nearly all the natural and artificial uranium isotopes, including ^{238}U, ^{234}U, ^{235}U, ^{236}U, ^{233}U, and ^{232}U. The measurements are affected by the presence of large quantities of inactive substances (silicates, sodium, potassium, iron, phosphates, etc.) and the potential α-emitting interferences [21,37,38]. Thus, the technique requires removal and preconcentration of the uranium from the sample matrix using precipitation, ion exchange, extraction, and liquid–liquid extraction. Typical detection limits in urine are in the order of 10–50 ng/l based on the preconcentration and counting times. The advantage of this method is the measurement of all U isotopes at the same time and the equipment's low maintenance cost. The major disadvantages are the long sample preparation time (1–2 d) and the long counting times, especially in order to obtain accurate measurements for ^{235}U at low total U concentrations.

Inductively coupled plasma mass spectrometry (ICP-MS) provides an excellent tool for rapid and simple analysis of low concentration of uranium in biological samples [53]. Schematic diagram of a typical ICP-MS instrument is shown in Figure 6.1. In ICP-MS,

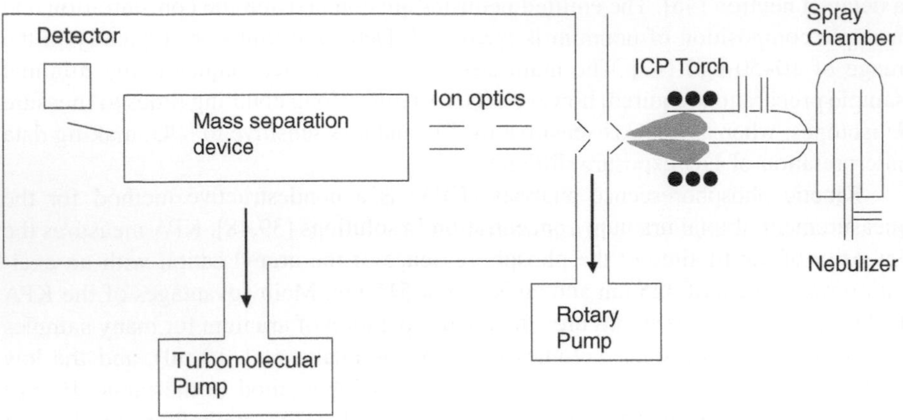

FIGURE 6.1 Schematic of an ICP-MS instrument.

the sample is introduced into the instrument in the form of a fine aerosol using a nebulizer. The aerosol is injected into a high-temperature plasma (5000–10,000 K) where positively charged ions are generated. An ion optics system guides the ions into the mass spectrometer, where they are separated according to mass-to-charge ratios. Finally, the ions are counted by an ion multiplier or Faraday detector. Types of mass spectrometers used in ICP-MS are quadrupole, magnetic sector and time of flight instruments [53]. Each of these mass spectrometers has its advantages and disadvantages. Quadrupole mass spectrometers (QMS) are the most common and consist of four which a radio frequency (RF) and a direct current (DC) are applied. By changing the RF and DC potentials, the quadrupole allows only ions with a single mass-to-charge ratio to pass toward the detector. The resolution (resolution = m/FW, where m is the mass of the selected ion and FW is width of the peak) of a quadrupole is about 350 for uranium isotopes. Magnetic sector field (SFMS) instruments use a magnetic field to disperse the ions according to their mass and energy. After passing through the magnet, the ions are focused, based on their energy, using an electrostatic sector. The combination of the magnetic and electrostatic sector provides the high resolution mass-to-charge separation. The resolution of double-focusing instruments is between 300–10000. Time-of-flight (TOF) mass spectrometers incorporate a flight tube in which ions are accelerated and travel at various speeds depending on their mass-to-charge ratios. A time-of-flight instrument overcomes the scanning nature of the QMS and SFMS in the respect that they provide mass spectrum almost instantaneously (in less than a 1ms). Resolutions measured for uranium isotopes on ICP-TOF-MS are about 2000.

ICP-MS is a multielement technique, which can provide information about the different isotopes in the sample [54]. Thus, it can measure all natural and artificial uranium isotopes and provide an accurate ratio analysis needed for the assessment of depleted uranium exposure. The detection limits for ICP-MS instruments are below 1 ng/l and the method can provide isotopic ratios on samples containing 10 ng/l of total uranium. The advantages of ICP-MS are short sample preparation

TABLE 6.1
Comparison between Instrumental Analysis Techniques Used for Analysis of Uranium in Biological Samples

Technique	Total U Detection Limit, ng/L	Type of Analysis Possible	Precision, RSD	Sample Throughput	Cost
KPA	10	Total U	5%	High	Low
α-Spectometry	40	Total and isotopic U	10%	Low	Moderate
TIMS	<1	Total and isotopic U	0.02%	Low	High
INAA	10	Total and isotopic U	10%	Low	High
DNC	10	Total and isotopic U	10%	Moderate	High
ICP-MS	<1	Total and isotopic U	0.1–5%	High	Moderate

and instrument analysis time, low sample volumes, and high sample throughput. The disadvantage of the techniques is the relatively high cost of the instrumentation and actual analysis.

Table 6.1 summarizes the differences between the above described techniques for the analysis of uranium in urine. Neutron activation analysis in both the form of INAA and DNC provides good detection limits for total U quantification. However, INAA analysis is based on the ^{238}U, and the DNC analysis is based on ^{235}U, which make measurements of isotopic composition of uranium difficult. Adding to that is the high cost of the equipment as it requires a nuclear reactor for the generation of neutrons. KPA, on the other hand, provides a low-cost analysis and high sample throughput but lacks the capability of discrimination between the different uranium isotopes, making the ascertaining of DU exposure impossible. α-Spectroscopy is capable of accurate isotopic uranium composition determination but lacks the detection limits of the rest of the techniques, and it has a low sample throughput. All of the above techniques suffer from one additional disadvantage: low precision of the analysis at total U levels below 100 ng/l in biological fluids. TIMS offers the probably the best sensitivity for total and isotopic uranium analysis but, because of its long sample preparation times and lack of automated analysis, is not suited for large-scale biomonitoring applications. ICP-MS does not have the sensitivity of TIMS (at least when single collector instruments are used) but offers the highest throughput and precise ratio analysis at uranium levels of 3–10 ng/l in urine.

URANIUM LEVELS IN URINE SPECIMEN

Uranium is widely distributed in the natural environment in soils, air, and food [55]. The toxicity of uranium has been the subject of many environmental and occupational health studies [56,57]. Earlier studies have been focused on uranium exposure from miners and mining communities. More recently, a renewed interest on the environmental health impacts of uranium has been developed based on potential exposure to depleted uranium during the Gulf War, Kosovo, and Operation Iraqi Freedom [3–6]. Recent toxicological and clinical studies have demonstrated that high levels of uranium can affect the kidney and bone tissues [55,58]. In addition, several biomonitoring programs have been developed to provide reference levels of uranium exposure to the general population. For example, the U.S. Center for Disease Control has added uranium to their 2nd and 3rd National Reports on Human Exposure to Environmental Chemicals (NHANES) [59]. In this NHANES survey, a sample of ~5000 individuals across the U.S. demonstrated a geometrical mean urinary uranium excretion of 9 ng/l, and 95% of the population showed levels below 46 ng/l. Studies in Canada, India, and Israel have shown uranium levels of the general population in the range of 1–50 ng/l [47,60,61]. Surveillance exposure assessment of soldiers and veterans from military conflicts where DU has been used have been performed by many laboratories, with one of the most comprehensive studies published by the Baltimore Veterans Administration (VA) Medical Center [3,4]. The VA study is related to a cohort of veterans that have been directly exposed to DU from metal fragments in "friendly fire" incidents and a control group of non-DU–exposed veterans. The last survey conducted by the VA research group showed uranium urinary exposure levels in the range of 1–50,000 ng/l [3]. From these biomonitoring investigations it can be concluded that the method detection limit for the determination of total uranium in urine should be in the range of 1–10 ng/l. In order to correctly identify depleted uranium exposure isotopic composition should be measured accurately on specimen containing 9 ng/l and above of total uranium.

URANIUM ANALYSIS IN URINE BY ICP-MS

The sample preparation for urine uranium analysis for ICP-MS vary greatly between the different laboratories based on the detection limits desired and sample introduction systems used for the analysis. The first step in the sample preparation is acidification of the urine in order to preserve it and to leach any uranium from the walls of the collection vessel. Often the urine is digested in a microwave using concentrated acid to break down the organic contents [29,30,34–36]. Another option is wet or dry ashing of the samples in the presence of acid [12,31]. The ashing step often results in a preconcentration of U, since most procedures start with high initial volume of urine and dissolve the resulting ash in small volume of diluted acid. Digestion/ashing is often followed by extraction chromatography using UTEVA and TRU resins [29,34–36]. This procedure separates uranium from other actinides as well as high abundant ions in urine such as Na, P, Cl, I, etc. Other studies have demonstrated direct extraction or ion exchange chromatography on a small volume urine specimen (1–5 ml) [11,62,63]. Results from direct dilution of urine for total and isotopic uranium measurements have also been reported [11,32,33,40,47,64–66].

The sample introduction system is inherent part of the ICP-MS instrumentation. Depending on the sample preparation chosen for the DU analysis, various sample introduction systems can be used. Most frequently ICP-MS instruments are equipped with either concentric/Meinhard or cross-flow nebulizers connected to a cyclonic or Scott double-pass spray chambers. These systems can handle high content of dissolved solids (as permitted by the ICP-MS) and are generally not susceptible to clogging. They provide good detection limits and are suitable for use with 0.2–2 ml/min flow rates. Using a quadrupole reaction cell ICP-MS equipped with a Meidhard type A nebulizer and a cyclonic spray chamber, detection limits of 0.3 ng/l for total uranium and 15 % accuracy in the $^{235}U/^{238}U$ ratios were determined when total uranium levels were above 10 ng/l [40]. In the study conducted by Ejnik et al. [40], the urine sample was diluted 2x with 2 % HNO_3 and was directly introduced in the nebulizer. To increase the precision and accuracy the in the measurements or to reduce the amount of sample used for the analysis other sample introduction systems can be used including quartz direct-injection high-efficiency nebulizers (DIHEN), membrane desolvation systems and ultrasonic nebulizers. DIHEN introduces the sample directly into the plasma without the need of a spray chamber. The flow rates from a DIHEN are in the order of 10–120 µl/min, resulting in a low sample consumption without a significant decrease in sensitivity [24,31]. Desolvation systems or ultrasonic nebulizers increase the sample transport efficiency which results in 5–15-fold enhancement of the signal intensity.

In studies where ICP-MS has been used to assess depleted uranium exposure, difficulties were encountered at low uranium levels (< 50 ng U L^{-1}) caused by increased $^{235}U/^{238}U$ ratios [12,30,32]. These higher ratios were explained either by (1) high background in urine at m/z 235 [31], (2) formation of isobaric interferences at mass to charge (m/z) 235 associated with biological matrices [11,67], (3) matrix-induced signal reduction [35,36,40], or (4) insufficient sensitivity for low abundance of ^{235}U isotope. On a direct analysis of urine using a high resolution magnetic sector field instrument, a polyatomic interference at m/z 234.81 that was not a combination of any two isotopes of any elements was observed [11]. A similar study on geological samples with high organic content indicated the presence of polyatomic interferences on all m/z between 229 and 237 [67]. Extensive sample preparations involving extraction chromatography were used to extract uranium from biological matrices and to separate the uranium isotopes from the polyatomic interferences at m/z 234.81 [11,29,34–36,62,63]. Extraction procedures are cumbersome, time consuming, and subject to uranium contamination from chemicals used in the extraction techniques. The polyatomic interference at m/z 234.81 has an additive effect on the ^{235}U signal causing the measured amount of ^{235}U acquired in the analysis to be greater than expected for low-level urine specimen (below 20 pg U ml^{-1}). The polyatomic interference results in samples that appear to contain uranium enriched in ^{235}U based on their measured $^{235}U/^{238}U$ ratios. This incorrect measurement can lead to false concerns that processed uranium from nuclear fuel or spent nuclear fuel is being released into the environment.

The types of ICP-MS instruments that are designed to minimize polyatomic interferences are high resolution magnetic sector field instruments and dynamic reaction cell (DRC) or collision cell instruments [53,54,68–70]. Magnetic sector instruments use a high current magnet in order to resolve the sought isotope from the interferences [53]. In collision cell or DRC system the interference's effects are minimized by tandem quadrupoles in which the first quadrupole can be pressurized with an external gas [53]. The principle behind the DRC is that (1) an external gas reacts with the interfering ions or (2) the external gas reacts with the ions of interest [53,68]. In the first case, this allows the reaction cell quadrupole to remove the interference while minimal reactions take place between the analyte ion and reaction gas. In the second case, the reaction gas and the analyte form a new chemical ion which moves the m/z ratio of the analyte away from the interference. For uranium isotopic analysis, oxygen was reacted with U^+ to form the chemical species UO_2^+, moving the analysis away from the m/z 234.81 interference to m/z 267.034. A second key feature of the reaction cell is that the pressurized quadrupole rods have rf and DC applied between the pole pairs allowing both for band pass tuning and kinetic energy discrimination. This provides a third mode of operation for this type of instrument, based on the changing the low and high band pass filters in order to minimize interferences. The ability to remove spectral interferences and achieve low detection limits is why the ICP-DRC-MS was determined to suitable for a DU bioassay.

In a recently published study from our laboratory, two parameters of the ICP-DRC-MS were utilized in order to eliminate the effect of polyatomic interferences, namely: (1) a cell reaction gas and (2) band pass mass cut-off ranges within the quadrupole in the reaction cell [40]. Utilizing the first technique, oxygen gas was used in the DRC to react with ionized uranium to form UO_2^+. Therefore, uranium signals were monitored at m/z 267 ($^{235}UO_2^+$) and m/z 270 ($^{238}UO_2^+$), away from m/z 235 where polyatomic interferences would otherwise interfere with the analysis. To evaluate the rate of UO_2^+ formation, the dependence of the intensities of m/z 238.05, 254.05, and 270.00 as a function of the oxygen flow rate was studied (Figure 6.2).

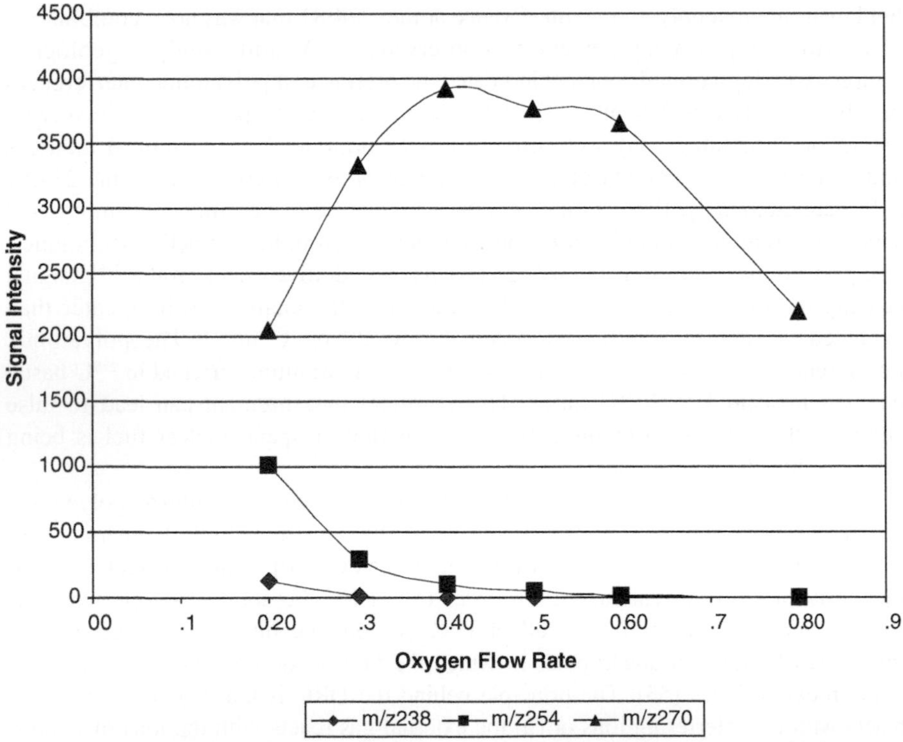

FIGURE 6.2 Optimization of DRC gas flow rate. Plot of signal intensity of m/z 238.05 ($^{238}U+$), 2540.05 ($^{238}UO+$), and 270.00 ($^{238}UO_2+$) vs. oxygen gas flow rate of the DRC.

The optimal oxygen gas flow for the DRC was determined by monitoring the signals of $^{238}U^+$, $^{238}UO^+$, and $^{238}UO_2^+$ with an increasing oxygen gas flow to the DRC. A gas flow of 0.5 ml min^{-1} was used because of the stability and high efficiency of UO_2^+ formation at this flow rate. The efficiency of $^{238}U^+$ conversion to $^{238}UO_2^+$ was measured to be greater than 90%. To further refine the analysis, the bandpass mass limits (RPa, RPq) were optimized at 0.0 and 0.7, respectively. At these parameters, no polyatomic interferences were observed, and the background signals were the same as the electronic noise at m/z 267 and 270. The bandpass mass cut off parameter was also investigated without the use of a reaction gas within the DRC. The band pass mass limits were not changed, and the m/z signals were monitored at 235.045 and 238.05. Interestingly, no polyatomic interferences were observed again, and the background signals were the same as the electronic noise.

An example of uranium total and isotopic analysis of 15 samples with known concentration is presented in Table 6.2 [40]. The total measured uranium in the urine sample was determined using an external calibration curve and ^{233}U as an internal standard. The ^{233}U signal corrects for all biological matrix effects and any instrument changes during the analysis. This correction is essential as the ionization of uranium decreases up to 60% from components within urine samples. Ionization is a chemical property governed by electron configurations; thus, all isotopes of uranium are affected exactly in the same manner by a sample's matrix. As ^{233}U, ^{235}U, and ^{238}U have the same ionization properties, ^{233}U as an internal standard corrects for any matrix effects from the sample and instrumentation variations during the analysis. The isotope of ^{238}U was used for quantification. The small change in ^{238}U composition

TABLE 6.2
Results of Urine Samples with Known DU Composition

Sample Number	Measured Concentration (ng/l)	Calculated Concentration (ng/L)	Measured $^{235}U/^{238}U$	Calculated^{235} $U/^{238}U$
1	3.5 (±0.1)	3.4	0.0073 (±0.0010)	0.0073
2	13.2 (±0.3)	13.4	0.0034 (±0.0006)	0.0038
3	13.4 (±0.5)	13.4	0.0054 (±0.0006)	0.0055
4	14.0 (±0.4)	13.4	0.0057 (±0.0006)	0.0063
5	13.9 (±0.6)	13.4	0.0068 (±0.0005)	0.0073
6	28.2 (±0.8)	28.4	0.0030 (±0.0003)	0.0032
7	29.3 (±1.2)	28.4	0.0050 (±0.0008)	0.0052
8	29.3 (±0.7)	28.4	0.0063 (±0.0004)	0.0064
9	27.7 (±2.0)	28.4	0.0072 (±0.0003)	0.0073
10	51.8 (±1.2)	53.4	0.0029 (±0.0002)	0.0029
11	54.2 (±1.6)	53.4	0.0051 (±0.0002)	0.005
12	55.3 (±1.3)	53.4	0.0065 (±0.0002)	0.0063
13	55.7 (±1.5)	53.4	0.0073 (±0.0002)	0.0073
14	99.1 (±1.4)	103	0.0026 (±0.0003)	0.0028
15	472 (±9)	503	0.0026 (±0.0001)	0.0026

from natural uranium's 99.274% to DU's 99.794% was determined to not significantly alter quantification in samples derived from clinical applications. The table also shows the measurement of the $^{235}U/^{238}U$ ratio and is compared to calculated values. The average relative errors for the 15 urine sample $^{235}U/^{238}U$ measurements were −2.1%.

An example of the uncertainties analysis associated with the measurements shown in Table 6.2 are shown in Table 6.3. The uncertainty associated with quantitative analysis of uranium in urine ranges between 1–3%. The uncertainty remains relatively constant from 5 ng U L^{-1} urine to 500 ng U L^{-1}. At urine uranium concentrations between 0.1–5 pg U ml^{-1}, the uncertainty does increase as the urine uranium concentrations decrease. The uncertainty associated with the measurement of $^{235}U/^{238}U$ ratios is illustrated in the second part of Table 6.3. The uncertainty was used to determine a ratio range in which a measured ratio can be considered statistical normal uranium (0.00725 $^{235}U/^{238}U$). Any measured ratio outside of the reported ranges means the urine sample has a 99.7% probability of not containing natural uranium. If the ratio is greater than the range, the urine sample contains enriched uranium. If the ratio is less than the range, the urine sample contains DU. This strategy is applied for the clinical application to determine if the urine sample does or does not contain DU with a defined level of certainty.

Table 6.4 displays a number of examples for the analysis of uranium in urine using ICP-MS. The figures of merit that were included in the summary are detection limit for total uranium determination, precision of total or ^{238}U quantification, and the precision of the $^{235}U/^{238}$ ratio measurement. From the table it can be seen that the total quantification is easily accomplished in most studies at levels around 1 ng/l or below using direct dilution, digestion, or extraction sample preparation methods.

TABLE 6.3
Method Uncertainties for Ranges of Urine Uranium Concentrations

	Quantitative Uranium Analysis		
	0.1–20 ng U l^{-1}	20–100 ng U l^{-1}	100–500 ng U l^{-1}
SD[a]	0.2	0.6	3
3 * (SD)	0.5	1.8	9
	Isotopic $^{235}U/^{238}U$ ratio		
	5–20 ng U l^{-1}	20 ng U l^{-1}	
SD[a]	0.0006	0.0004	
3 * (SD)	0.0017	0.0011	
Normal uranium range (99.7% CL[b])	0.0056–0.0090	0.0062–0.0084	

Note: All pooled standard deviations are in ng U l^{-1}.

[a] SD: Pooled standard deviation.
[b] CL: Confidence level.

TABLE 6.4
Figures of Merit for ICP-MS

	Sample Preparation	Instrument Used	Limit of Detection Total, ng/L	Precision (%RDS) of Total U Measurement	Precision (%RDS) of Total Isotopic ($^{235}u/^{238}u$)
Mohagheghi et al. [62]	Direct dilution	Q-ICP-MS	15		
Becker et al. [71]	Laser ablation	SF-ICP-MS	0.1	7	
Westphal et al. [31]	Ashing	Q-ICP-MS			16–100
Hang et al. [65]	Online extraction	Q-ICP-MS	2		10
Tomalchyov et al. [72]	Digestion, extraction	Q-ICP-MS	0.05	4.9	1.8
Benkhdda et al. [63]	Online extraction	SF-ICP-MS	0.02	3.0–9.2	0.8–1
Tresl et al. [34]	Digestion, extraction	SF-ICP-MS			1.3
Gwiazda et al. [11]	Direct dilution	SF-ICP-MS	0.017		2–8.6
Vanhaecke et al. [73]	Direct dilution	SF-ICP-MS	0.2	5	
Becker et al. [74]	Digestion	SF-ICP-MS	0.1		
Krystek et al. [66]	Direct dilution	SF-ICP-MS	0.2		
Haldiman et al. [32]	Direct dilution, isotope dilution	Q-ICP-MS	0.8		
Rodushkin et al. [75]	Direct dilution	SF-ICP-MS	<1		
Ejnik et al. [12]	Ashing	Q-ICP-MS	0.1		15
Ejnik et al. [40]	Direct dilution	Q-ICP-MS	0.3	1	8
Caddia et al. [33]	Direct dilution	Q-ICP-MS	0.3	3.9	
Karpas et al. [16]	Direct dilution	Q-ICP-MS	3		
Lorber et al. [61]	Direct dilution	Q-ICP-MS	1.5		
Ting et al. [64]	Direct dilution	SF-ICP-MS	1	2–5.5	
Schaumloeffel et al. [29]	Digestion, preconcentration	SF-ICP-MS	1.6		2.5
Pappas et al. [36]	Digestion, extraction	SF-ICP-MS			<1

Results from urine direct dilution analysis shows very similar precision as more laborious sample preparation approaches. Thus, the ICP-MS technique is well suited for analysis of uranium at levels below the range for the average population.

Determination of the source of exposure is more difficult as it can be observed from the decreased precision of the $^{235}U/^{238}U$ measurements, especially at low ng/l total U levels. From Table 6.4, it can be concluded the most accurate analysis on isotopic results are obtained using extraction chromatography coupled to a high resolution instrument magnetic sector instruments. In most cases such a method decreases the sample throughput as it introduces an extra sample preparation step. In a study by Pappas et al. [36], urine was first digested and uranium was extracted using a TRU resin column. The analysis was performed on a magnetic sector instrument and resulted in high precision $^{235}U/^{238}U$ ratio measurements (below 1%

RSD within run) at total uranium contents below 5 ng/l. Benkhdda et al. further improved the extraction analysis by developing a flow injection system for online uranium extraction from the urine matrix and when coupled with a magnetic sector instrument they were able to achieve very precise $^{235}U/^{238}$ analysis (below 1% RSD) in samples below 5 ng/l [63]. The analysis time in this case was shortened to about 10 min sample making such an analysis suitable for large-scale biological monitoring of depleted uranium exposure.

ELECTRON MICROSCOPIC ANALYSIS OF DU FRAGMENTS

Ultrastructural and microprobe methods, such as scanning electron microscopy (SEM) combined with energy dispersive x-ray analysis (EDS), is another useful method for identification of metals in solid samples (i.e., tissues) [71]. However, despite the wide use of these electron-based optical systems, there are disadvantages of the SEM and EDS techniques. Both techniques require extensive sample preparation (e.g., paraffin sections mounted on pure carbon discs and coated with a layer of gold or carbon to avoid electron beam charging effects).

Ultrastructural analysis allows metal fragments to be analyzed in an efficient and sensitive manner. The scanning electron microscope is a versatile approach for the study of microinclusions in tissues, with the capability for examining objects at low and high magnification with substantial depth of field and with a three-dimensional appearance. Scanning electron microscopy combined with energy dispersive x-ray analysis is a very powerful tool for the identification of mineral substances and to establish the elemental composition of foreign materials in tissues.

Standard sample preparation methods for SEM/EDS analysis of tissue sections require 4–6 micron thick sections that are then placed on pure carbon disks. Prior to mounting the section on the disk, the carbon disk is washed with concentrated H_2SO_4 (to remove iron), thoroughly rinsed with distilled deionized water, placed in acetone, and ultrasonicated for 2 h. Then the carbon disk is washed with distilled deionized water again and placed in a vacuum oven to dry for 12 h. Prior to SEM-EDS, tissue sections mounted on the disks are deparaffinized with xylene and absolute ethyl alcohol. In our laboratory, a Hitachi S-3500N scanning electron microscope interfaced to an energy dispersive spectrometer (EDS), and KEVEX software were used to examine the tissue sections discussed here. An accelerating voltage of 20 KeV and 90 nA current was used for most of the measurements. Backscattered electron images at magnifications from 60–6000× were used to observe morphologic characteristics of tissue and to record the composition of the inorganic material.

Figure 6.3 demonstrate the use of SEM/EDXA for the identification of uranium inclusion in a case of a DU fragment in subcutaneous soft tissue. The SEM-EDS was obtained at two magnifications, at 20× and 200×. At the lower magnification (Figure 6.3C), the x-ray microanalysis demonstrated the presence of not only uranium, but also Al and Fe (Figure 6.3B). At higher magnification, the SEM-EDAX analysis identified the metal fragment as uranium (Figure 6.3D). Accordingly, SEM

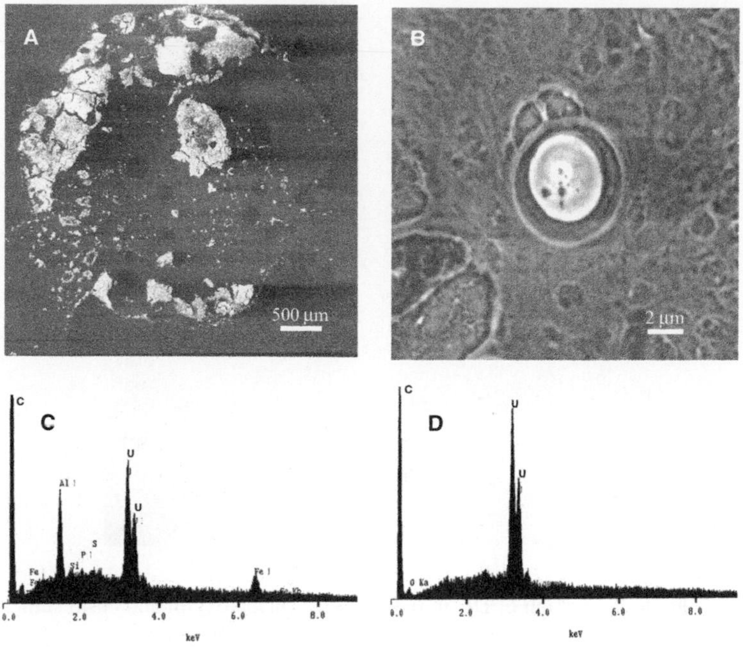

FIGURE 6.3 Scanning electron photomicrograph of uranium in a subcutaneous soft tissue biopsy. Panel B demonstrates the energy dispersive x-ray spectrum of uranium at an energy of approximately 3.8 KeV, as well as other elements including aluminum and iron (magnification 20×). By using a higher magnification (200×) only the presence of uranium was detected within the subcutaneous soft tissue section.

coupled with energy dispersive x-ray microanalysis allows us to examine the elemental composition of solid materials in tissues and identify whether a material is of foreign or natural origin.

HISTOPATHOLOGIC EVALUATION OF DU IN TISSUES

DU fragments in tissues induced a chronic inflammatory reaction with reactive fibrosis and foreign body giant cell formation (Figure 6.4 to Figure 6.6) [77]. Figure 6.4 displays a typical reactive soft tissue with foci characteristic of foreign material surrounded by syncytial cells, lymphocytes, and prominent hemosiderin-laden macrophages. The fibrovascular tissue has focally hyperchromatic irregular nuclei consistent with radiation effect. Although the DU foreign material does not polarize, the material has a variegated coloration by hemotoxylin and eosin stain, from a light to dark brown tan as demonstrated in Figure 6.5. Iron stain (Prussian blue) demonstrates positivity with the foreign material as well as prominent positivity of hemosiderin in surrounding tissue as shown in Figure 6.6.

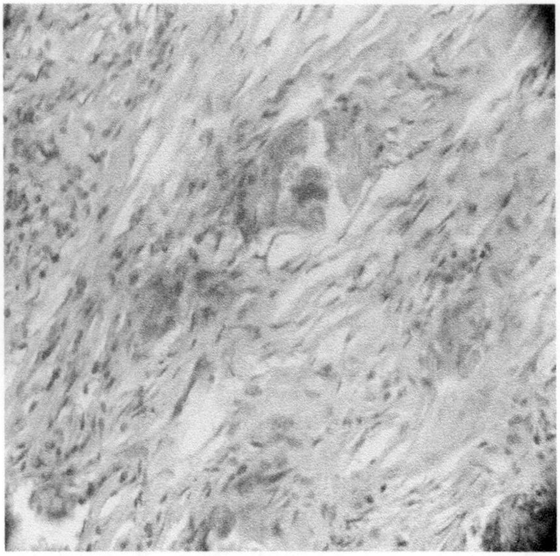

FIGURE 6.4 Photomicrograph of DU in tissue (skin) (hematoxylin and eosin stain; original magnification, 20×).

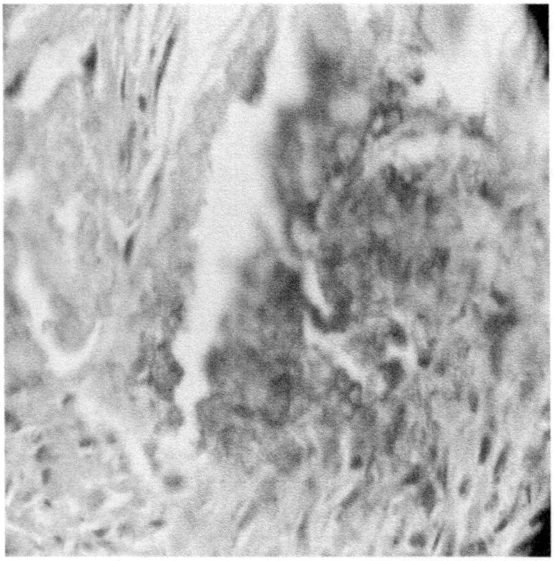

FIGURE 6.5 Photomicrograph of DU in tissue (skin) (hematoxylin and eosin stain; original magnification, 40×).

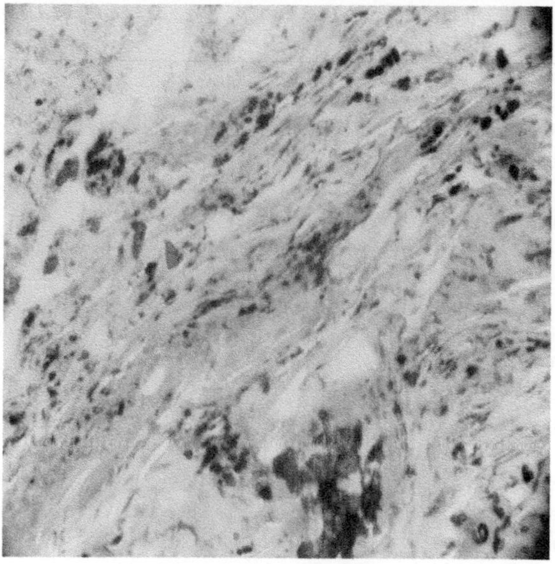

FIGURE 6.6 Photomicrograph of DU in tissue (skin) (Prussian blue stain; original magnification, 20×).

REFERENCES

1. Army Environmental Policy Institute, Health and Environmental Consequences of Depleted Uranium Use in the U.S. Army: Technical Report, Atlanta, 1995.
2. Uijt de Haag, P.A., Smetsers, R.C., Witlox, H.W., Krus, H.W., and Eisenga, A.H., Evaluating the risk from depleted uranium after the Boeing 747-258F crash in Amsterdam, 1992, *J. Hazard. Mater.,* 76, 39, 2000.
3. McDiarmid, M.A., Engelhardt, S.M., Oliver, M., Gucer, P., Wilson, P.D., Kane, R., Kabat, M., Kaup, B., Anderson, L., Hoover, D., Brown, L., Albertini, R.J., Gudi, R., Jacobson-Kram, D., Thorne, C.D., and Squibb, K.S., Biological monitoring and surveillance results of Gulf War I veterans exposed to depleted uranium, *Int. Arch. Occup. Environ. Health,* 79, 11, 2006.
4. McDiarmid, M.A., Engelhardt, S., Oliver, M., Gucer, P., Wilson, P.D., Kane, R., Kabat, M., Kaup, B., Anderson, L., Hoover, D., Brown, L., Handwerger, B., Albertini, R.J., Jacobson-Kram, D., Thorne, and C.D., and Squibb, K.S., Health effects of depleted uranium on exposed Gulf War veterans: a 10-year follow-up, *J. Toxicol. Environ. Health A,* 67, 277, 2004.
5. McDiarmid, M.A., Keogh, J.P., Hooper, F.J., McPhaul, K., Squibb, K., Kane, R., DiPino, R., Kabat, M., Kaup, B., Anderson, L., Hoover, D., Brown, L., Hamilton, M., Jacobson-Kram, D., Burrows, B., and Walsh M., Health effects of depleted uranium on exposed Gulf War veterans, *Environ. Res.,* 82, 168, 2000.
6. McDiarmid, M.A., Squibb, K., Engelhardt, S., Oliver, M., Gucer, P., Wilson, P.D., Kane, R., Kabat, M., Kaup, B., Anderson, L., Hoover, D., Brown, L., and Jacobson-Kram, D., Surveillance of depleted uranium exposed Gulf War veterans: health effects observed in an enlarged "friendly fire" cohort, *J. Occup. Environ. Med.,* 43, 991, 2001.

7. Voegtlin, C. and Hodge, H.C., *Pharmacology and Toxicology of Uranium Compounds, I and II*, McGraw-Hill, New York, 1949.
8. Voegtlin, C. and Hodge, H.C., *Pharmacology and Toxicology of Uranium Compounds, III and IV*, McGraw-Hill, New York, 1953.
9. Miller, A.C., Blakely, W.F., Livengood, D.L., Whitaker, T., Xu, J., Ejnik, J.W., and Hamilton, M.M., Parlette E., St. John T., Gerstenburg H.M., Hsu H., Transformation of human osteoblast cells to the tumorigenic phenotype by depleted uranium-uranyl chloride, *Environ. Health Perspect.*, 106, 465, 1998.
10. Hooper, F.S., Squibb, K.S., Siegel, E.L., McPhaul, K., and Keogh, J.P., Elevated urine uranium excretion by soldiers with retained uranium shrapnel, *Health Phys.*, 77, 512, 1999.
11. Gwiazda, R.H., Squibb, K., McDiarmid, M., and Smith, D., Detection of depleted uranium in urine of veterans from the 1991 Gulf War, *Health Phys.*, 86, 12, 2004.
12. Ejnik, J.W., Carmichael, A.J., Hamilton, M.M., McDiarmid, M.A., Squibb, K.S., Boyd, P., and Tardiff, W., Determination of the isotopic composition of uranium in urine by inductively coupled plasma mass spectrometry, *Health. Phys.*, 78,143, 1999.
13. Horan, P., Dietz, L., and Durakovic, A. The quantitative analysis of depleted uranium isotopes in British, Canadian, and U.S. Gulf War veterans, *Mil. Med.*, 167, 620, 2002.
14. Brina, R. and Miller, A.G., Direct detection of trace levels of uranium by laser-induced kinetic phosphorimetry, *Anal. Chem.*, 64, 1413, 1992.
15. Caravajal, G.S. and Mahan, K.I., The determination of uranium in natural waters at ppb levels by thin-film x-ray fluorescence spectrometry after coprecipitation with an iron dibenzyldithiocarbamate carrier complex, *Anal. Chim. Acta.*, 135, 205, 1982.
16. Karpas, Z., Halicz, L., Roiz, J., Marko, R., Katorza, E., Lorber, A., and Goldbart, Z., Inductively coupled plasma mass spectrometry as a simple, rapid, and inexpensive method for determination of uranium in urine and fresh water: comparison with LIF, *Health Phys.*, 71, 879, 1996.
17. Kressin, I.K., Spectrophotometric method for the determination of uranium in urine, *Anal. Chem.*, 56, 2269, 1984.
18. Lang, S. and Raunemaa, T., Behaviour of neutron-activated UO2-dust particles in the gastrointestinal tract of the rat, *Radiat. Res.*, 126, 273, 1991.
19. Pleskach, S.D., Determination of U and Th in urine by neutron activation, *Health Phys.*, 48, 303, 1985.
20. Rendl, J., Seybold, S., and Borner, W. Urinary iodide determined by paired-ion reversed-phase HPLC with electrochemical detection, *Clin. Chem.*, 40, 908, 1994.
21. Sachett, I.A., Nobrega, A.W., and Lauria, D.C. Determination of uranium isotopes by chemical stripping and alpha-spectrometry, *Health. Phys.*, 46, 133, 1984.
22. Allain, P., Berre, S., and Premel-Cabic, A.Y. Investigation of the direct determination of uranium in plasma and urine by inductively coupled plasma mass spectrometry, *Anal. Chim. Acta*, 251, 183, 1991.
23. Desideri, D., Meli, M.A., Roselli, C., Testa, C., Boulyga, S.F., and Becker, J.S., Determination of (236)U and transuranium elements in depleted uranium ammunition by alpha-spectrometry and ICP-MS, *Anal. Bioanal. Chem.*, 374, 1091, 2002.
24. Boulyga, S.F., Matusevich, J.L., Mironov, V.P., Kudjashov, V.P., Halicz, L., Segal, I., McLean, J.A., Montaser, A., and Becker, J.S., Determination of $^{236}U/^{238}U$ isotope ratio in contaminated environmental samples using different ICP-MS instruments, *J. Anal. At. Spectrom.*, 17, 958, 2002.
25. Uchida, S., Garcia-Tenorio, R., Tagami, K., and Garcia-Leon, M., Determination of U isotopic ratios in environmental samples by ICP-MS, *J. Anal. At. Spectrom.*, 15, 889, 2000.
26. Boulyga, S.F., Testa, C., Desideri, D., and Becker, J.S., Optimization and application of ICP-MS and alpha-spectrometry for determination of isotopic ratios of depleted

uranium and plutonium in samples collected in Kosovo, *J. Anal. At. Spectrom.,* 16, 1283, 2001.

27. Boulyga, S.F. and Becker, J.S., Isotopic analysis of uranium and plutonium using ICP-MS and estimation of burn-up of spent uranium in contaminated environmental samples, *J. Anal. At. Spectrom.,* 17, 1143, 2002.
28. Zoriy, M.V., Halicz, L., Ketterer, M.E., Pickhardt, C., Ostapczuk, P., and Becker, J.S., Reduction of UH^+ formation for $^{236}U/^{238}U$ isotope ratio measurements at ultratrace level in double focusing sector field ICP-MS using D_2O as solvent, *J. Anal. At. Spectrom.,* 19, 362, 2004.
29. Schaumloeffel, D., Giusti, P., Zoriy, M.V., Pickhardt, C., Szupnar, J., Lobinski, R., and Becker, J.S., Ultratrace determination of uranium and plutonium by nano-volume flow injection double-focusing sector field inductively coupled plasma mass spectrometry (nFI-ICP-SFMS), *J. Anal. At. Spectrom.,* 20, 17, 2005.
30. Hodge, S.J., Ejnik, J., Squibb, K.S., McDiarmid, M.A., Morris, E.R., Landauer, M.R., and McClain, D.E., Detection of depleted uranium in biological samples from Gulf War veterans, *Mil. Med.* 166, Suppl. 2, 69, 2001.
31. Westphal, C.S., McLean, J.A., Hakspiel, S.J., Jackson, W.E., McClain, D.E., and Montaser, A., Determination of depleted uranium in urine via isotope ratio measurements using a large-bore direct injection high efficiency nebulizer — inductively coupled plasma mass spectrometry, *Appl. Spectrosc.,* 58, 1044, 2004.
32. Halidiman, M., Barudaux, M., Eastgate, A., Froidevaux, P., O'onovan, S., Von Gunten, D., and Zoller, O., Determining pictogram quantities of uranium in urine by isotope dilution inductively coupled plasma mass spectrometry: comparison with alpha-spectrometry, *J. Anal. At. Spectrom.,* 16, 1364, 2001.
33. Caddia, M. and Iversen, B.S., Determination of uranium in urine by inductively coupled plasma mass spectrometry with pneumatic nebulization, *J. Anal. At. Spectrom.,* 13, 309, 1998.
34. Tresl, I., DeWannemacker, G., Quetel, C.E., Petrov, I., Vanhaecke, F., Moens, L., and Taylor, P.D., Validated measurements of the uranium isotopic signature in human urine samples using magnetic sector-field inductively coupled plasma mass spectrometry, *Environ. Sci. Technol.,* 38, 581, 2004.
35. Pappas, R.S., Ting, B.G., Jarret, J.M., Paschal, D.C., Caudill, S.P., and Miller, D.T., Determination of uranium-235, uranium-238 and thorium-232 in urine by magnetic sector inductively couple plasma mass spectrometry, *J. Anal. At. Spectrom.,* 17, 131, 2002.
36. Pappas, R.S., Ting, B.G., and Paschal, D.C. A practical approach to determination of low concentration uranium isotope ratios in small volumes of urine, *J. Anal. At. Spectrom.,* 18, 1289, 2003.
37. Jia, G., Belli, M., Sansone, U., Rosamilia, S., Ocone, R., and Gaudino, S., Determination of uranium isotopes in environmental samples by alpha-spectrometry, *J. Radioanal. Nucl. Chem.,* 253, 395, 2002.
38. Jia, G., Torri, G., and Innocenzi, P., An improved method for the determination of uranium isotopes in environmental samples by alpha-spectrometry, *J. Radioanal. Nucl. Chem.,* 262, 433, 2004.
39. Hedaya, M.A., Birkenfeld, H.P., and Kathren, R.L., A sensitive method for the determination of uranium in biological samples ulitizing kinetic phosphorescence analysis (KPA), *J. Pharm. Biomed. Anal.,* 15, 1157, 1997.
40. Ejnik, J.W., Todorov, T.I., Mullick, F.G., Squibb, K., McDiarmid, M.A., and Centeno, J.A., Uranium analysis in urine by inductively coupled plasma dynamic reaction cell mass spectrometry, *Anal. Bioanal. Chem.,* 382, 73, 2005.

41. Othman, I., The relationship between uranium in blood and the number of working years in the Syrian phosphate mines, *J. Environ. Radioact.* 18, 151, 1993.
42. Byrne, A.R. and Benedik, L., Uranium content of blood, urine and hair of exposed and non-exposed persons determined by radiochemical neutron-activation analysis with emphasis on quality control, *Sci. Total Environ.*, 107, 143, 1991.
43. Ide, H.M., Moss, W.D., Minor, M.M., and Campbell, E.E., Analysis of uranium by delayed neutrons, *Health Phys.*, 37, 405, 1979.
44. Brits, R.J.N. and Holemans, E.A., Determination of uranium in urine by delayed neutrons, *Health Phys.* 36, 65, 1979.
45. Alfassi, Z., Introduction — principles to activation analysis, in *Activation Analysis*, Vol. 1, Alfassi, Z., Ed., CRC Press, Boca Raton, FL, 1990, chap. 1.
46. Alfassi, Z., Use of delayed neutrons in activation analysis, in *Activation Analysis*, Vol. 1, Alfassi, Z., Ed., CRC Press, Boca Raton, FL, 1990, chap. 6.
47. Ough, E.A., Lewis, B.J., Andrews, W.S., Bennett, L.G.I., Hancock, R.G.V., and Scott, K., An examination of the uranium levels in Canadian forces personnel who served in the Gulf War and Kosovo, *Heath Phys.*, 82, 527, 2002.
48. Ejnik, J.W., Hamilton, M.M., Adams, P.R., and Carmichael, A.J., Optimal sample preparation conditions for the determination of uranium in biological samples by kinetic phosphorescence analysis (KPA), *J. Pharm. Biomed. Anal.*, 24, 227, 2000.
49. Elliston, J.T., Glover, S.E., and Filby, R.H., The determination of natural uranium in human tissues by recovery corrected kinetic phosphorescence analysis, *J. Radioanal. Nucl. Chem.*, 248, 487, 2001.
50. Elliston, J.T., Glover, S.E., and Filby, R.H., Comparison of direct kinetic phosphorescence analysis and recovery corrected kinetic phosphorescence analysis for the determination of natural uranium in human tissues, *J. Radioanal. Nucl. Chem.*, 263, 301, 2005.
51. Smith, D., Thermal ionization mass spectrometry, in *Inorganic Mass Spectrometry: Fundamentals and Applications*, Barshick C., Ed., Marcel Dekker, New York, 2000, chap. 1.
52. Ide, H.M., Moss, W.D., and Gautier, M.A., Bioassay alpha spectrometry: energy resolution as a function of sample source preparation and counting geometry, *Health Phys.*, 56, 71, 1989.
53. Thomas, R., *Practical guide to ICP-MS*, 1st ed., Marcel Dekker, New York, 2004.
54. Becker, J.S., Recent developments in isotope analysis by advanced mass spectrometry techniques, *J. Anal. At. Spectrosc.*, 20, 1173, 2005.
55. Agency for toxic substances and disease registry, Toxicological Profile for Uranium, U.S. Department of Health and Human Services, Public Health Service, Atlanta, 1999.
56. Fisher, D.R., Jackson, P.O., Brodaczynski, G.G., and Scherpelz, R.I., Levels of ^{234}U, ^{238}U, and ^{230}Th in excreta of uranium mill crushermen, *Health Phys.*, 45, 617, 1983.
57. Thun, M.J., Baker, D.B., Steenland, K., Smith, A.B., Halperin, W., and Berl, T., Renal toxicity in uranium mill workers, *Scand. J. Work. Environ. Health*, 11, 83, 1985.
58. Squibb, K.S., Leggett, R.W., and McDiarmid, M.A., Prediction of renal concentration of depleted uranium and radiation dose in Gulf War 1 veterans with embedded shrapnel, *Health Phys.*, 89, 267, 2005.
59. Centers for Disease Control, Third National Report on Human Exposure to Environmental Chemicals, U.S. Department of Health and Human Services, Public Health Service, Atlanta, 2005.
60. Dang, H.S., Pullat, V.R., and Pillai, K.C., Determining the normal concentrations of uranium in urine and applications of the data to its biokinetics, *Health Phys.*, 62, 562, 1992.
61. Lorber, A., Karpas, Z, and Halicz, L., Flow injection method for the determination of uranium in urine and serum by inductively coupled plasma mass spectrometry, *Anal. Chim. Acta*, 334, 295, 1996.

62. Hang, W., Zhu, L., Zhong, W., and Mahan, C., Separation of actinides at ultra-trace level from urine matrix using extraction chromatography-inductively coupled plasma mass spectrometry, *J. Anal. At. Spectrosc.,* 19, 966, 2004.
63. Benkhedda, K., Epov, V.N., and Evans, R.D., Flow-injection technique for determination of uranium and thorium isotopes in urine by inductively coupled plasma mass spectrometry, *Anal. Bioanal. Chem.,* 381, 1596, 2005.
64. Ting, B.G., Paschal, D.C., Jarrett, J.M., Pirkle, J.L., Jackson, R.J., Sampson, E.J., Miller, D.T., and Caudill, S.P., Uranium and thorium in urine of United States residents: reference range concentrations, *Environ. Res.,* 81, 45, 1999.
65. Mohagheghi, A.H., Shanks, S.T., Zigmond, J.A., Simmons, G.L., and Ward, S.L.A., A survey of uranium and thorium background levels in water, urine, and hair and determination of uranium enrichments by ICP-MS, *J. Radioanal. Nucl. Chem.* 263, 189, 2005.
66. Krystek, P. and Ritsema, R., Determination of uranium in urine — measurement of isotope ratios and quantification by use of inductively coupled plasma mass spectrometry, *Anal. Bioanal. Chem.,* 374, 226, 2002.
67. Chen, C.C., Edwards, R.L., Cheng, H., Dorale, J.A., Thomas, R.B., Moran, S.B., Weinstein, S.E., and Edmonds, H.N., Uranium and thorium isotopic and concentration measurements by magnetic sector inductively coupled plasma mass spectrometry, *Chem. Geol.*, 185, 165, 2002.
68. Tanner, S.D., Baranov, V.I., and Bandura, D.R., Reaction cells and collision cells for ICP-MS: a tutorial review, *Spectrochim. Acta Part B,* 57, 1361, 2002.
69. Becker, J.S., Applications of inductively coupled plasma mass spectrometry and laser ablation inductively coupled plasma mass spectrometry in materials science, *Spectrochim. Acta Part B* 57, 1805, 2002.
70. Vanhaecke, F. and Moens, L., Recent trends in trace element determination and speciation using inductively coupled plasma mass spectrometry, *Fresenius J. Anal. Chem.,* 364, 440, 1999.
71. Becker, J.S., Burow, M., Zoriy, M.V., Pickhardt, C., Ostapczuk, P., and Hille, R., Determination of uranium and thorium at trace and ultratrace levels in urine by laser ablation ICP-MS, *At. Spectrosc.,* 25, 197, 2004.
72. Tolmachyov, S.Y., Kuwabara, J., and Noguchi, H., Flow injection extraction chromatography with ICP-MS for thorium and uranium determination in human body fluids, *J. Radioanal. Nucl. Chem.,* 261, 125, 2004.
73. Vanhaecke, F., Stevens, G., De Wannemacker, G., and Moens, L., Sector field ICP-mass spectrometry for the routine determination of uranium in urine, *Can. J. Anal. Sci. Spectrosc.,* 48, 251, 2003.
74. Becker, J.S., Burow, M., Boulyga, S.F., Pickhardt, C., Hille, R., and Ostapczuk, P., ICP-MS determination of uranium and thorium concentrations and $^{235}U/^{238}U$ isotopic ratios at trace and ultratrace levels in urine, *At. Spectrosc.,* 23, 177, 2002.
75. Rodushkin, I. and Odman, F., Application of inductively coupled plasma sector field mass spectrometry for elemental analysis of urine, *J. Trace Elem. Med. Biol.,* 14, 241, 2001.
76. Goldstein, J.I., Newbury, D., Joy, D., Lyman, C., Echlin, P., Lifshin, E., Sawyer, L., and Michael, J., *Scanning Electron Microscopy and X-Ray Microanalysis*, 1st ed, Plenum Publishing Corporation, New York, 2002.
77. Katzin, W.E., Centeno, J.A., Feng, L.U., Kiley, M., and Mullick, F.G., Health effects of depleted uranium exposure, *Histopathology* 41(Suppl. 2), 327, 2002.

7 Exposure and Health Surveillance in Gulf War Veterans Exposed to Depleted Uranium

Katherine S. Squibb and Melissa A. McDiarmid

CONTENTS

Despite the wealth of knowledge available on uranium (U) metabolism, excretion, and toxicity in animals and humans exposed to natural and enriched U in 1991 [1–3], there was minimal information available on depleted uranium (DU) toxicity when U.S. Gulf War soldiers were exposed to DU for the first time under battlefield conditions. In multiple friendly-fire incidents involving munitions containing DU penetrators and DU armored tanks, U.S. soldiers were exposed by inhalation, ingestion, and wound contamination to aerosolized DU oxides formed when DU penetrators struck U.S. Bradley and Abrams tanks. In addition, some soldiers in or on the U.S. tanks were hit by fine metal fragments that still remain embedded in their muscle tissue. Precise predictions of the short- and long-term health effects of DU under these unique exposure conditions were not fully possible despite the substantial toxicological database available, due to the limited information available on the size and solubility of the DU oxides formed under these battlefield conditions. In addition, knowledge gaps existed regarding the local and systemic effects of the DU embedded fragments.

In this context, the U.S. Veterans Administration (VA) in 1993 established a health surveillance program for the Gulf War soldiers involved in these friendly-fire incidents to assess the health status of this cohort, and to better inform future decisions regarding the surgical handling of DU shrapnel and wound management for these soldiers and those involved in future incidents. Also to be addressed were the concerns voiced regarding the long-term health effects of inhaled, insoluble DU oxide particles, which could produce local tissue damage if retained for long periods of time. To address concerns among other Gulf War veterans about their exposures to DU, though they were not involved directly in friendly-fire incidents, a second VA activity emerged which offers any Gulf War veteran an opportunity to have a urine specimen analyzed for uranium [4]. The results of these two VA programs have considerably improved our understanding of the potential health effects of military DU exposures and the range of DU exposures that occurred during the war.

BALTIMORE VA DEPLETED URANIUM (DU) HEALTH SURVEILLANCE PROGRAM

Since its initiation in 1993, the Baltimore VA DU health surveillance program has enrolled 74 of the approximately 104 tank crew members who survived the Gulf War friendly fire incidents that occurred within a 48 h period in February 1991. These incidents involved 6 Abrams tanks equipped with DU armor for protection against enemy fire and 14 Bradley fighting vehicles mistakenly fired upon by U.S. forces using munitions containing DU penetrators [5]. DU penetrators are very effective weapons because the DU metal sharpens and burns as it penetrates tank armor, giving rise to aerosols of fine DU-oxide particles, which consist of a range of sizes and solubility [6]. Both of these parameters are critical determinants of toxicity when exposure occurs by inhalation, ingestion, or wound contamination, all of which occurred in soldiers present in or on these tanks when they were hit and in those that immediately entered the tanks as rescuers. In addition to the formation of fine DU dust particles, small shards of DU metal were also produced and became embedded in muscle tissue [7]. Due to the small size of many of these shards (from 1 mm to 2 mm in diameter or larger) and the fact that soldiers were hurt by multiple fragments (from 10 to more than 30) [8], these fragments were not immediately surgically removed due to concern for excessive soft tissue damage.

The Baltimore VA health surveillance program was established to follow this group of soldiers, which was defined as the "DU-Exposed" cohort, based on the high potential for significant exposure to DU in these soldiers. As a part of this program, all members of the cohort are invited to the Baltimore Veterans Affairs Medical Center (VAMC) every 2 years for a 3-d clinical assessment program. Of the total number of soldiers that survived these incidents (104), some have been lost to follow-up despite an active recruiting effort conducted by the Office of the Special Assistant for Gulf War Illness (OSAGWI) or have chosen not to participate in the program for personal reasons.

TABLE 7.1
Summary of the Number of GWI Soldiers Attending Health Surveillance Visits

Year	DU-Exposed	Nonexposed	Total*
1993–4	33		33
1997	29	38	67
1999	21+ 29 new		50
2001	31+ 8 new		39
2003	32		32
2005	30 + 4 new		34

Note: 74 participants have attended at least 1 inpatient evaluation, 30 have attended 3–5 inpatient evaluations, and 9 have attended all 6 inpatient evaluations.

Table 7.1 provides a summary of the number of Gulf War soldiers in the DU-exposed cohort who have participated in the six in-patient visits to date. A total of 74 soldiers have participated in at least one of the biennial visits, which started in 1993–1994. Many (30) of the DU-exposed cohort have participated in most of the visits (3–5), whereas 9 have participated in all six in-patient visits. Of the 74 who are participating in the surveillance program, 19 have evidence of retained shrapnel as indicated by skeletal x-ray analysis. In one visit (1997) a group of non-DU-exposed Gulf War soldiers were invited to participate as a control group to provide a better basis for interpreting the clinical information gathered from the DU–Exposed soldiers (Table 7.1). Gulf War–deployed soldiers with no known opportunities for DU exposure during their tour of duty were selected for the non-DU–exposed group based on their answers to a DU-exposure questionnaire. This questionnaire included information on their locations and activities during the war and is more fully described in McDiarmid et al. [9].

DU CHEMICAL AND RADIATION EXPOSURE ASSESSMENT

DU exposure assessment was one of the most critical issues that needed to be addressed in the early surveillance visits for the Gulf War DU-exposed cohort. Because DU is a radioactive metal (though less so than natural U by 40%) [7], different body-burden measurements were selected, based on both DU's radioactive and chemical nature. These included whole body radiation counting, urine–uranium excretion, and tissue radiation dose estimates.

Whole Body Radiation Counting

Whole body radioactivity counting (WBRC) was conducted at the Boston VA Medical Center on DU-Exposed soldiers, including those with shrapnel [9]. Although the Boston VA counting facility is ideal for detection of exposure to low doses of radioactive material, due the low background levels provided by the World War II steel used in its construction, results using this approach showed that whole body radiation measurements for DU were not a sensitive measure of DU body burden. Only 9 of the 29 DU-exposed soldiers for whom WBRC was conducted as part of the 1997 visit had detectable scores above the background provided by the counting chamber. All nine were soldiers with shrapnel. The lack of sensitivity of this method is due, in large part, to the low radioactivity of DU and the tissue absorption of DU radiation that was measured using a tissue-equivalent phantom containing known amounts of DU at different depths [8].

This work was useful for characterizing differences within the DU-exposed cohort and for establishing that all WBRC scores for the non-DU-exposed group were below the detection level of this method. The annual radiation dose estimate calculated from this work also indicated that the dose received by the soldier with the highest WBRC score was only slightly above the NRC's annual exposure allowance for the general public (0.11 compared to 0.1 rem, respectively) (NRC Regulation 20, Title 10, Chapter 1, Code of Federal Regulations-Energy 20.1301).

Urine Uranium Excretion

Urine U excretion has long been used as a measure of recent uranium exposure in the workplace [1,2], based on knowledge that uranyl ion (UO_2^{+2}), the most stable species of U in solution, is the most likely U species in blood [10]. About 40% of uranyl ions in blood form a complex with transferrin, which does not filter through the glomerulus, whereas about 60% form complexes with small anions, mainly carbonate, which do pass through the glomerulus into the tubular lumen. Association with $HCO_3^=$ decreases the interaction of the uranyl ion with the renal proximal tubule cell brush border membrane, thereby limiting its reabsorption and increasing its urinary excretion [10]. An occupational exposure decision level of 0.8 μg/l is used by the Department of Energy's Fernald Environmental Management Project (FEMP) [11] as a trigger for investigating work areas for unsuspected elevated exposure to U. Mean urine U concentrations in uranium mill workers reported by Thun et al. [12] were 65.2 μg/L of urine in 1975, with 39.1% of the workers exceeding the action level of 30 μg/L.

Urine excretion of U has been measured in all six of the surveillance visits of the DU Health Surveillance Program as a measure of systemic U exposure. The advantage of using urine U concentrations as a biomarker of exposure is the availability of well-established analysis methods for total U concentrations in urine samples and the availability of comparison values that help in the interpretation of the data.

As early as the first visit in 1993–1994, data indicated that DU-exposed soldiers with shrapnel were excreting high concentrations of U in their urine compared to DU-exposed soldiers without shrapnel (Figure 7.1). In succeeding years through 2005, urine U excretion has remained high in the soldiers with shrapnel. These data indicate that chronic, systemic exposure to DU in soldiers with shrapnel has been occurring for over 14 y due to the slow release of DU from embedded fragments oxidizing *in situ*. The mean for the soldiers with shrapnel is well above an upper limit value that could occur in a normal population due to intake of natural U from drinking water sources (0.365 μg/l) [13] and the occupational exposure decision

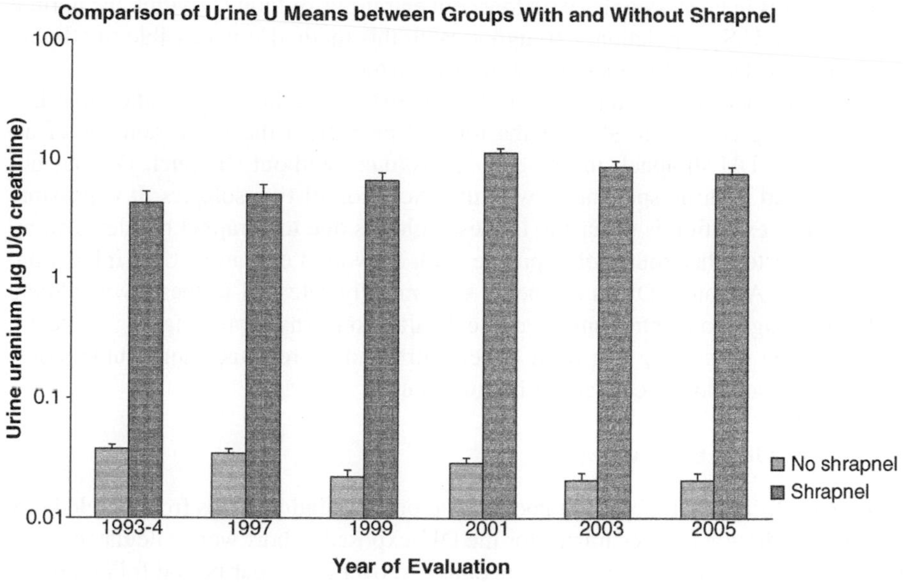

FIGURE 7.1 Comparison of mean urine uranium concentrations (µg U/g creatinine) between DU-exposed Gulf War soldiers who do (dark shaded bars) or do not (light shaded bars) have evidence of shrapnel as determined by x-ray analysis. Mean concentrations have not changed significantly between visits for those with shrapnel who are excreting uranium in their urine at concentrations well above the NHANES [14] 95th percentile range for the normal U.S. population aged 6 years and older (0.034 µg/g creatinine).

level of 0.8 µg/L is used by the Department of Energy's Fernald Environmental Management Project (FEMP) [11].

The mean for DU-exposed soldiers without shrapnel each year has remained close to the 95th percentile range reported by NHANES [14] for the normal U.S. population aged 6 years and older (0.034 µg/g creatinine). These data indicate that current systemic exposure to U in these soldiers, stemming from their exposure to DU in February 1991, is sufficiently low as to be within the normal range of U excretion. Slow release of DU remaining in tissues such as the lung and pulmonary lymph nodes may be occurring but it is not detectable against the natural background of total U in urine.

DU isotopic analysis: In order to better address the question of whether DU body burden can be detected using urine even at low exposures, new analytical methods have been developed to specifically measure DU in urine samples, based on determination of the U^{235}/U^{238} isotopic ratio of the total U present in a urine sample. Because DU is produced by the extraction of U^{235} and U^{234} from natural U during the U^{235} enrichment process, the U^{235}/U^{238} ratio in DU is 0.2 while that of natural U is 0.7.

An inductively coupled plasma dynamic reaction cell mass spectrometer (ICP/DRC/MS) technique developed recently [15] has the ability to measure U isotopic ratios in urine samples with low concentrations of total U (within the normal range for the U.S. population, ~10 ng/L). With this method it is possible to reliably detect DU as distinct from natural U in urine samples.

Isotopic analysis of samples from the DU-exposed cohort has verified that DU accounts for greater than 99% of the total U present in the urine samples of all soldiers with DU shrapnel. In DU-exposed soldiers without shrapnel, DU has not been detected in urine specimens, with the exception of two soldiers. It is possible that low DU excretion is occurring in these soldiers due to shrapnel not detected by x-ray or due to other routes of exposure such as wound contamination, inhalation, or ingestion. Although DU is not being systemically released in these two soldiers at high enough concentrations to cause health concerns from ongoing systemic exposure; the possibility that local effects surrounding residual, nonsoluble stores of DU in these soldiers cannot not be ruled out.

Radiation Dose Estimates

Due to the difficulty in obtaining good whole body radiation doses from whole body counting, radiation dose estimates for the DU-exposed cohort were calculated using each soldier's urinary U excretion data obtained over a 10-year period following the Gulf War (1993 through 2003) and the ICRP [16] biokinetic model for U [17]. The upper bound estimated lifetime (50-year) radiation dose for the DU-exposed soldier excreting the highest urine U concentration was 0.06 Sv, which is very close to the NCRP [18] allowable radiation dose for the public of 0.05 Sv and more than a factor of 10 less than the occupational lifetime limit of 1 Sv. The upper-bound estimate of 50-year tissue committed dose equivalents for the kidney and bone (1.3×10^{-5} and 3.6×10^{-5}, respectively) were well below the NCRP [18] occupational tissue committed effective dose standard of 25 Sv.

CLINICAL SURVEILLANCE RESULTS

Members of the DU-exposed cohort are invited to the Baltimore VAMC every 2 years for a 3-d clinical medical assessment program. The clinical assessments conducted during these visits have included a detailed medical history, including an extensive exposure history, a thorough physical examination, laboratory studies, and radiological surveys for retained DU fragments. Twenty-four hour urine samples are obtained for total uranium and DU analysis, and for clinical chemistry parameters related to renal function. Soldiers also complete a battery of neurocognitive tests and all soldiers are clinically evaluated during their first visit for posttraumatic stress disorder (PTSD). Laboratory studies include hematologic and blood clinical chemistry measures, as well as semen quality, neuroendocrine, immunologic, and genotoxicity parameters.

Careful surveillance of the function of specific target organs included in this program are based on the known toxicity of U as a heavy metal and known effects of other heavy metals such as lead (Pb). As a naturally radioactive metal, potential health effects caused by U due to its radiological and/or chemical nature have been considered. Animal studies conducted since the 1940s provide a strong base of knowledge regarding the most likely target organs for DU [1,10,19–23]. Damage to lung tissue is most likely to occur following exposure to insoluble U oxides, depending on dose, whereas the primary, rate limiting health effect following inhalation and oral exposures to soluble forms of U involves the kidney [3,23,24]. Thus, for example, urinary low molecular weight protein β_2-microglobulin, known to be increased in cadmium (Cd) exposed workers [25,26] has been measured since the first surveillance visit (1993–1994). Since 1999, additional low molecular weight proteins have been analyzed in urine samples for a more complete surveillance of renal tubular function, together with enzymes indicative of renal cell damage (such as N-acetyl-6-glucosaminidase (NAG) and intestinal alkaline phosphatase (IAP). Similarly, additional genotoxicity parameters have been measured over the course of the six surveillance visits. Sister chromatid exchange (SCE) and chromosomal aberration (CA) assays have been conducted since 1997. The hypoxanthine guanine phosphoribosyl transferase (HPGRT) mutation frequency test was added in 2001, and fluorescent *in situ* hybridization (FISH) was added in 2005. Whereas standard hematological parameters are measured during every visit, a detailed study of effects of U on specific lymphocyte cell populations was conducted in 2001, which showed no evidence a DU-exposure related changes. More detailed information on the special parameters included in each visit have been reported [9,27–30]. Standard hematological, renal, hepatic, and neuroendocrine parameters are measured every visit.

Approach to Data Analysis

Data gathered during each surveillance visit are examined for evidence of DU exposure–related health effects by comparing results based on the soldiers' ongoing systemic exposure to U as measured by their urine U concentrations. For each surveillance visit, participating soldiers are placed in one of two groups: low U exposure (urine U concentration < 0.1 μg/g creatinine) or high U exposure (urine U concentration ≥ 0.1 μg/g creatinine). Table 7.2 provides the mean urine U concentrations for the low U exposure vs. high U exposure groups for each of the

TABLE 7.2
Mean Urine Uranium for Low and High Urine Uranium Groups for Each Evaluation

Year of Evaluation	Low Uranium Group (mean ± sd)	High Uranium Group (mean + sd)
1994	0.022 ± 0.021	4.489 ± 6.568
1997	0.025 ± 0.023	5.350 ± 8.041
1999	0.013 ± 0.013	7.005 ± 11.854
2001	0.018 ± 0.020	11.296 ± 23.347
2003	0.015 ± 0.022	8.626 ± 13.527
2005	0.021 ± 0.028	9.827 ± 14.404

six surveillance visits. Although the specific soldiers participating in each visit are not exactly the same from year to year, the mean urine U concentrations of the groups have remained very similar. The means within the low U groups and those within the high U groups are not significantly different from each other ($p > 0.05$) from year to year. Thus, the groups are similar in their U exposure over the different years.

Surveillance Results

Hematological and hepatic parameters

Over the six surveillance visits conducted from 1994 through 2005, there have been no consistent, significant differences between the low vs. high U exposure groups in most of the clinical parameters examined. To be conservative, the level of significance has been set at $p = 0.2$ so as to not miss subtle changes that might be occurring. Even so, hematological parameters showed significant differences between the groups in at most two of the five surveillance visits as shown in Table 7.3; thus, observed differences have not have not been consistent over time. Serum proteins indicative of liver and other organ damage also showed minimal, nonconsistent changes.

Renal Function Parameters

Standard clinical measurements of renal function (serum creatinine, serum uric acid, and urine creatinine) have shown no evidence of acute or progressive renal damage in the low or high U exposed groups since 1994. Serum creatinine was higher in the low U group only in 2001, a change that was not in the expected direction for decreased renal function and was not accompanied by a change in serum uric acid. Urine creatinine was higher in the low group only in 1999. Although serum PO_4 was higher in the high U group in 2003, this was not accompanied by a change in urine PO_4 and was not observed in the 2005 visit (Table 7.3).

Changes in other serum and urine parameters also have shown no pattern of progressive changes except for a consistent, though nonsignificant, increase in the urinary excretion of the low molecular weight retinol-binding protein (RBP) since 2001. This could be an indication of early changes in renal proximal tubule function. These results have been accompanied by inconsistent changes in total urine protein, however, and no differences have been observed in beta$_2$-microglobulin or two other markers of renal cell damage, NAG and IAP. RBP excretion will continue to be carefully monitored, based on work by Squibb et al. [17], which reports that predicted renal concentrations of U in the DU-exposed soldiers excreting the highest concentrations of U are likely to have already accumulated U in their kidneys to concentrations approaching 1 ppm. Although a critical threshold of 3 ppm was established for workers in U industries in 1960 by the ICRP [31], more recent work suggests that a 10-fold lower concentration (0.3 ppm) may be more protective [10,32,33].

Reproductive Health Parameters

Significant differences in the neuroendocrine hormones FSH, LH, and TSH did not occur between the two groups in any of the visits. Though prolactin and free thyroxine were both decreased in the high U group in 2001, neither of these differences

TABLE 7.3
Differences between Low and High Urine Uranium Groups for Selected Clinical Parameters

Clinical Parameter	Differences Observed (and Year)
Renal Function	
Urine creatinine	l > h (p = 0.07) 1999
Urine calcium	
Urine PO_4	
Urine β-2 microglobulin	
Urine intestinal alkaline phosphatase (IAP)	
Urine N acetyl-ß-glucosaminidase (NAG)	
Urine total protein	h > l (p = 0.06) 2001; l > h (p = 0.21) 2003
Urine microalbumin	
Retinol binding protein (RBP)	h > l (p = 0.06) 2001; h > l (ns) 2003
Serum creatinine	L > H 2001
Serum calcium	H > L 1999
Serum PO_4	H > L 2003
Serum uric acid	
Other Serum Values	
GPT	
GOT	l > h (p = 0.1) 2003
CPK	
LDH	L > H 1999, 2001
Alkaline phosphatase	
Semen Characteristics	
Days abstinence (2–5)	
Semen volume (2–5 ml)	
Sperm concentration (>20 million/ml)	H > L 1997; h > l (p = 0.09) 1999; h > l (ns) 2001, 2003
Total sperm count (>40 million)	H > L 1999; h > l (p = 0.06) 2001; h > l (ns) 2003
Percent motile sperm (>50%)	
Total progressive sperm (>20 million)	H > L 1999
Percent progressive sperm (>50%)	
Total rapid progressive sperm (>10 million)	H > L 1999
Percent rapid progressive sperm (>25%)	
Neuroendocrine Hormones	
FSH	
LH	
Prolactin	l > h (p = 0.06) 2001
Testosterone	
TSH	
Free thyroxine	L > H 2001

(continued)

TABLE 7.3 (CONTINUED)
Differences between Low and High Urine Uranium Groups for Selected Clinical Parameters

Clinical Parameter	Differences Observed (and Year)
	Hematologic Function
White blood cells	H > L 1994
Hematocrit	L > H 2001
Hemoglobin	L > H 2001
Platelets	
Lymphocytes	L > H 1994,1999
Neutrophils	H > L 1994,1999
Basophils	
Eosinophils	H > L 1997
Monocytes	L > H 1994,1999

Note: L = low urine uranium group (U < 0.1 μg/g creatinine) and H = high urine uranium group (U ≥ 0.1 μg/g creatinine).

Upper case letters (H, L) = significant findings. ($p < 0.05$)
Lower case letters (h, l) = nonsignificant findings.

was observed in earlier or later years. Although blood testosterone concentrations were not different between groups in previous years, in 2005 testosterone concentrations were slightly lower in the high U group. This may be related to age and will be carefully monitored (Table 7.3)

The primary differences observed in semen characteristics over the six surveillance visits have been related to sperm count. Total sperm count, sperm concentration, and progressive sperm have been higher in the high U group since 1997; however, this does not represent an adverse health effect.

Neurocognitive Parameters

Neurocognitive parameters have been measured by a battery of computerized and paper-and-pencil neurocognitive tests from which four impairment indices are constructed and compared to military norms as described by McDiarmid et al. [4]. Scores on these tests have generally been within normal ranges for the high and low U groups. A small but consistent decrease in mean accuracy scores obtained from a computerized test have been reported over time in the high U group. The may be related in part to complex comorbid conditions present in the two soldiers driving this difference. Continued analysis of this test will help provide insight to its cause.

Genotoxicity

Sister chromatid exchange (SCE) and chromosomal aberrations (CA) have been measured in blood lymphocytes since 1997. These have given mixed results, with

TABLE 7.4
Summary of Differences in Genotoxicity Parameters across Evaluations

Genotoxicity Parameter	Evaluation Year 1994	1997	1999	2001	2003	2005
Sister chromatid exchange		l>h	H>L	l>h		
(SCE)		ns		ns	ns	
Chromosomal aberrations						
(CA)		ns	ns	H>L	ns	ns
Hypoxanthine-guanine						
Phosphoribosyl transferase				h>l	h>l	h>l
(HPRT) mutation frequency				(p = 0.1)	ns	ns

Note: L = low urine uranium group (U < 0.1 μg/g creatinine); H = high urine uranium group (U ≥ 0.1 μg/g creatinine); ns = no significant differences between groups.

Lower case letters (h, l) = nonsignificant findings.
Upper case letters (H, L) = significant findings. ($p < 0.05$)

SCEs higher but not significantly so in the low U groups in 1997 and 2001, but significantly different in the opposite direction (H > L) in 1999 (Table 7.4). No difference was observed in 2003. Chromosomal aberrations were not different between groups in any year except 2001, in which CAs were significantly higher in the high U group. Due to the possible concern for U-related genotoxic effects based on *in vitro* work by [34–41] and *in vivo* animal carcinogenicity studies [42–46], the HPGRT mutation frequency assay was added to the surveillance protocol in 2001. In both 2001 and 2003, the high U group showed a greater frequency of mutations in blood lymphocytes; however, this difference has not been consistently significant at the $p < .05$ level.

SUMMARY

The Baltimore VA Health Surveillance Program established in 1993 has conducted six biennial in-patient clinical surveillance visits for 1991 Gulf War soldiers exposed to DU during friendly-fire incidents involving DU-munitions and DU armored tanks. Exposure to DU occurred through inhalation, ingestion, and/or wound contamination due to the formation of fine DU oxide particles when DU penetrators sharpen and burn as they penetrate tank armor. Some soldiers were also struck by small fragments of DU metal and retain multiple DU fragments in their soft tissues. Exposure assessment using urine U concentrations measured at each surveillance visit have shown that soldiers with embedded fragments continue to excrete elevated concentrations of U, indicating that they have been experiencing chronic systemic exposure to U since the Gulf War.

Radiation dose estimates determined from urine U excretion measures over time suggest that risks from radiation induced tissue damage in the DU-exposed cohort are low. Potential for chemical-induced renal effects of DU exposure, however, remain of concern. Surveillance visit results have shown evidence of a change in an early marker of renal damage, which is consistent with the well known effects of U on kidney proximal tubule cells. Data from genotoxicity tests also indicate a need for close follow-up and will be examined in more detail in future visits. With the exception of the continuously elevated urine U excretion in the soldiers with DU embedded fragments, however, no clinically significant U-related health effects have been identified in this cohort of soldiers. The potential for health impacts due to the chronic exposure to U being experienced by these soldiers indicates the need for continued surveillance of this cohort of Gulf War soldiers and others who may be involved in similar battlefield exposures to DU.

REFERENCES

1. ATSDR (Agency for Toxic Substances and Disease Registry), *Toxicological Profile for Uranium.* U.S. Department of Health and Human Services, Washington, D.C., 1999.
2. IOM (Institute of Medicine), *Gulf War and Health, Volume 1. Depleted Uranium, Pyridostigmine Bromide, Sarin, Vaccines,* Eds., C.E. Fulco, C.T. Liverman, and H.C. Sox, Washington, DC, National Academy Press, 2000.
3. The Royal Society, The Health Hazards of Depleted Uranium Munitions Part I. Policy document 6/01. Royal Society, London, 2001, report available at www.royalsociety.ac.uk.
4. McDiarmid, M.A., Squibb, K., and Engelhardt, S.M., Biologic monitoring for urinary uranium in Gulf War I veterans. *Health. Phys.* 87: 51–56, 2004.
5. OSAGWI (Office of the Special Assistant for Gulf War Illness), Environmental Exposure Report — Depleted uranium in the Gulf (II), Interim Report, U.S. Department of Defense, Washington, DC, 2000, http://www.deploymentlink.osd.mil/du_library/du_ii/.
6. Guilmette, R.A., Parkhurst, M.A., Miller, G., Hahn, F.F., Roszell, L.E., Daxon, E.G., Little, T.T., Whicker, J.J., Cheng,, Y.S, Traub, R.J., Lodde, G.M., Szrom, F., Bihl, D.E., Creek, K.L., and McKee, C.B., Human Health Risk Assessment of Capstone Depleted Uranium Aerosols, Columbus, OH: Battelle Press, 2005, http://www.deploymentlink.osd.mil/du_library/du_capstone/index.pdf
7. Army Environmental Policy Institute (AEPI), Health and Environmental Consequences of Depleted Uranium Use in the U.S. Army. Atlanta, GA: AEPI Technical Report, 1995.
8. Toohey, R.E., Excretion of depleted uranium by Gulf War veteran. *Radiat. Prot. Dosimetry* 105, 171–174, 2003.
9. McDiarmid, M.A., Hooper, F.J., Keogh, J.P., McPhaul, K., Squibb, K., Kane, R., DiPino, R., Kabat, M., Kaup, B., Anderson, L., Hoover, D., Brown, L., Hamilton, M., Jacobson-Kram, D., Burrrows, B., and Walsh, M., Health effects of depleted uranium on exposed Gulf War veterans, *Environ. Res.* 82, 168–180, 2000.
10. Leggett, R.W., The behavior and chemical toxicity of U in the kidney: a reassessment. *Health Phys.* 57, 365–383, 1989.
11. Fernald Environmental Management Project (FEMP), Technical Basis for Internal Dosimetry at the Fernald Environmental Management Project, Revision 3 dated December 23, 1997, Fernald, OH: FEMP Publication SD 2008, 1997.

12. Thun, M.J., Baker, D.B., Steenland, K., Smith, A.B., Halperin, W., and Berl, T., Renal toxicity in uranium mill workers. *Scand. J. Work Environ. Health* 11, 83–90, 1985.
13. ICRP (International Commission on Radiological Protection), *Report of the Task Groups on Reference Man,* Vol. 23, Elmsford, NY: Pergamon Press, 1974.
14. NHANES (National Health and Nutrition Examination Survey), Second National Report on Human Exposure to Environmental Chemicals, NCEH Publication No. 02-0716. Atlanta, GA: Centers for Disease Control and Prevention, 2003.
15. Ejnik, J.W., Todorov, T.I., Mullick, F.G., Squibb, K., McDiarmid, M.A., and Centeno, J.A., Uranium analysis in urine by inductively coupled plasma dynamic reaction cell mass spectrometry, *Anal. Bioanal. Chem.* 382, 73–79, 2005.
16. ICRP (International Commission on Radiological Protection), Age-Dependent Doses to Members of the Public from Intake of Radionuclides, Part 3. Ingestion Dose Coefficients, Oxford: Pergamon Press, ICRP Publication 69, 1995.
17. Squibb, K.S., Leggett, R.W., and McDiarmid, M.A., Prediction of renal concentrations of depleted uranium and radiation dose in Gulf War veterans with embedded shrapnel, *Health. Phys.* 89: 267–273, 2005.
18. NCRP (National Council on Radiation Protection and Measurements), Limitation of Exposure to Ionizing Radiation. Bethesda, MD: National Council on Radiation Protection and Measurements, Report Number 116, 1993.
19. Voegtlin, I.C. and Hodge, H.C., Eds., *Pharmacology and Toxicology of Uranium Compounds,* National Nuclear Energy Series (VI). New York: McGraw-Hill, 1949.
20. Voegtlin, C., and Hodge, H.C., Eds., *Pharmacology and Toxicology of Uranium Compounds,* New York: McGraw-Hill, 1953.
21. Tannenbaum, A., Ed., *Toxicology of Uranium.* New York: McGraw-Hill, 1951.
22. Hodge, H.C., Stannard, N., and Hursh, J.B., Eds., *Uranium, Plutonium, Transplutonic Elements: Handbook of Experimental Pharmacology,* Vol. 36. New York: Springer-Verlag, 1973.
23. Parkhurst, M.A., Daxon, E.G., Lodde, G.M., Szrom, F., Guilmette, R.A., Roszell, L.E., Falo, G.A., and McKee, C.B., Depleted Uranium Aerosol Doses and Risks: Summary of U.S. Assessments. Columbus, OH: Battelle Press, 2005, http://www.deploymentlink.osd.mil/du_library/du_capstone/index.pdf.
24. The Royal Society, The Health Effects of Depleted Uranium Munitions. Summary, Policy Document 6/02. Royal Society, London, 2002, report available at www.royal society.ac.uk.
25. Bernard, A.M. and Lauwerys, R.R., Dose response relations between urinary cadmium and tubular proteinuria in adult workers. *Am. J. Ind. Med.* 31: 116–118, 1977.
26. Jarup, L. and Elinder, C.G., Letter to the editor re: Commentary on Dose-response relations between urinary cadmium and tubular proteinuria in adult workers. *Am. J. Ind. Med.* 31: 119–120, 1997.
27. Hooper, F.J., Squibb, K.S., Siegel E.L., McPhaul, K., and Keogh, J.P., Elevated urine uranium excretion by soldiers with retained uranium shrapnel. *Health Phys.* 77: 512–519, 1999.
28. McDiarmid, M.A., Squibb, K., Engelhardt, S., Oliver, M., Gucer, P., Wilson, D., Kane, R, Kabat, M., Kaup, B., Anderson, L., Hoover, D., Brown, L., and Jacobson-Kram, D., Surveillance of depleted uranium exposed Gulf War veterans: health effects observed in an enlarged "friendly fire" cohort. *J. Occup. Environ. Med.* 43, 991–1000, 2001.

29. McDiarmid, M.A., Engelhardt, S., Oliver, M., Gucer, P., Wilson, D., Kane, R., Kabat, M., Kaup, B., Anderson, L., Hoover, D., Brown, L., Handwerger, B., Albertini, R.J., Jacobson-Kram, D., Thorne, C.D., and Squibb, K., Health effects of depleted uranium on exposed Gulf War veterans: a ten-year follow-up. *J. Toxicol. Environ. Health Part A,* 67, 277–296, 2004.
30. McDiarmid, M.A., Engelhardt, S., Oliver, M., Gucer, P., Wilson, P.D., Kane, R., Kabat, M., Kaup, B., Anderson, L., Hoover, D., Brown, L., Albertini, R.J., Gudi, R., Jacobson-Kram, D., Thorne, C.D., and Squibb, K.S., Biological monitoring and surveillance results of Gulf War I veterans exposed to depleted uranium. *Int. Arch. Occup. Environ. Health* 79: 11–21, 2006.
31. ICRP (International Commission on Radiological Protection), Report of committee II on permissible dose for internal radiation. *Health Phys* 3: 1–380, 1960.
32. Kathren, R.L. and Weber, J. Eds., *Ultrasensitive Techniques for Measurement of Uranium in Biological Samples and the Nephrotoxicity of Uranium.* U.S. Nuclear Regulatory Commission, Proceedings of the Meeting. Nuclear Regulatory Commission staff-originated report NUREG/CP-0093, Washington, DC, 1988.
33. Diamond, G., Biological consequences of exposure to soluble forms of natural uranium. *Radiat. Prot. Dosimetry* 26: 23–33, 1989.
34. Lin, R.H., Wu, L.J., Lee, C.H., and Lin-Shiau, S.Y., Cytogenetic toxicity of uranyl nitrate in Chinese hamster ovary cells. *Mutation Res.* 319: 197–203, 1993.
35. Miller, A.C., Fuciarelli, A.F., Jackson, W.E., Ejnik, J., Edmond, C., Strocko, S., Hogan, J, Page, N., and Pellmar, T., Urinary and serum mutagenicity studies with rats implanted with depleted uranium or tantalum pellets. *Mutagenesis* 13, 643–648, 1998.
36. Miller, A.C., Blakely, W.F., Livengood, D., Whittaker, T., Xu, J., Ejnik, J.W., Hamilton, M.M., Parlette, E., St. John Gerstenberg, H.M., and Hsu, H., Transformation of human osteoblast cells to the tumorigenic phenotype by depleted uranium-uranyl chloride, *Environ. Health Perspect.* 106, 465–471, 1998.
37. Miller, A.C., Stewart, M., Brooks, K., Shi, L., and Page, N., Depleted uranium-catalyzed oxidative DNA damage: Absence of significant alpha particle decay. *J. Inorganic. Biochem.* 91, 246-252, 2002.
38. Miller, A.C., Xu, J., Stewart, M., Brooks, K., Hodge, S., Shi, L., Page, N., and McClain, D., Observation of radiation–specific damage in human cells exposed to depleted uranium: dicentric frequency and neoplastic transformation as endpoints, *Radiat. Prot. Dosimetry* 99, 275–278, 2002.
39. Miller, A.C., Brooks, K., Stewart, M., Anderson, B., Shi, L., McLain, D., and Page, L., Genomic instability in human osteoblast cells after exposure to depleted uranium: delayed lethality and micronuclei formation*, J Environ Radiol* 64, 247–259, 2003.
40. Yazzie, M., Gamble, S.L., Civitello, E.R., and Stearns, D.M., Uranyl acetate causes DNA single strand breaks in vitro in the presence of ascorbate (vitamin C), *Chem. Res. Toxicol.* 16, 524–530, 2003.
41. Stearns, D., Yazzie, M., Bradley, A.S., Coryell, V.H., Shelley, T., Ashby, A., Asplund, C.S., and Lantz, R.C. Uranyl acetate induces hprt mutations and uranium — DNA adducts in Chinese hamster ovary EM9 cells, *Mutagenesis* 20: 417–423, 2005.
42. Hueper, W.C., Zuefle, J.H., Link, A.M, and Johnson, M.G., Experimental studies in metal carcinogenesis. II. Experimental uranium cancers in rats, *J. Natl Cancer Inst.* 13, 291–305, 1952.

43. Leach, L.J., Yuile, C.L., Hodge, G.C., Sylvester, G.E., and Wilson, H.B., A five-year inhalation study with natural uranium dioxide (UO_2) Dust-II. Postexposure retention and biologic effect in the monkey, dog, and rat, *Health. Phys.* 25, 239–258, 1973.
44. Cross, F.T., Palmer, R.F., Busch, R.H., Filipy, R.E., and Stuart, B.O., Development of lesions in Syrian Golden hamsters following exposure to radon daughters and uranium ore dust, *Health Phys.* 41, 135–153, 1981.
45. Mitchell, R.E.J., Jackson, J.S., and Heinmiller, B., Inhaled uranium ore dust and lung cancer risk in rats, *Health Phys.* 76, 145–155, 1999.
46. Hahn, F.F., Guilmette, R.A., and Hoover, M.D., Implanted depleted uranium fragments cause soft tissue sarcomas in the muscles of rats, *Environ. Health Perspect.* 110, 51–59, 2002.

8 Health Hazards of Depleted Uranium Munitions: Estimates of Exposures and Risks in the Gulf War, the Balkans, and Iraq

Brian G. Spratt

CONTENTS

INTRODUCTION

Concerns about the health hazards associated with the use of depleted uranium (DU) munitions in the conflicts in the Balkans (1994–1999), and previously in the 1991 Gulf War, led to the formation by the United Kingdom Royal Society of a DU working group that undertook an independent assessment of the risks to health and the environment. Further data have become available since the publication of the two parts of the Royal Society report (Royal Society, 2001, 2002); new test firings of DU munitions have allowed more secure estimates of the likely intakes of DU on the battlefield (Capstone Report, 2004), and studies of the individual levels of exposure to DU of soldiers from these conflicts are now available. Additionally, DU munitions were deployed in the Iraq conflict of 2003 and, unlike in the earlier conflicts, timely monitoring for exposure of soldiers to DU has been carried out. This chapter briefly reviews the conclusions of the Royal Society report and of two subsequent major studies that estimate intakes of DU in the Gulf War. It discusses the consequent risks to health (Capstone Report, 2004; Sandia Report, 2005) and the available data on exposure levels derived from measuring the concentrations of DU in the urine of veterans from the three conflicts in which DU munitions have been used. Detailed reviews of the military uses of DU, and of the methods used to estimate intakes of DU and the risks to health, can be found in the Royal Society Capstone and Sandia reports and elsewhere in this volume.

DU EXPOSURES ON AND BEYOND THE BATTLEFIELD

The risks of exposure to DU are twofold. DU is weakly radioactive and a toxic heavy metal with its main effect on the kidney. The relative nature of the risk depends on the extent to which DU is retained in the body, providing a radiation risk, or is excreted through the kidney, increasing the risk of renal toxicity. Exposures on the battlefield mostly arise from DU penetrator strikes, which can generate impact aerosols containing high concentrations of DU oxides that can be inhaled, and DU fragments, which can cause shrapnel wounds. Ingestion of DU particulates in contaminated vehicles may also occur but is generally not considered to be a very significant exposure route and is not discussed further.

Following an impact, the proportion of the mass of a DU penetrator rod that is converted into an aerosol containing DU oxide particulates depends on a number of factors, particularly the nature of the struck target. For "soft targets," DU rounds are likely to enter the vehicle and pass straight through, with little conversion of the penetrator rod into DU oxides. Conversely, the greatest proportion of the penetrator mass that is converted into DU oxides is believed to occur when a penetrator strikes the DU armor of a modern battle tank, such as during "friendly-fire" incidents involving an M1A1 Abrams tank.

A range of DU concentrations have been measured within struck vehicles in test firings of large-caliber DU rounds (Capstone Report, 2004). Inhalation intakes of DU oxides for surviving crew have been estimated to be about 10–20% greater in an Abrams tank with DU armor, compared to one with conventional armor, and about three times greater in a struck tank than in a less heavily armored Bradley fighting vehicle. In one test firing, the activation of the tank's ventilation system reduced the concentration and hence the estimated inhalation intake of DU by about 90% (Capstone Report, 2004), but in friendly-fire incidents in the Gulf War, it is not clear whether ventilation systems were active, and assessment of risks in most studies are based on the cautious assumption of no ventilation. Those surviving within, or close to, a tank struck with a DU round will inhale DU oxides in the impact aerosols and may also receive DU shrapnel wounds. Therefore, they are believed to be at highest risk from DU, although a broader view of the risks to soldiers on the battlefield is required. Radiation risks to military personnel of handling DU rounds, and to tank crews from the DU rounds stored within tanks, are considered to be very low as exposure levels can easily be monitored and managed.

Most studies of the hazards of DU munitions have focused on battlefield exposures, but there are also concerns about longer-term exposures to local populations who live in, or return to, areas where DU munitions were deployed. In the Gulf War, the conflicts in the Balkans, and in Iraq, the great majority of the total mass of DU that was deployed was small-caliber DU rounds (30 mm; ~300 gram of DU), fired in strafing runs by attack aircraft, with only a small proportion being large-caliber tank rounds (mostly 120 mm; ~4.5 kg of DU). Most of the DU penetrators in strafing runs miss their intended targets and end up embedded several feet in the ground, where they corrode and could lead to elevated uranium levels in soil or local water supplies. In the years following a conflict, the local population may also be exposed to DU by inhalation of resuspended soil containing DU oxides. Understanding the exposure pathways of most significance for soldiers and local populations, and the resulting risks to health, requires estimates of intakes of DU for a number of different scenarios.

Exposures on the battlefield have usually been considered in one of three categories (OSAGWI Report, 2000; CHHPM Report, 2000), an approach that provides a useful framework for considering the potential intakes and risks and which was used in the Royal Society, Capstone, and Sandia reports, summarized here:

- *Level I* includes soldiers surviving in a vehicle struck by a DU penetrator(s) or first responders entering the struck vehicle to rescue the occupants. Intakes are likely to be dominated by inhalation of DU particulates and, for surviving crew, by DU shrapnel wounds.
- *Level II* includes those who work within struck vehicles for substantial periods, at some time after the impact, and who may be exposed to inhalation of DU oxides resuspended within the contaminated vehicle.
- *Level III* includes all others on the battlefield who may be exposed from being downwind of struck vehicles or from briefly entering struck vehicles.

Identification and health monitoring of soldiers who are believed to have received Level I and II exposures has been the main priority for veterans of the Gulf War and subsequent conflicts. The number of soldiers who are known to have received Level I exposures is relatively small. For example, after the 1991 Gulf War, approximately 100 U.S. soldiers known to have been involved in friendly fire incidents were assessed as being in this category (McDiarmid et al., 2005), and 14 U.K. soldiers were assessed as having received Level I exposures in the Iraq conflict of 2003 (Ministry of Defence, personal communication). However, there are likely to be larger numbers of Iraqi soldiers from the Gulf War and the conflict in Iraq who had Level I exposures, but no data appear to be available. Exposures for soldiers categorized as Level III are likely to be much lower than for those in Level I or II but, as this group is potentially very much larger, estimates are required of the range of exposures (and health risks) for soldiers and support personnel who were on or near the battlefield where DU munitions were deployed.

ESTIMATING INTAKES OF DU AND RISKS TO HEALTH

The effects of exposure to DU can be assessed by long-term health surveillance of veterans and exposed populations, or by using modeling approaches to predict the health effects (excess cancers, kidney damage, etc.) resulting from known intakes of DU. Both approaches have serious limitations when applied to the Gulf War veterans. Surveillance of morbidity and mortality among Gulf War Veterans is being carried out in the U.S. and the U.K. but, except for those with retained DU shrapnel, and perhaps others with Level I exposures, this approach is unlikely to provide compelling evidence of health effects attributable to DU exposure, as any observed effects could be due to a number of other potentially toxic exposures that occurred in this war. The health effects from known intakes of DU can be predicted, but this approach requires a knowledge of the levels of exposure to DU of soldiers and support personnel, as well as the local population, during or after the conflict. Surprisingly, for 10 years after the Gulf War, little effort appears to have been put into developing procedures to estimate the levels of exposure of individual soldiers to DU. Similarly, there were almost no timely data obtained on exposures to DU of individual soldiers and peacekeepers following their service in the Balkans, and there is still no data on intakes among any of the local populations who live in areas where DU munitions were used. The likely intakes of DU for various categories of soldiers on the battlefield, and for individuals living or working in areas where DU munitions were deployed, therefore, have to be estimated. Recently, measurements of DU in urine have been obtained for soldiers from the Gulf War, the Balkans, and Iraq, which can be used to estimate the original intakes of individual soldiers and these studies are discussed below.

Occupational exposures to uranium have occurred since the 1940s and increasingly sophisticated models have been developed to predict the distribution of uranium in the body following inhalation of uranium containing dusts (Chapter 11, this book). The International Commission on Radiological Protection (ICRP) human respiratory model for radiological protection describes the processes of inhalation, lung deposition, clearance, and absoption to blood in the respiratory tract (ICRP, 1994), and the systemic model for uranium (ICRP, 1995) describes the uptake to organs from the blood and subsequent excretion through the kidneys to urine. Estimates of the intakes of DU are required as source terms for these models and, together with estimates of the key parameters of the respiratory tract model, including the size distribution and absorption profile of the inhaled DU oxides, allow the concentrations of DU in different tissues and organs of the body to be predicted at any time after the estimated intake. These concentrations can then be related to the available information on the chemical toxicity of uranium and, for radiological effects, the doses to tissues and organs can be calculated and the resulting cancer risks derived. These models can also be run in reverse to estimate the original inhalation intake from the amount of DU excreted in urine at any time after the exposure.

The modeling approach is subject to considerable uncertainties, both in the intakes and the model parameters. The Royal Society study therefore used central estimates of the range of concentrations of DU oxides in impact aerosols from test firings and other sources, and of the absorption parameter values, size distribution, etc., of the DU oxides, to produce a central estimate of intakes and risks for a number of exposure scenarios. To account for uncertainties in the intakes, and the variation in the observed values of the key model parameters, worst-case estimates were also provided by using values at the upper end of the likely range. The chosen parameter values for the worst-case estimates are those that increase uranium concentrations in the kidney when considering chemical toxicity, or which increase retention of DU particulates in the respiratory tract or associated lymph nodes for the radiological effects. Worst-case scenarios provide estimates of intakes and risks that could possibly apply to a few individuals under exceptional circumstances and that are unlikely to be exceeded. They are useful for prioritizing further studies as exposure scenarios that lead to very low-risk estimates using the worst-case are unlikely to be of concern.

Test firings of DU rounds and test fires involving DU munitions provide the major source of data on the air concentrations of DU oxides in and around struck or burning tanks, and their size distribution, chemical speciation, and absorption characteristics. The most relevant test firings are those that use armored vehicles, but measuring the amount of DU oxides in the aerosols generated following a DU-penetrator impact is problematic, as air samplers within the struck vehicle have to continue working after the impact. The data from test firings that were available in 2001 provided only limited information on the concentrations of DU oxides within struck tanks as air samplers typically failed to function after impact. The Royal Society report used the best data then available of initial air concentrations of DU oxides within a struck tank to provide the central and worst-case estimates of the inhalation intakes for those with Level I exposures. Similarly, estimates were made of the amounts of DU inhaled from resuspension of DU particulates for those working in struck tanks at some time after the initial impact (Level II exposures). Level III exposures were also estimated for those briefly entering a struck tank at some time after the impact and, by dispersion modeling, for those exposed to DU particulates downwind of vehicles strikes by inhalation of the resulting plumes. Chronic exposures of residents to DU in the years following a conflict were also considered.

The estimates of intakes were combined with the available data on the properties of the resulting DU oxides. For inhalation, the key parameters include the size distribution, which determines the proportion of inhaled particulates that are retained in the lungs, and the absorption characteristics of the DU oxides, which determines the kinetics of release of DU into the blood and then to other tissues and to urine (Chapter 11, this book). The inadequacy of the available data from test firings for providing reliable estimates of intakes for soldiers with Level I exposures in the Gulf War was highlighted in 2002 by the U.S. General Accounting Office. These deficiencies were addressed by 13 new test firings of large-caliber DU rounds against Abrams tanks and Bradley fighting vehicles in the Capstone DU Aerosols Study which, by careful design, determined the DU air concentration and size distribution within the struck vehicle as a function of time after impact (Capstone Report, 2004). The Capstone project included test firings into the crew compartment of tanks with conventional or DU armor, with and without activation of the tank's ventilation system. They provide improved data on the air concentrations of DU oxides in struck vehicles and of the composition, particle size, and *in vitro* solubility of particulates in the impact aerosols, and these allow revised estimates of the likely intakes of DU, as well as health risks, for those surviving in struck vehicles or first responders (Level I exposures). Two test firings by the French army have also been briefly reported (Chazel, 2003), but these give very limited data compared to those in the detailed report from the Capstone project.

The published report of the Capstone project provides detailed estimates of intakes and risks to health for a range of exposure scenarios in the Gulf War, and the Capstone test firing data have been independently analyzed by the Sandia National Laboratories, who estimate intakes for soldiers on the battlefield in the Gulf War, and also for Iraqi civilians, and the predicted risks to health (Sandia Report, 2005). Comparing the estimates of intakes and risks between the three recent studies is complicated by differences in methodology and terminology. The Capstone and Sandia reports produce typical intakes and risks for each exposure scenario that are broadly equivalent to the central estimates in the Royal Society report. They also provide maximum intakes and risks, based on the highest observed values in test firings, that are generally considerably lower than those using the more cautious worst-case approach of the Royal Society report, as the latter assumes the most unfavorable values of intakes and all model parameters. The differences in the methodologies, the availability of improved data from the Capstone trials on the concentrations of DU particulates in vehicles after an impact (Level I), and the data on resuspension of deposited DU particulates on entering a tank at some time after the strike (Level II), together with new data on the properties of DU particulates in impact aerosols, inevitably lead to differences in the estimated intakes and risks to health in the Royal Society, Capstone, and Sandia reports. However, these differences are not great, and it is reassuring that all three reports agree on the general nature and extent of the risks to health from the use of DU munitions.

LEVEL I AND II EXPOSURES AND RISKS

Level I exposures will often involve both DU shrapnel wounds and inhalation of DU oxides, and the health of these soldiers needs to be carefully monitored. In the Gulf War there were 6 Abrams tanks and 15 Bradley fighting vehicles involved in friendly-fire incidents. The veterans involved in these incidents are classed as having Level I exposures and are likely to be among those who received the greatest exposures to DU in the Gulf War. A cohort of these veterans, some of whom have retained DU shrapnel, has been monitored at intervals since 1993. Several continue to excrete high levels of uranium, and all of these individuals are believed to have retained DU shrapnel. In the latest report on the health of this cohort (McDiarmid et al., 2005), kidney function tests for the latter group are mostly similar to those of a comparator group from the friendly-fire cohort who have normal levels of uranium excretion and who do not have retained DU shrapnel. However, levels of retinol binding protein, a marker for proximal tubular dysfunction, are elevated in the high uranium excretion group, although the mean difference between the two groups does not reach statistical significance, and none of the individual values are above the normal clinical range. The amount of uranium excreted in some of the veterans with retained shrapnel appears to be still increasing, and in two individuals the current levels of excretion are predicted to result in uranium concentrations of 0.63 and 0.95 μg per gram kidney (Squibb et al., 2005). In uranium workers and individuals with high levels of uranium in their water supply, these levels of uranium in the kidney have been associated with evidence of proximal tubular dysfunction (Royal Society, 2003), which is consistent with the elevated retinol binding protein levels observed in recent health monitoring of the individuals with retained DU shrapnel. Adverse effects on the kidney from chronic excretion of these levels of uranium are possible, and continued long-term monitoring of kidney function is required for those who have retained DU shrapnel.

Squibb et al. (2005) have also estimated the committed effective dose from the DU solubilized from retained shrapnel. It indicates that the upper-bound dose (obtained using the assumption that 10 times more DU will be released to blood than has been released during the first 10 years) for the veteran with the highest uranium excretion (0.06 Sv) is similar to the dose received from a lifetime's exposure at the dose limit for members of the general public (0.05 Sv) and is less than 10% of the dose received from a lifetime's exposure at the dose limit for workers (1 Sv). The uranium solubilized and released to blood from retained DU shrapnel is therefore considered to be more a threat to the kidney than a radiological risk.

The chronic irradiation of tissues immediately surrounding the retained DU shrapnel is of more concern, and soft tissue sarcomas have been observed in animals around DU implants (Hahn et al., 2002). Continued monitoring of kidney function, abnormalities in tissues surrounding retained shrapnel, and cancers in general are important for those veterans with retained DU shrapnel. Furthermore, these individuals may have inhaled substantial amounts of DU oxides, which cannot be quantified by urinary analysis due to the release to urine of DU from the retained shrapnel, and they may be at increased risk of lung cancer.

For those in the Gulf War with Level I or II exposures but who do not have retained DU shrapnel, the inhalation intakes are still unclear as relatively insensitive methods of uranium isotope analysis have been used to detect DU in the urine of this category of veteran from the friendly-fire incidents (McDiarmid et al., 2004). Inhalation intakes for Level I and II exposures, therefore, have had to be estimated by modelling using data from test firings (Table 8.1). The Royal Society central estimate of the amount of DU inhaled by a survivor in a struck tank (Level I exposure) was 250 mg, with a worst-case intake of 5 gram (Royal Society, 2001). The data from the Capstone test firings indicate a range of inhalation intakes of 250–710 mg of DU for the surviving crew of an Abrams tank, struck in the crew compartment by a single large-caliber DU penetrator (without activation of the ventilation system), and of 160–200 mg for first responders, with a maximum intake for Level I exposures of 1 gram (Capstone Report, 2004). The values of the best estimate and maximum estimate of inhalation intakes for Level I in the Sandia study (250 mg and 4 gram; Sandia Report, 2005), based on the Capstone test firing data, are almost identical to those in the Royal Society report.

The predicted increase in the lifetime risk of death from lung cancer from Level I inhalation exposures is about 0.1% in the Royal Society and Sandia reports, and 0.06–0.3% in the Capstone report), which can be compared to the overall lifetime

TABLE 1
Comparison of Estimated Inhalation Intakes of DU in Royal Society Report and Those Based on New Data from the Capstone DU Aerosols Study

	Level I	Level II	Level III (Vehicles)	Level III (Other)
Royal Society	250 mg (5 gram)	10 mg (2 gram)	1 mg (200 mg)	< 1 mg (80 mg)
Capstone	160–710 mg[a] (1 gram)	5 mg (1.45 gram)	0.5 mg (145 mg)	—
Sandia	250 mg (4 gm)	40 mg (600 mg)	6 mg (60 mg)	0.004 mg (6 mg)

Note: Central or best estimates of inhalation intakes are shown; values in parentheses are estimated worst-case intakes (Royal Society report) or maximum intakes (Capstone and Sandia reports).

[a]Capstone provide a range of values for Level I; those given are the range for tanks with conventional or DU armor without activation of the ventilation system. For Level II, estimates of intakes are given for an individual working for 10 h or 100 h (parentheses) within contaminated vehicles without respiratory protection. Estimated intakes for Level III are given for working in contaminated vehicles for 1 h or 10 h (parentheses), and for inhalation intakes of DU in plumes and from resuspended soil (other). The Royal Society Report used data available prior to the Capstone Trials; the other two estimates used Capstone data.

risk of fatal lung cancer for young adults (about 6–7%) (Table 8.2). These estimates are based on a single large-caliber impact and, as documented in the Gulf War, vehicles may be struck by more than one DU round; the risks would increase proportionately in such cases. The total cancer risk is dominated by the risk to the lung, and the increased lifetime risk of other fatal cancers, including leukemia, is

TABLE 2
Excess Lifetime Risk of Fatal Lung Cancer and Leukemia from Inhalational Intakes of DU

	Level I	Level II	Level III (Vehicles)
Royal Society (lung cancer)	0.12% (6.5%)	0.0025% (2.4%)	< 0.0001% (0.02%)
Capstone (lung cancer)	0.06 – 0.3% (0.4%)	0.001% (0.4%)	0.0001% (0.04%)
Sandia (lung cancer)	0.085% (1.4%)[a]	0.014% (0.21%)	< 0.0001% (0.002%)
Royal Society (leukemia)	0.0005% (0.005%)	< 0.00001% (0.001%)	< 0.00001 (0.00001%)
Capstone (leukemia)	—	—	—
Sandia (leukemia)	0.0004% (0.007%)	0.0001% (0.001%)	< 0.00001% (0.00001%)

Note: Central or best estimates of excess risk are shown; values in parentheses are worst-case estimates (Royal Society) or maximum estimates (Capstone and Sandia). Values are based on the estimates of inhalation intakes in Table 1. Capstone did not specifically consider leukemia risks.

[a] Sandia estimates a 3.5% excess risk as an upper-bound value. Level III excess risks from inhalation of plumes and resuspended soil are lower than those from entry into vehicles and are not shown.

very much lower for Level I exposures ($\leq$ 0.0005%). It remains low even for the maximum or worst-case risk estimates (< 0.01%). For lung cancer, the maximum or worst-case estimates indicate the possibility of much more substantial risks to health: an increased risk of 6.5%, corresponding to a doubling of the lifetime risk of fatal lung cancer using the cautious worst-case approach of the Royal Society report, and an upper-bound value of 3.5% in the Sandia report. These levels of risk would only be expected in exceptional circumstances, but the maximum increased risks of 0.4-1.4% in the Capstone and Sandia reports indicate risks that could apply to a few soldiers with Level I exposures.

Level II intakes depend on the amount of time spent working in contaminated tanks (and assuming a lack of adequate respiratory protection). Central or best estimates that assume 10 h working within contaminated vehicles give risks of lung cancer that are at least fivefold less than the corresponding Level I estimates, and the risks of other cancers are low (Table 8.2). Maximum and worst-case Level II estimates that assume 100 h working within contaminated vehicles and higher air concentrations of DU oxides indicate risks of lung cancer that are 0.2–0.4% for the Capstone and Sandia studies, and 2.4% for the worst-case Royal Society estimate.

The three studies estimate similar peak kidney uranium concentrations for Level I exposures, and the central or best estimates of up to 6 μg uranium per gram kidney (Table 8.3) suggest the possibility of some transient kidney dysfunction (Royal Society, 2002). The worst-case kidney concentration in the Royal Society study (400 μg per gram), and the maximum value in the Sandia study (53 μg per gram), suggest the possibility of kidney failure under extreme circumstances. Level II exposures are of less concern, and the central or best estimates are unlikely to lead to anything other than the possibility of transient minor kidney dysfunction, although the worst-case estimate from the Royal Society study suggests serious kidney effects under exceptional circumstances.

TABLE 3
Peak Kidney Uranium Concentrations (μg Uranium per Gram Kidney)

	Level I	Level II	Level III (Vehicles)	Residents[a]
Royal Society	4 (400)	0.05 (96)	0.005 (10)	0.002 (0.2)
Capstone	0.7–6.5 (16)	0.03 (14)	0.003 (1.4)	—
Sandia	3 (53)	0.5 (8)	Negligible (0.08)	0.001 (0.01)[b]

Note: Central or best estimates of peak kidney uranium concentrations are shown; those in parentheses are worst-case estimates (Royal Society) or maximum estimates (Capstone and Sandia). Values are based on the estimates of inhalation intakes in Table 8.1.

[a] From inhalation of DU particulates in resuspended soil.

[b] The highest estimated central value (or maximum value) for chronic exposures of civilians are shown, corresponding to children at play.

LEVEL III EXPOSURES AND RISKS

The intakes of DU estimated for Level III inhalation exposures resulting from briefly entering struck tanks, and from exposure to plumes downwind of penetrator impacts and resuspended soil, are shown in Table 8.1 and Table 8.2. The risks to health are predicted to be very low for these scenarios (Table 8.2), and thus for most soldiers on the battlefield, whose exposure will be at or less than that of Level III. Peak kidney uranium concentrations for a range of Level III exposure scenarios are also very low although the Royal Society worst-case value could lead to some kidney dysfunction (Table 8.3).

EXPOSURES TO RESIDENTS OF AREAS WHERE DU MUNITIONS WERE DEPLOYED

The Royal Society and Sandia reports estimate intakes and risks for civilian populations after a conflict (Royal Society, 2002; Sandia Report, 2005). These risks differ from military scenarios as exposures may continue for many years and, as well as involving exposures analogous to those in the military situation (such as children entering and playing within struck vehicles), they include alternative exposure pathways (including the possibility of contaminated food and water). Additionally, they will involve children and the elderly, who may be more susceptible to health effects than soldiers. Both reports highlight the great variability of possible intakes as, although no widespread contamination from deposited DU particulates has been observed, local concentrations of DU can be high around penetrator impact sites (Danesi et al., 2003; Sansone et al., 2001; UNEP, 2001, 2002, 2003; Uytenhove et al., 2002). However, so far, almost all data on environmental contamination have been from surveys by the United Nations Environmental Programme in the Balkans (UNEP, 2001, 2002, 2003) where relatively small amounts of DU munitions were deployed (about 13 t), compared to those used in the Gulf War (about 320 t), and presumably in Iraq where the amounts used by the U.S. forces in Iraq are still not available.

Cancer risks for local populations from long-term inhalation of resuspended DU particles deposited in soil are estimated in both the Royal Society and Sandia studies to be extremely low, even in the worst-case risk. Kidney concentrations from resuspended DU are also likely to be low, although the Royal Society worst-case estimate of 0.1–0.2 μg uranium per gram kidney suggests the possibility of minor effects on the kidney. The most significant risk to residents identified in the Sandia report was for children playing for long periods in DU-contaminated vehicles (based on 300 h within and 700 h outside the vehicles), where they estimated an increased lifetime risks of fatal lung cancer of 0.035% (maximum of 0.14%) and an increased total cancer risk of about 0.1% (maximum of 0.3%).

Contamination of soil (and food produce) from deposition of DU particulates will be insignificant compared to the natural levels of uranium in soil, but high concentrations of DU around sites of penetrator impacts may present some risks to children playing in these areas. The Royal Society report also highlighted the longer term possibility of contamination of local water supplies in areas where large numbers of corroding penetrators leach uranium into the soil, although the Sandia report considered significant contamination of both food and water unlikely. At present, studies in the Balkans have identified DU contamination at low levels in water from only one well (in Bosnia and Herzegovina; UNEP, 2003), in an area where DU penetrators from strafing runs are buried in the soil.

RADIOLOGICAL RISKS OF INTERNALLY DEPOSITED DU PARTICULATES

Discussion of radiological risks from internally deposited radionuclides is inevitably intertwined with politics, and a range of views exist that either greatly inflate or reduce cancer risks from small intakes of radionuclides from those proposed by the ICRP. Inevitably, a number of criticisms of the published reports on DU have been raised by those who believe exposures to DU are more hazardous than suggested by the application of standard ICRP risk models. These criticisms tend to focus on the perceived inappropriateness of using cancer risks derived from long-term epidemiological studies of populations exposed to external radiation (e.g., atomic bomb survivors) to derive risk estimates for alpha-emitting radioactive particles retained within the body. Could DU particles retained in the lung, or translocated to tracheo-bronchial lymph nodes, be substantially more hazardous that predicted from ICRP models? The risks of internal radionuclides has been reviewed in recent years and, at present, the most robust studies of internally exposed cohorts suggest that the risks of internal radiation are not substantially underestimated (or overestimated) by ICRP models, although the uncertainties associated with the inferred risks are considered to be greater than for external radiation (BEIR VII Phase 2 Report, 2005; CERRIE Report, 2004; Harrison and Muirhead, 2003).

One possible reason why internally deposited DU particulates could be more dangerous than predicted from standard radiological risk models is the evidence from studies of immortalized human osteoblasts that indicate (depleted) uranium salts may induce DNA damage by chemical, as well as radiochemical processes, and may result in neoplastic transformation (Miller et al., 2002) and genomic instability in the progeny of the exposed cells (Miller et al., 2003). Chemical damage to DNA from high local uranium concentrations around DU particles retained in the lung or tracheo-bronchial lymph nodes could enhance the cellular damage produced by alpha-particles and increase the risk of cancer. The possibility of synergistic effects was considered in the Royal Society report (2001) and has recently been raised by others (e.g., Baverstock, 2005), and adds to the uncertainties surrounding the risks from retained DU oxides or DU shrapnel. However, any synergistic effects would also be expected to apply to inhalation exposures to uranium dusts in milling and fabrication workers, but there is no evidence of unexpectedly high rates of cancer or kidney disease in long-term epidemiological studies of these occupational cohorts (Royal Society Report 2001, 2002).

A recent study reports that DU particulates recovered from a test firing at an armored vehicle and instilled into the rat lung were more toxic to the kidney than natural uranium particulates (Mitchel and Sunder, 2004). This was tentatively attributed to the DU particulates containing other metals from the impacted tank, which could increase toxicity. The Capstone test firings show the presence of about 10% aluminum and 5% iron in recovered DU particulates, and other metals in lesser amounts, as well as titanium, which is present at 1% in the DU alloy used in munitions (Capstone Report, 2004).

The possibility that retained DU in the lungs, or DU shrapnel, could be more toxic than expected leads to continuing caution about the health of those who may have been exposed to substantial levels of DU (Level I or II). However, for the small intakes predicted for Level III exposures, the estimated risks to health are sufficiently low that these uncertainties are unlikely to be of consequence.

EXPOSURES TO DU IN THE GULF WAR AND THE BALKANS CONFLICTS

The Royal Society report highlighted the fact that there were almost no measurements of DU concentrations in urine taken from soldiers potentially exposed to DU in the Gulf War or the conflicts in the Balkans. Timely measurements of DU concentrations in urine would have provided estimates of the intakes of individual soldiers in these conflicts and a better assessment of the risks to health. In 2001, the United Kingdom Ministry of Defence (MOD) set up the Depleted Uranium Oversight Board (DUOB), which includes independent expert scientists and representatives of the Red Cross, the British Legion, the National Gulf Veterans and Families Association and their scientific advisors, and the MOD, and is responsible for overseeing the development and implementation of a testing program for DU excretion in the urine of U.K. veterans from the Gulf War and the conflicts in the Balkans.

Provided there has been no significant exposure to enriched uranium, the fraction of the total amount of uranium excreted in urine that is DU can be obtained from the $^{238}U/^{235}U$ ratio, which is approximately 500 for the DU used in penetrators (0.2% ^{235}U; Mitchel and Sunder, 2004; Pöllänen et al., 2003; UNEP, 2003), and 137.9 for natural uranium. The proportion of DU, obtained from the inflation of the isotopic ratio above that of natural uranium, together with the total uranium concentration, provides the concentration of DU in the urine. The amount of DU excreted in urine per day at a specified interval after exposure can be used to estimate the amount of an earlier inhalation intake. The presence of ^{236}U in urine samples with elevated $^{238}U/^{235}U$ ratios can provide additional confirmation that DU is present as ^{236}U is a minor component of DU used for penetrators (about 0.0028% in penetrators recovered in the Balkans; UNEP, 2003) but should not be detected in natural uranium from the diet.

DU is expected to be excreted in urine for many years since a fraction of any inhaled DU particulates are relatively insoluble, and small particles of DU oxides retained in the lung, or translocated to the tracheo-bronchial lymph nodes, will release DU into the bloodstream, and thence to urine, over many years. Additionally, there is a slow recycling of uranium from organs (particularly the skeleton) back to blood, and then to urine, which contributes to the long-term excretion of DU. If a $^{238}U/^{235}U$ ratio of 142 can be shown to be significantly greater than the normal isotope ratio, as appears to be the case, and the isotope ratio of DU used in penetrators is 500, it should be possible to detect an excretion of 0.42 ng DU per day in the presence of 10 ng per day of natural uranium derived from the diet. Calculations based on central estimates of the model parameters (Royal Society report 2001) indicate that this level of sensitivity would allow estimated inhalation intakes of $\geq$ 4.3 mg of DU to be detectable after 14 years, corresponding to a committed effective dose of 0.25 mSv. Individuals with committed doses (i.e., the total radiation dose received over 50 years) from DU inhalation that are above the annual dose limit for the general public (1 mSv) should therefore be readily identified by uranium isotope analysis of urine, using the most sensitive mass spectrometers. At this level of sensitivity in measuring isotope ratios, a negative test for DU implies a predicted intake of < 5 mg DU, and a central estimate of excess lifetime risk of fatal lung cancer of < 0.002% (compared to the general lifetime risk of 6–7%). Risks of other cancers would be extremely low and significant effects on the kidney would be very unlikely.

Detecting and quantifying residual DU from small exposures that occurred 14 years ago in the Gulf War is, however, challenging as it requires accurate measurement of slight increases in the normal $^{238}U/^{235}U$ ratio of 137.9 in urine containing a few ng of uranium per liter, and is susceptible to errors due to contamination of samples with environmental sources of natural uranium and possibly with DU present in analytical laboratories. Several laboratories have recently published protocols for detecting DU in urine using thermal ionization mass spectrometry (TIMS) or inductively coupled plasma mass spectrometry (ICP-MS). Most of these studies have not been capable of detecting small amounts of DU in urine containing normal levels of uranium and have mainly been used to analyze urine from veterans with retained shrapnel and others involved in friendly-fire incidents. For example, Ejnik et al. (2005) published the uranium isotope analysis of urine samples from U.S. veterans of the Gulf War using ICP-MS and reported no evidence of DU in veterans without retained shrapnel but, with their protocol, urine containing 5–20 ng uranium per liter would be considered to contain natural uranium unless the $^{238}U/^{235}U$ ratio was outside the 95% confidence limits of approximately 108–188. Similarly, a study by Westphal et al. (2005) using ICP-MS could only measure accurately uranium isotope ratios in samples containing > 100 ng uranium per liter (far above the 95th percentile of normal urinary uranium concentrations in the U.S.). McDiarmid et al. (2004) in their analysis of veterans of the Gulf War have also used ICP-MS and consider that $^{238}U/^{235}U$ ratios between 128 and 151 are consistent with natural uranium, although refinement of the methodology appears to have improved their ability to detect smaller amounts of DU (Gwiazda et al., 2004). Methods of isotope analysis used for detecting DU in the urine of Canadian veterans from the Gulf War and Kosovo were also too insensitive (Ough et al., 2002). More sensitive mass spectrometric approaches than those used in the above studies are required to be able to establish whether veterans of the Gulf War have been exposed to small amounts of DU and, if so, to estimate the intake.

A recent study using instruments more advanced than quadrapole ICP-MS has been coordinated by the DUOB to evaluate the ability of laboratories to accurately measure small amount of DU in urine (Parrish et al., 2006). Urine samples from volunteers containing 1.6–6 ng of natural uranium per liter were spiked with different amounts of DU, increasing the $^{238}U/^{235}U$ ratio from the normal value of 137.9 to between 141.5 and 151, and these were provided (blinded) to three independent laboratories expressing an interest in the DU testing program. All three laboratories were able to measure $^{238}U/^{235}U$ ratios in samples containing these low amounts of total uranium to within ± 4%, and mostly within ± 2.5%. The two laboratories using the more sensitive instruments (multicollector [MC]-ICP-MS) measured isotope ratios to within about ±1.5% and, as expected, outperformed the laboratory using less sensitive sector field (SF)-ICP-MS.

Two of the laboratories involved in this validation exercise (National Environment Research Council Isotope Geosciences Laboratory [NIGL], Keyworth, U.K, and Scientifics Ltd., Harwell, U.K.) were chosen by the DUOB to offer a sensitive test for DU exposure to any U.K. veterans of the Gulf War and the conflicts in the Balkans. In this ongoing testing program, the NIGL laboratory uses MC-ICP-MS, has lower throughput than the Harwell laboratory, and requires larger volumes of urine (400 ml) and uranium separation and purification, but provides a more precise measure of the $^{238}U/^{235}U$ ratio and a better ability to measure the amount of ^{236}U. The Harwell laboratory uses SF-ICP-MS with an ultrasonic nebulizer and a single-collector ion counting detector system, without uranium separation and purification, and thus requires small volumes of urine (50 ml), and has a high throughput, but lower sensitivity. From a number of clinics around the U.K., 24-h urine samples from veterans are collected, coordinated by an independent health service provider, using thoroughly cleaned containers provided by one of the analysis laboratories. Most of the urine samples have been analyzed by both independent laboratories, with good agreement. Unannounced spiked samples containing small amounts of DU (inflating the normal isotope ratio to about 140–150) are sent at intervals to each laboratory to ensure continued reliability in the values of the reported isotope ratios and their ability to detect small amounts of DU in urine.

The estimated errors in the isotope ratios provided by the laboratories, and the performance of the laboratories in analyzing spiked samples, indicate that DU is present in a veteran's urine sample if the $^{238}U/^{235}U$ ratio is $\geq$ 142, using the more sensitive MC-ICP-MS. With the higher throughput, SF-ICP-MS, the Harwell laboratory considers that urine samples with ratios of >145 contain DU, and those with ratios of 142–145 possibly do so. Any sample reported by the Harwell laboratory to have an isotope ratio of 140 or higher is also analyzed by NIGL and (assuming the laboratory does not report a technical problem with the assay), a ratio of $\geq$142 is considered by the DUOB to be indicative of DU. The ability of NIGL to detect ^{236}U in those urines with elevated $^{238}U/^{235}U$ ratios provides additional evidence for the presence of DU (Parrish et al., 2006).

Any veteran with an isotope ratio measurement $\geq$ 142 would be informed that DU is present in their urine and given an estimate of the amount he is excreting and its possible consequences for his health. The proposed health advice for any veteran found to be excreting a small amount of DU is complicated by the wide divergence of views on the DUOB about the risks of internal alpha-emitting radionuclides, which has resulted in advice that is based on the majority view of health professionals — intakes of a few mg of DU oxides are very unlikely to significantly increase the risk of cancer or kidney disease — together with a statement that a minority of scientists consider that the standard radiation risk model substantially underestimates the hazards.

As of September 1, 2005, apart from a single urine sample that one of the two laboratories believes may contain a small amount of DU, none of the 229 urine samples analyzed by both laboratories have shown evidence of DU ($^{238}U/^{235}U$ ratios are in all cases < 142), suggesting that for most veterans the intakes of DU during the Gulf War and the conflicts in the Balkans were small and unlikely to lead to any health consequences. The DUOB results conflict with those of Horan et al. (2002), which indicated that urine samples from 14 of 27 British, Canadian, and U.S. Gulf War veterans, who considered they had been exposed to DU, had detectable levels of DU in their urine. Half of the positive samples from this latter study had $^{238}U/^{235}U$ ratios > 200, and one veteran had a ratio of 426, suggesting that 94% of the uranium excreted in urine was DU. The urine samples analyzed by Horan et al. (2002) were obtained 9 years after exposure, compared with 13–14 years for those studied by the DUOB, but this difference is unlikely to explain the high percentage of urine with greatly elevated isotope ratios in the former study and the absence of any urine with clear evidence of DU in the latter. It should be noted that, in contrast to the studies carried out under the auspices of the DUOB, the analysis carried out by Horan et al. (2002) included no control group of unexposed individuals and no blinded evaluation of the ability of their procedures to accurately measure small amounts of DU in urine and to avoid problems with sample contamination.

One study has suggested that British veterans from the Gulf War who were considered to have been exposed to DU had elevated levels of chromosome aberrations compared to a group of laboratory workers in Bremen (Schroder et al. 2003). These veterans were probably from the group studied by Horan et al. (2002) and, if so, their DU urine tests are likely to be suspect. The significance of this work is therefore difficult to assess as the levels of exposure (if any) of these veterans to DU is unknown, and a more rigorous assessment is required using a group of veterans who have independently validated exposures to DU and an appropriate control group.

EXPOSURES TO DU IN THE 2003 IRAQ CONFLICT

One of the recommendations of the Royal Society report was that there should be timely assessment of exposures to DU in military personnel in any future conflict in which DU rounds were deployed. After the 2003 Iraq conflict, this occurred in the U.K. through implementation of MOD guidelines (http://www.mod.uk/issues/depleted_uranium/du_biomonitoring.htm) to assess exposure in soldiers involved in friendly-fire incidents and in those concerned that they may have been exposed to DU. Additionally, the MOD has commissioned independent research from King's College, London, to assess exposure to DU in a sample of soldiers from Iraq, including some who might be expected to be at most risk of exposure (e.g., those involved in examining and removing battle-damaged vehicles).

The ongoing independent research study involves the direct measurement of DU in urine using uranium isotope analysis. For their own testing of military personnel, the MOD initially proposed to measure total uranium in urine, with those having high levels of uranium being tested for exposure to DU by uranium isotope analysis. The MOD sought independent advice from the Royal Society DU working group on a suitable threshold level of total uranium that warranted further investigation by isotope analysis. According to central estimates of model parameters, an inhalation exposure that would result in a committed radiation dose of 1mSv (the annual exposure limit for members of the public or ~50% of annual natural background in the U.K.) would lead to about 350 ng uranium per liter of urine at 6 months after intake and a maximum kidney uranium concentration well below 1 μg per gram of kidney. A value of 35 ng per liter of urine at 6 months (adjusted appropriately if sampling was earlier or later) was chosen as the action threshold since, even if all of the uranium detected in urine was DU (a cautious assumption as intakes of natural uranium will contribute), the committed radiation dose and kidney uranium concentration would be 10-fold less than the above values using central estimates and would still be below them using worst-case estimates of the model parameters.

The use of a threshold value of total uranium in urine, above which isotope analysis is warranted, provides a useful and cost-effective approach for screening in future conflicts. However, although this type of approach has been used for U.S. veterans of the Iraq conflict of 2003, for health monitoring of U.K. soldiers from Iraq, laboratory accreditation issues resulted in virtually all the samples being sent for uranium isotope analysis to one of the independent laboratories used in the DUOB testing program. Uranium isotope analysis has been used to examine intakes in soldiers who were concerned about exposures to DU and to investigate potential exposures in two incidents in Iraq in which about 40 U.K. personnel were identified as being at risk from DU intakes by inhalation, ingestion, or contamination of wounds by DU aerosols or from DU metal fragments. This group includes the Level I personnel mentioned above and a further group of about 25 personnel who were on the scene but not directly involved in the immediate rescue operations. The first incident involved severe damage to a Challenger 2 tank loaded with DU munitions, and the second concerned the suspected impact of small caliber DU rounds on two lightly armored vehicles known as Scimitars.

About 50% of the troops that were in or near these vehicles or involved in subsequent recovery operations agreed to be monitored. A small number are excreting DU (average $^{238}U/^{235}U$ ratio of 315) in their urine. However, as all of these latter individuals were wounded in the attack, there is uncertainty about the route(s) of exposure. Those with DU in their urine have total uranium in urine levels that vary by orders of magnitude but most individuals have levels not very different from those found amongst the general population (MOD, personal communication). Some injured and uninjured individuals that were in the two lightly armored vehicles when they were struck by small caliber DU munitions have normal uranium levels and isotope ratios, suggesting that, in the absence of DU shrapnel wounds, such incidents may not inevitably lead to substantial exposures to DU.

The absence of any detectable DU in the urine of some personnel assessed as having received Level 1 exposures is surprising. However, it is important to treat the finding with some caution as only localized and rather limited DU contamination, with a maximum level of the order of 500Bq cm^{-2} (MOD, personal communication), was found on the three vehicles involved in these incidents. It is possible that the small caliber DU rounds passed through the lightly armored vehicles with little generation of DU oxides and/or that ventilation systems were activated, resulting in low concentrations of DU inside the vehicles. It is also important to remember that

Level I exposures include those who were on, as well as in vehicles. Personnel on top of a struck vehicle may be thrown clear or may evacuate immediately after an incident so that only very limited exposure to DU would occur. In particular, the absence of any apparent DU exposure for some soldiers involved in these incidents should not be extrapolated to those in which heavily armored tanks are struck by large caliber rounds where very high concentrations of DU oxides in the struck vehicle can occur (Capstone Report, 2004).

Excluding the soldiers involved in the above incidents, none of the ~350 people tested (as of July 1, 2005) had $^{238}U/^{235}U$ ratios > 142. The results of the MOD monitoring must be interpreted with some caution as the program provides testing for individuals who are concerned about DU exposures but would not be expected to be at any great risk of DU exposure. Indeed, the MOD suggests that this may be the case for about 70% of those tested so far. However, the MOD assessment is that the remaining 30% (about 100 personnel) are more likely to have been exposed to DU than most soldiers serving in Iraq. The availability of urine samples taken within a few months of the 2003 Iraq conflict greatly increases the ability to detect small intakes of DU, compared to the analysis by the DUOB of urines from veterans of the 1991 Gulf War and the Balkans conflicts, and the absence of any isotope ratios > 142 suggests that there may have been very little exposure to DU for most U.K. soldiers in the 2003 Iraq conflict.

The U.S. Department of Defense has implemented guidelines for monitoring uranium concentrations in soldiers (http://www.pdhealth.mil/downloads/OTSG_DU_policy_memo_05-003.pdf). These guidelines state that "DU bioassays will be administered to all personnel with imbedded metal fragments … [who] were in, on, or near (less than 50 meters) an armored vehicle at the time (or shortly after) it was struck with a DU munition (Level I exposure category)" and to those who "routinely enter damaged vehicles as part of their military occupation or who fight fires involving DU munitions (Level II exposure category)." The "DU bioassay" involves measurement of total uranium in urine, with uranium isotope analysis for those who have > 50 ng uranium per liter of urine, a threshold that is just above the upper 95th percentile for the urinary uranium concentrations found in a U.S. study of individuals aged 20 years and older (46 ng per liter; Centers for Disease Control, 2005).

As of September 2004, 1607 U.S. soldiers from Iraq 2003 have been tested, and 111 had uranium concentrations above 50 ng per liter, but only six of these were detected as excreting DU, and these were all known or suspected to have retained DU shrapnel (http://www.pdhealth.mil/downloads/TAB_ref_OIF_DU_v1.pdf). The finding that 93% of those tested had total uranium levels < 50 ng per liter puts an upper limit on their intakes of DU (see previously mentioned research) and, as this group should include those most likely to have been exposed to DU intakes, the results suggest that intakes for most U.S. soldiers in the 2003 conflict in Iraq (i.e., Level III exposures or less) were low. The sensitivity of the uranium isotope analysis in the U.S. testing program for 1991 Gulf War veterans (McDiarmid et al., 2004) appears to be considerably lower than that used by the DUOB. Unless the sensitivity has been increased for testing soldiers of the 2003 Iraq conflict, the apparent lack of any DU in the great majority of soldiers excreting > 50 ng total uranium per liter should be treated with caution as a small proportion of DU would probably not be detected.

The results from the most recent conflict in Iraq support the view derived from isotope analysis of urine from soldiers in the Gulf War and the Balkans that exposures to DU for the majority of soldiers in these conflicts were low and probably less than those estimated in the published studies of the predicted intakes and hazards of DU munitions.

The lack of any evidence of DU in the urine of U.K. veterans is surprising, particularly for those from the 2003 Iraq conflict where urine samples were taken soon after potential exposure, given the high sensitivity of the uranium isotope measurements. Are there reasons to believe that there may have been substantial exposures, but they are not being detected by urinary analysis as an elevated $^{238}U/^{235}U$ ratio? One claim is that DU may not be detected in urine because the particles inhaled on the battlefield are totally insoluble "ceramic" DU. This is highly implausible as it is inconsistent with the *in vitro* absorption characteristics determined for DU oxides produced under realistic conditions in test firings. Furthermore, a substantial proportion of DU particulates from impacts is readily or moderately soluble (4–30% of the DU from the test firings in the Capstone project dissolved *in vitro* within 1 d; Capstone Report, 2004) and, following translocation to blood, about 10% of the solubilized DU that enters the bloodstream will be deposited in bone, from which it will be slowly released to appear in urine (ICRP, 1995). If large inhalation intakes of DU oxides occurred in the Gulf War or the more recent conflicts, DU should still be readily detected in urine using highly sensitive MC-ICP-MS, both from the very slow dissolution of relatively insoluble particulates retained in the lungs or associated lymph nodes and from the slow release of DU from the soluble components deposited in bone. Additionally, the uranium isotope measurements from Iraq were taken soon after the conflict and any substantial intakes of DU should have been very evident

CONCLUSIONS

The Royal Society report, and subsequent reports, all estimate that intakes of DU for the great majority of exposed soldiers on the battlefield (Level III exposures) are likely to be low, and the consensus view is that these intakes would not be expected to lead to any significant increase in a veteran's risk of lung cancer (or any other malignant disease) or kidney disease. The Royal Society report reflected the uncertainties in intakes and risks by taking a cautious approach that considered the consequences for health if risks were 100 greater than those predicted by ICRP models, and concluded that, even with this level of uncertainty, the increased risks of lung cancer or other cancers would be very small for the great majority of veterans who, if they were exposed to DU, would be categorized as Level III. As shown in Table 8.1, even if underestimated 100-fold, the central or best estimates of the increased cancer risks in all three reports discussed in this chapter remain $\leq 0.01\%$ for Level III exposures.

The available information on individual levels of exposure, derived from uranium isotope measurements, strongly support the view that exposures for most veterans of those conflicts in which DU rounds were deployed are likely to have been low and reinforce the view that the risks of cancer or kidney damage are slight for most veterans.

There remain considerable uncertainties about the risks of cancer and kidney disease for any soldiers who received large inhalation intakes of DU particulates (Level I and II) and/or have retained DU shrapnel. Long-term health monitoring is essential for this group of veterans and is ongoing in the U.S. and U.K., but presumably is not being carried out for Iraqi soldiers who survived incidents involving DU munitions in the 1991 Gulf War or the conflict in 2003 (or soldiers in the Balkans).

No widespread contamination with DU has been found in the Balkans and, given the evidence for low intakes of DU by most soldiers in the Gulf War, the Balkans and Iraq, it seems likely that inhalation exposures for those returning to live in areas where DU munitions were deployed will also be low. An exception is for children playing for long periods within contaminated vehicles where the cancer risks could be as high as 0.3%. However, some caution is required as the estimated exposure levels for residents of areas where DU munitions were used are subject to considerable uncertainties, and measurements of the amounts of DU in the urine of residents of these areas are needed to establish whether or not their exposures, and risks to health, from the use of DU munitions are low. If intakes of DU are shown to be low it would make repeated claims that the use of DU has led to a marked increase in birth defects and cancers in Iraq highly implausible (Hindin et al., 2005). Environmental concerns would then focus largely on areas of local DU contamination around penetrator impact sites, and the possibility that corrosion of large numbers of buried penetrators from strafing attacks may contaminate local water or food supplies.

ACKNOWLEDGMENTS

I thank the U.K. Ministry of Defence for providing a summary of the uranium isotope measurement for U.K. troops from Iraq, David Coggon (University of Southampton) for permission to summarize results from the DU testing program conducted through the DUOB, Randy Parrish (National Environment Research Council Isotope Geosciences Laboratory) for unpublished information of urine testing of individuals potentially exposed to DU, and Ron Brown (Ministry of Defence), Mike Bailey, and George Etherington (Health Protection Agency Radiation Protection Division) for comments on the manuscript. The views expressed in this article are my own and not necessarily those of the Royal Society DU Working Group or the DUOB.

REFERENCES

Baverstock, K. 2005. Science, politics and ethics in the low dose debate. *Med. Confl. Surviv.* **21:** 88–100.

BEIR VII — Phase 2 report. 2005. Health Risks from Exposure to Low Levels of Ionizing Radiation: BEIR VII Phase 2. Committee to Assess Health Risks from Exposure to Low Levels of Ionizing Radiation. National Academies Press, Washington, DC (http://www.nap.edu/openbook/030909156X/html/).

Capstone Report. 2004. Depleted Uranium Aerosol Doses and Risks; Summary of U.S. Assessments. (http://www.deploymentlink.osd.mil/du-library/du_capstone/index.pdf).

CERRIE Report. 2004. Report of the Committee Examining Radiation Risks of Internal Emitters (CERRIE) (http://www.cerrie.org/pdfs/cerrie_report_e-book.pdf).

Chazel, V., Gerasimo, P., Dabouis, V., Laroche, P., and Paquet, F. 2003. Characterization and dissolution of depleted uranium aerosols produced during impacts of kinetic energy penetrators against a tank. *Radiat. Prot. Dosimetry* **105:** 163–166.

CHPPM Report. 2000. U.S. Army Center for Health Promotion and Preventive Medicine: Depleted Uranium – Human Exposure Assessment and Health Risk Characterization. Health risk assessment consultation number 26-MF-7555-00D. (http://www.gulflink.osd.mil/chppm_du_rpt_index.html).

Danesi, P.R., Markowicz, A., Chinea-Cano, E., Burkart, W., Salbu, B., Donohue, D., Ruedenauer, F., Vogt, S., Zahradnik, P., and Ciurapinski, A. 2003. Depleted uranium particles in selected Kosovo samples. *J. Environ. Radioact.* **64:** 143–154.

Ej, J.W., Todorov, T.I., Mullick, F.G., Squibb, K., McDiarmid, M.A., and Centeno, J.A. 2005. Uranium analysis in urine by inductively coupled plasma dynamic reaction cell mass spectrometry. *Anal. Bioanal. Chem.* **382:** 73–79.

Hahn, F.F., Guilmette, R.A., and Hoover, M.D. 2002. Implanted depleted uranium fragments cause soft tissue sarcomas in the muscles of rats. *Environ. Health Perspect.* **110:** 51–59.

J.D. and Muirhead, C.R. 2003. Quantitative comparisons of cancer induction in humans by internally depositied radionuclides and external radiation. *Int. J. Radiat. Biol.* 79: 1–13.

P., Dietz, L., and Durakovic, A. 2002. The quantitative analysis of depleted uranium isotopes in British, Canadian, and U.S. Gulf War veterans. *Mil. Med.* **167:** 620–627.

Gwiazda, R., Brugge, D., and Pannikar, B. 2005. Teratogenicity of depleted uranium aerosols: a review from an epidemiological perspective. *Environ. Health* **4:** 17.

Gwiazda, R.H., Squibb, K., McDiarmid, M., and Smith, D. 2004. Detection of depleted uranium in urine of veterans from the 1991 Gulf War. *Health Phys.* **86:** 12–18.

ICRP 1994. Human respiratory tract model for radiological protection. International Committee on Radiological Protection (ICRP) Publication 66. *Annals of the ICRP* 24, No. 1-3, Oxford, Pergamon Press.

ICRP 1995. Age-dependent doses to members of the public from intake of radionuclides: Part 3 ingestion dose coefficients. International Committee on Radiological Protection (ICRP) Publication 69. *Annals of the ICRP* 25, No. 1, Oxford, Pergamon Press.

McDiarmid, M.A., Engelhardt, S.M., Oliver, M., Gucer, P., Wilson, P.D., Kane, R., Kabat, M., Kaup, B., Anderson, L., Hoover, D., Brown, L., Albertini, R.J., Gudi, R., Jacobson-Kram, D., Thorne, C.D., and Squibb, K.S. 2005. Biological monitoring and surveillance results of Gulf War I veterans exposed to depleted uranium. *Int. Arch. Occup. Environ. Health* August 2 (*epub ahead of print edition*).

McDiarmid, M.A., Squibb, K., and Engelhardt, S.M. 2004. Biologic monitoring for urinary uranium in Gulf War I veterans. *Health Phys.* **87:** 51–56.

Miller, A.C., Brooks, K., Stewart, M., Anderson, B., Shi, L., McClain, D., and Page, N. 2003. Genomic instability in human osteoblast cells after exposure to depleted uranium: delayed lethality and micronuclei formation. *J. Environ. Radioact.* and Sunder, S. 2004. Depleted uranium dust from fired munitions: physical, chemical and biological properties. *Health Phys.* **87:** 57–67.

Miller, A.C., Xu, J., Stewart, M., Brooks, K., Hodge, S., Shi, L., Page, N., and McClain, D. 2002. Observations of radiation-sepcific damage to human cells exposed to depleted uranium: dicentric frequency and neoplastic transformation as endpoints. *Radiat. Prot. Dosimetry* **99:** 275–278.

National Health and Nutrition Examination Survey, 1999–2002. (http://www.cdc.gov/exposurereport/3rd/pdf/thirdreport.pdf).

OSAGWI Report. 2000. Office of the Deputy Secretary of Defense for Gulf War Illnesses, DOD. Exposure Investigation Report: Depleted Uranium in the Gulf (II). (http://www.gulflink.osd.mil/du_ii).

Ough, E.A., Lewis, B.J., Andrews, W.S., Bennett, L.G., Hancock, R.G., and Scott, K. 2002. An examination of uranium levels in Canadian forces personnel who served in the Gulf War and Kosovo. *Health Phys.* **82:** 527–532.

Parrish, R.R., Thirlwall, M.F., Pickford, C., Horstwood, M., Gerdes, A., Anderson, J., and Coggon, D. 2006. Determination of 238U/235U, 236U/238U and uranium concentration in urine using SF-ICP-MS and MC-ICP-MS: An inter-laboratory comparison. *Health Phys*, Vol. 90:127–138.

Pöllänen, R., Ikäheimonen, T.K., Klemola, S., Vartti, V-P., Vesterbacka, K., Ristonmaa, S., Honkamaa, T., Sipilä, P., Jokelainen, I., Kosunen, A., Zilliacus, R., Kettunen, M., and Hokkanen, M. 2003. Characterization of projectiles composed of depleted uranium. *J. Environ. Radioact.* **64:** 133–142.

Royal Society Report. 2001. The health hazards of depleted uranium munitions, part 1. (http://www.royalsoc.ac.uk/displaypagedoc.asp?id=11496).

Royal Society Report. 2002. The health hazards of depleted uranium munitions, part 2. (http://www.royalsoc.ac.uk/displaypagedoc.asp?id=11498).

Sandia Report. 2005. An Analysis of Uranium Dispersal and Health Effects Using a Gulf War Case Study. Sandia National Laboratories. (http://www.sandia.gov/news-center/news-releases/2005/def-nonprolif-sec/snl-dusand.pdf).

Sansone, U., Stellato, L., Jia, G., Rosamilia, S., Gaudino, S., Barbizzi, S., and Belli, M. 2001. Levels of depleted uranium in Kosovo soils. *Radiat. Prot. Dosimetry* **97:** 317–320.

Schroder, H., Heimers, A., Frentzel-Beyme, R., Schott, A., and Hoffmann, W. 2003. Chromosome aberration analysis in peripheral lymphocytes of Gulf War and Balkans War veterans. *Radiat. Prot. Dosimetry* **103:** 211–219.

Squibb, K.S., Leggett, R.W., and McDiarmid, M.A. 2005. Prediction of renal concentrations of depleted uranium and radiation dose in gulf war veterans with embedded shrapnel. *Heath Phys.* **89:** 267–273.

UNEP. 2001. Depleted Uranium in Kosovo: Post-Conflict Environmental Assessment. United Nations Environmental Programme. (http://postconflict.unep.ch/publications/uranium.pdf).

UNEP. 2002. Depleted Uranium in Serbia and Montenegro: Post-Conflict Environmental Assessment in the Federal Republic of Yugoslavia. United Nations Environmental Programme. (http://postconflict.unep.ch/publications/duserbiamont.pdf).

UNEP. 2003. Depleted Uranium in Bosnia and Herzegovina: Post-Conflict Environmental Assessment. United Nations Environmental Programme. (http://postconflict.unep.ch/publications/BiH_DU_report.pdf).

Uyttenhove, J., Lemmens, M., and Zizi, M. 2002. Depleted uranium in Kosovo: results of a survey by gamma spectrometry on soil samples. *Health Phys.* **83:** 543–548.

Westphal, C.S., McLean, J.A., Hakspiel, S.J., Jackson, W.E., McClain, D.E., and Montaser, A. 2004. Determination of depleted uranium in urine via isotope ratio measurements using large-bore direct injection high efficiency nebulizer-inductively coupled plasma mass spectrometry. *Appl. Spectrosc.* **58:** 1044–1050.

9 Canadian Forces Uranium Testing Program

Edward A. Ough, Ph.D.

CONTENTS

INTRODUCTION

The Canadian Forces (CF) voluntary uranium testing program is currently in its 5th year of operation. The program was initiated about 10 years after the 1990 Gulf War. Following the war, a number of military personnel suffered lingering health problems thought to be due to their wartime service. Concern was raised in the public media about the role of depleted uranium in these illnesses.

During the war, the U.S. and U.K. expended 300 tons of depleted uranium munitions [1]. Since the war, American forces have fired-off 11 tons of DU munitions in Bosnia (1994–1995) and Kosovo (1999) [2] and an estimated 68 tons in the 2003 Gulf War [3]. Although Canada did have DU munitions in its military arsenal (Phalanx Close-In Weapons Support System) aboard ships between 1990–1998, no Canadian DU was ever fired in combat.

Canadian forces were generally well removed from the conflict areas where DU was used during the 1990 Gulf War. (The one exception would be the Canadian soldiers stationed at Camp Doha during the 1991 ammunition explosion.) Although well removed from areas where DU munitions were expended, Canadian veterans expressed concern about long-distance exposure to the aerosols produced when DU rounds impacted on armored targets. In order to ease veteran's fears, the Canadian Department of National Defence (DND), through the Directorate of Medical Policy, initiated a voluntary uranium testing program. This program was one of the first voluntary uranium testing programs to be initiated by a national government.

The setup and day-to-day operation of the uranium testing program was an interesting and complicated task. Decisions had to be made on what analytical tests needed to be performed, what type of biological sample to test, and what laboratories would be selected to perform the analyses. To actually test an individual for past exposure — often 10 years or more in the past — to depleted uranium is daunting due, in part, to the ubiquitous presence of natural uranium in the environment and to its low (i.e., subparts per billion [ppb]) concentration in biological samples. It was decided early on that testing for both total uranium and uranium-238/uranium-235 isotopic ratio measurements were required in order to determine the source of the uranium. Urine was selected as the testing media because of the ease of collection and that an invasive medical procedure was not required for its collection. Private commercial laboratories, with the required analytical expertise, were contracted to perform the urine bioassays in order for the government to be completely hands-off in the analysis of all samples. Those laboratories selected met a series of criteria, which included the capability to perform both total uranium and isotope ratio measurements.

Although DND went to considerable effort to be removed from the testing, there was much public scrutiny over the test results, and some skepticism over the reliability of the results. This public skepticism prompted DND, through Defence Research and Development Canada (DRDC), to fund an interlaboratory analytical exercise where the commercial laboratories contracted by DND were compared against other laboratories, utilizing different analytical instruments, to test the reliability of the CF test laboratories and to determine whether the CF program could be improved through use of some or all of these instruments.

In this chapter, we will describe those analytical instruments and procedures suited for total uranium and $^{238}U/^{235}U$ isotopic ratio bioassays. We will look at the up-to-date results from the CF uranium bioassay program [4] and the results from the DRDC interlaboratory comparison study [5,6] to see how the CF testing laboratories rated against other analytical laboratories. We will briefly compare the CF program and test results with those from other countries and international organizations. Finally, some recommendations will be proposed on how to improve the CF program and other uranium bioassay programs.

ISOTOPIC URANIUM ANALYTICAL PROCEDURES

The natural background concentrations of uranium in biological samples have to be taken into account when attempts are made to test for depleted uranium. Depleted uranium (DU) differs from natural uranium (NU) in the ratio of the three naturally occurring isotopes: 99.2739% ^{238}U, 0.7204% ^{235}U, and 0.0057% ^{234}U in natural uranium and 99.799% ^{238}U, 0.2001% ^{235}U, and 0.0009% ^{234}U in depleted uranium.

Consequently, a measure of the $^{238}U/^{235}U$ isotope ratio is required to determine the source of the uranium present in any given biological or environmental sample. A $^{238}U/^{235}U$ isotopic ratio of 138 is indicative of NU and a $^{238}U/^{235}U$ isotopic ratio of 500 is indicative of DU. A $^{238}U/^{235}U$ isotopic ratio between these values (i.e., a mixture of natural and depleted uranium) is expected for biological samples contaminated with DU.

A number of analytical techniques are available for the measurement of uranium in test samples. Kinetic phosphorescence analysis (KPA) is a standard analytical procedure employed in industrial uranium bioassay programs [7], but the technique only measures total uranium and is not suitable for DU bioassays. Neutron activation (instrumental and delayed), alpha spectrometry, thermal ionisation mass spectrometry (TIMS), and inductively coupled plasma mass spectrometry (ICP-MS) are four analytical procedures that have the capability to measure $^{238}U/^{235}U$ isotopic ratios.

Neutron Activation Analysis

When exposed to a thermal neutron source, a fraction of the ^{238}U present in a sample of uranium can be converted to ^{239}U. Instrumental neutron activation analysis (INAA) takes advantage of the β decay (half-life of 23.5 min) of ^{239}U to the short-lived ^{239}Np to measure concentrations of ^{238}U in samples. A low-energy gamma ray (~74.7 keV) is released during the decay of ^{239}U. The following equation:

$$A = N\,\sigma\,\phi(1 - e^{-ll})$$

relates the activity (A), after an irradiation time t, of a sample to a number (N) of ^{239}U atoms in the sample. The other variables and constants in the equation are: the radioactivity decay constant for ^{239}U ($\lambda = 4.92 \times 10^{-4}\ s^{-1}$), the neutron capture cross-section ($\sigma = 2.73 \times 10^{-24} cm^2$) and the neutron flux ($\phi$). Intereference from activated species such as sodium and potassium in biological matrices limits INAA detection of ^{238}U in biological samples to about 1 ppb (part per billion; $\mu g\ L^{-1}$). Anion exchange separation of uranium from interfering species can lower this detection limit to 1 ppt (part per trillion; $ng\ L^{-1}$) for neutron sources with sufficiently high thermal neutron flux [8].

Delayed neutron activation analysis (DNAA) is a more sensitive technique for determining ^{235}U in samples. Fission occurs when a sample containing ^{235}U is irradiated with thermal neutrons. In approximately 1.6% of the fissions, fission products or daughter species emit neutrons as part of their decay mechanism. This neutron emission is delayed or later than neutrons born in fission. A DNAA detection limit of 6.9 pg L^{1} ^{235}U (4 ng L^{-1} DU) in a urine sample has been reported in the literature [9].

Delayed NAA has two significant analytical advantages over INAA. These are the significantly lower detection limit (approximately 200 times) and that there is no need to remove the ^{235}U from the biological matrix. When it comes to DU bioassays for DU, these advantages are for naught due to the fact that direct measurements of both ^{235}U and ^{238}U are required to measure the isotope ratio. As such, INAA is the "weakest link" in neutron activation analysis, and the detection limit for isotope ratio measurements depends on this technique's detection limit.

Alpha (α) Particle Spectrometry

In alpha particle spectrometry (APS) the natural radioactivity of the uranium isotopes can be used to determine total and isotopic ratios in a sample. The detection limits in APS are determined by the detector efficiency, chemical recovery (uranium has to be removed from the sample matrix), background radioactivity, sample size, and counting time. The uranium detection limit is related to the specific activity (S), the alpha particle yield per disintegration (k) constant, and the number of counts (n) such that:

$$L_D = \frac{n}{259 * k * S}$$

The estimated uranium isotope detection limits for 10 counts are listed in Table 9.1. It can be seen that APS is 5 times more sensitive to ^{235}U compared to ^{238}U, but the technique also requires the removal of the uranium from the sample matrix and has relatively high uranium isotopic composition detection limit (1 mBq L^{-1} per isotope (3-d counting period) or 1 μg L^{-1} for ^{238}U [10]).

Mass Spectroscopy

Mass spectroscopy (MS) is fast becoming the preferred technique for ultra-trace level metal analysis. A mass spectrometer can be broken down into three basic components the ion source, the mass analyzer, and the ion detector system. Several methods, including resonance ionization, thermal ionization, and inductively coupled plasma (ICP), are available for the ionization of a sample prior to introduction into the mass spectrometer. Mass spectrometers equipped with ICP front ends can be found in most commercial and university analytical laboratories. The mass analyzer is responsible for the separation of the ions based on their mass and charge. Some of the more common mass analyzers include time of flight, quadrupole, magnetic, and electrostatic. Many commercial instruments combine magnetic and electrostatic mass analyzers (double focusing) in the increasingly popular high-resolution sector field instruments [11]. An ion detector system is normally comprised of either a

TABLE 9.1
Estimated Uranium Isotope Detection Limits (Silicon Surface-Barrier Detectors) for Alpha (α) Particle Spectrometry [7]

Isotope	Energy (MeV)	k (yield per Disintegration)	Activity (Bq ng^{-1})	Detection Limit (ng L^{-1})
^{234}U	4.77	0.72	0.232	2×10^3
^{235}U	4.40	0.57	7.98×10^{-5}	8
^{238}U	4.19	0.77	1.24×10^{-5}	40

single ion collector or multiple ion collectors. Instruments with a single ion collector are only able to measure the intensity of one ion peak at a time and have to "hop" from peak to peak when isotopic ratios are required. Any variability in the intensity of the ion peak, as normally observed for ICP-MS, will affect the precision of these measurements. A multicollector system can measure two or more ion peaks simultaneously thus negating any signal variability and providing higher precision for isotopic ratio measurements than a single collector instrument. The most common mass spectrometers for uranium in urine analysis are thermal ionization MS (TIMS), inductively coupled plasma quadrupole MS (ICP-Q-MS), and inductively coupled plasma sector field MS (ICP-SF-MS) with either single collector or multicollector detectors.

Thermal Ionization Mass Spectrometry (TIMS)

Some of the first media reports of DU contamination in Gulf War veterans were based on TIMS analyses carried out at Memorial University (St. John's, Newfound land, Canada) [12]. In TIMS, the uranium is removed from the sample matrix (i.e., urine), preconcentrated and a small sample (~6µL) evaporated on a platinum or rhenium filament. The filament is loaded into the sample compartment of the TIMS and placed under a high vacuum. Sufficient current is then run through the filament to thermally ionization the uranium. The uranium ions are then focused through a sector field mass analyzer with the ions of select isotopes (i.e., ^{238}U and ^{235}U) directed to individual cups of a multidetector array for quantitative analysis.

The advantages of TIMS are the low detection limit (0.1 picogram for both ^{235}U and ^{238}U) and the precision in the isotopic ratios (0.05–0.2%). At their best, ICP front ends deliver approximately 10% of a sample into the mass spectrometer, whereas approximately 100% of a sample will be thermally ionized in a TIMS. As a result, TIMS is approximately 10 times more sensitive than any ICP-MS. The high precision for isotopic ratios is due to the simultaneous measurement of the ion intensity from the two isotopes of interest, which negates any instrument instability. The only ICP-MS instrument with comparable precision is a sector field instrument equipped with a multicollector array.

The main disadvantages of TIMS are the need for the removal of uranium from the sample matrix and extensive preconcentration. Both disadvantages significantly increase the turnaround time (up to 1 week) for a sample and can be problematic in an operational setting where fast (less than 24 h) turnaround times are required.

Inductively Coupled Plasma Mass Spectrometry

Inductively coupled plasma mass spectrometry (ICP-MS) has been around for over 25 years and is widely used in trace analysis of environmental and biological samples. The inductively coupled plasma is an electrical flame produced by the inductive coupling of high-frequency energy to a gas flow, which is normally argon. The liquid sample is nebulized and injected as an aerosol into the plasma where it is first vaporized and then ionized. The ionized gas is directed into the mass spectrometer where it passes through the mass analyzer; ions are separated based on their mass-to-charge ratio and then directed to the detector for qualitative and quantitative analysis.

Some advantages of ICP-MS are the user's ability to perform online "hands-off" analysis, to measure most elements (elements with high ionization potentials are difficult to measure with TIMS), to measure uranium in urine with minimal sample preparation, and to process large numbers of samples in short periods of time. Some disadvantages of ICP-MS are the lower precision in isotopic ratio measurements (as compared to TIMS) when using a single collector instrument, the need to reduce the percentage of total dissolved solid to < 1% which necessitates an approximate 10 times dilution of urine samples, and the presence of spectral (e.g., isobaric, poly-atomic) interferences.

The choice of mass analyzer, either quadrupole or sector-field, and the choice of detector, either single-collector or multicollector, will have a significant effect on an instrument's capability to detect trace quantities of uranium, to distinguish isotopes of uranium from spectral interference, and on the precision of isotope ratio measurements. Sector-field instruments have relatively low (< 0.2 counts per second [cps]) background noise whereas the best quadrupole instruments have relatively high (~10 cps) background noise. As an instrument's detection limits (DL) are based on the signal to noise (s/n) ratio, a sector-field instrument will have limits at least 10 times lower than quadrupole instruments. Sector field instruments, unlike quadrupole instruments, can also be operated at higher resolution ($m/\Delta m > 300$), and at these settings it is possible to separate isotopes of interest from any spectral interference. If the interference is polyatomic, an ICP-Q-MS spectrometer can be equipped with a collision cell, which is positioned before the mass analyzer, to effectively remove the interference from the ion beam. A multicollector instrument, as previously mentioned, will have better precision on isotope ratio measurements than one with a single collector, therefore data acquired on a multicollector ICP-SF-MS will have precision comparable to that acquired on a TIMS.

TIMS vs. ICP-MS

Typical detection limits (total and isotopic) and precision on isotope ratio measurements for the four MS techniques have been listed in Table 9.2, along with indications

TABLE 9.2
Typical Uranium Detection Limits (Total and $^{238}U/^{235}U$ Isotope Ratio) and Precision for Several Mass Spectrometer Systems

	Detection Limits (ng L^{-1})			
Method	Total U	$^{238}U/^{235}U$ [b]	Isolation from Matrix[a]	Precision (2σ)
TIMS[c]	0.001	0.14	Yes	0.05–0.2%
ICP-Q-MS	0.1	14	No	5%
ICP-SF-MS	0.01	1.4	No	0.1–1.0%
ICP-SF-MC-MS	0.01	1.4	Some	0.05–0.2%

[a] For the measurement of total uranium.

[b] Uranium-238 concentration.

[c] 100 ml urine sample reduced to 6 μl sample.

as to whether or not the technique requires the removal of uranium from the sample matrix. Based on both detection limits and precision, TIMS is ranked first for isotope ratio measurements, but it may not be the ideal instrument for uranium bioassays. If the environmental levels of uranium are significantly higher than the TIMS limit, then the other techniques become as reliable (if a slight reduction in the precision of $^{238}U/^{235}U$ isotope ratios is acceptable). If factors such as the cost to acquire and run an instrument, the cost to analysis a sample, and the time to process a sample are more important, then ICP-Q-MS becomes the instrument of choice. These were some of the issues that were addressed when the CF testing program was first initiated.

RESULTS FROM CF TESTING PROGRAM

Following reports in the media of elevated uranium concentrations in Canadian veterans of the Gulf War conflict, the Canadian Minister of National Defence announced, in February 2000, a uranium testing initiative for active and retired CF personnel. At this time, enquiries were made to a number of commercial and university laboratories (private and university) across Canada with regard to their ability to measure total isotopic uranium concentrations. The ability to perform isotopic ($^{238}U/^{235}U$) measurements, suitable instrument (method) detection limits, sample processing cost, sample turnaround time, and the ability to handle the volume of samples were some of the selection criteria. Two laboratories, one employing ICP-MS and the other I/DNAA, met the criteria and were selected for testing program.

To date, 228 active or retired CF personnel have submitted 24-h urine samples for uranium assay. The samples were from concerned individuals and therefore represent a directed, rather than a random, sampling. A control group (e.g., nondeployed CF personnel) was not included in the testing program. Studies on nonoccupationally exposed individuals (Table 9.3) provide data for the expected natural background levels of uranium in urine. The uranium distribution for the entire sample cohort, as measured by both ICP-MS (217 samples) and INAA (163 samples), are presented in Figure 9.1. The range, mean, and median values for the entire study population are listed in Table 9.3. The total uranium concentration ranges were 0.11 ng L^{-1} to 49.5 ng L^{-1} for ICP-MS and 1 ng L^{-1} to 81 ng L^{-1} for INAA). From the data, the mean and median uranium concentrations were calculated to be 4.5 ng L^{-1} and 2.8 ng L^{-1}, respectively, for ICP-MS analysis and 17 ng L^{-1} and 15 ng L^{-1}, respectively, for INAA. For duplicate analyses of individuals, the INAA results tended to be consistently higher than those for ICP-MS.

The reported analytical differences were attributed to sample in-homogeneity and from differing analytical sensitivities (instrument detection limits). The laboratories received 24 h urine samples collected on separate days; therefore, variations in people's daily intake of food and water might affect the amount of uranium excreted in their urine. In the German MoD study, 24 h urine samples were collected from two individuals, 57 and 55 years of age, for a period of 10 d and 15 d, respectively [13]. The individuals were shown to have mean daily urinary uranium concentrations of 19.5 ng (15–30 ng daily range; 57 y) and of 27.16 ng (10–65 ng daily range; 55 y). These results demonstrate the daily variation that can occur in urinary uranium concentrations. The I/DNAA laboratory reported a significantly

TABLE 9.3
Reported Uranium Concentrations (ng L^{-1}) in Urine Samples

Group	Tech.	Country	Samples	Range	Mean	Median
Canadian Soldiers						
Ough et al. [4]	ICP-MS	Canada	217	0.11–49.5	4.5	2.8
Non-DU–Exposed American Veterans						
McDiarmid et al. [15]	KPA	U.S.	23	<10–130	<10	<10
Hooper et al. [14]	KPA	U.S.	10		70	
Gwiadza et al. [17][a]	ICP-MS[c]	U.S.	12	1.2–88		17
DU Exposed American Veterans						
Hooper et al. [14]	KPA	U.S.	15		10,080	
McDiarmid et al. [15]	KPA	U.S.	29	10–30,740	5800	210
Ejnik et al. [16]	KPA	U.S.	7	150–45,000		9200
Gwiadza et al. [17][a]	ICP-MS[c]	U.S.	28	1.8–200		66
Gwiadza et al. [17][a,b]	ICP-MS[c]	U.S.	17	5–71,600		2610
German Soldiers Deployed in Kosovo (May and September 2000)						
Roth et al. [13]	ICP-MS	Germany	43	1–27	12.8	
Roth et al. [13]	ICP-MS	Germany	34	1–27	13.8	
Control Group for German Deployment in Kosovo (May and September 2000)						
Roth et al. [13]	ICP-MS	Germany	50		12.5	
Roth et al. [13]	ICP-MS	Germany	50		8.6	
Swedish Soldiers in Kosovo [18]						
		Sweden	200		9.9	
Swedish Soldiers Control (in Sweden) [18]						
		Sweden	200		38.3	
International Red Cross						
Meddings et al. [19]	ICP-MS	Kosovo[d]	31	3.5–26.9		8.9
Non-Occupation–Exposed Individuals						
Leuong [20]	ICP-MSc	Canada	37	1.4–11	4.3	
Lorber et al. [21]	ICP-MS	Israel	200	1–40		10
Dang et al. [22]	INAA	India	27	2.9–40	12.8	9.4
Rodushkin [23]	ICP-MSc	Sweden	19	1.4–17	9.3	
Galletti et al. [24]	ICP-MS	Italy	38	3–26	10	

[a] Nanograms uranium per gram creatinine.
[b] With shrapnel.
[c] ICP-SF-MS with single collector.
[d] Individuals resided for at least 3 months in Kosovo (range: 3–11 months).

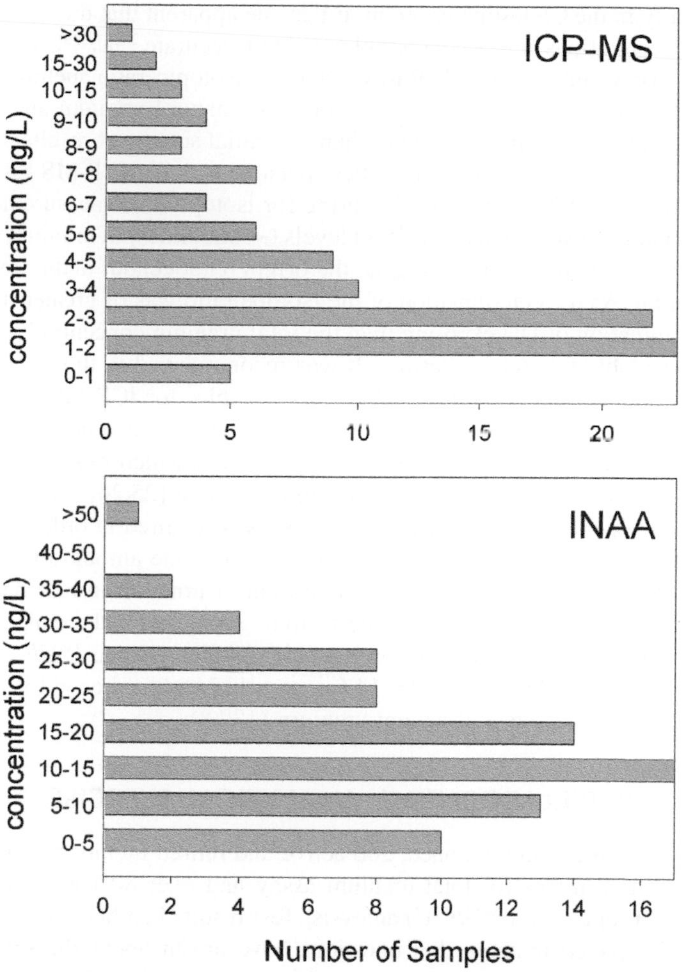

FIGURE 9.1 A histogram of the distribution for total uranium, in urine, as measured by ICP-Q-MS and INAA.

higher uranium detection limit for INAA (~50 ng L^{-1}) compared to the ICP-MS laboratory (~0.5 ng L^{-1}). As a result, the reported INAA values were are all in the vicinity of detection limit levels. This interpretation was confirmed by the reported sizes of the percent relative standard deviations for individual INAA data (20–60%) as compared to ICP-MS data (2–32%), for uranium concentrations less than 1 ng L^{-1}.

Early in the testing program, it became apparent that, at the very low levels of uranium present in the urine samples analyzed, the ICP-MS data were more accurate whereas the INAA data were less sensitive as a result of the higher detection limit for this technique. As a result, the use of INAA was discontinued and current CF uranium testing is by ICP-MS.

Also early in the CF testing program, it became apparent that due to low urinary uranium concentrations, it was not possible to obtain accurate $^{238}U/^{235}U$ isotope ratios from 24-h urine samples. The ICP-MS $^{238}U/^{235}U$ isotopic ratio method detection limits (MDL) of 14 ng L^{-1} total uranium for 100% natural uranium and 50 ng L^{-1} for 100% DU, were significantly higher than the initial set of test results. Due to the difficulty in measuring such low quantities, personnel at the ICP-MS testing laboratory suggested that hair could replace urine for isotope ratio measurements. Hair has the advantage of significantly elevated levels (~1000 times) of uranium compared to urine, but there may be concerns about the origin (endogenous and/or exogenous) of the uranium. As the determination of the isotope ratio was of greater importance in the determination of DU exposure than the total uranium concentration, hair was considered suitable for isotopic testing. It was recommended to those veterans still requesting isotopic assays that they submit hair samples for ICP-MS analysis.

To date, 102 hair samples have been analyzed for total uranium and 87 for $^{238}U/^{235}U$ isotopic ratio. The concentration of uranium, which ranged from 0.7 μg kg^{-1} to 354 μg kg^{-1}, was consistent with published data [25,26]. The selection of hair samples close to the skin (nape of the neck) was required in order to minimize contamination by exogenous uranium and to maximize the amount of endogenous uranium. It has been shown that the concentration of uranium in hair decreases as a function of the sample distance from the scalp (e.g., highest at the scalp and lowest at the tip of the hair) [25]. Analyses on the hair samples yielded $^{238}U/^{235}U$ isotopic ratios ranging from 122 ± 17 to 147 ± 14 (mean = 137 and median = 137) that were consistent with the presence of natural uranium (137.8).

INTERLABORATORY ANALYTICAL EXERCISE

When this study was commissioned, 200 active and retired members of the CF had submitted urine samples for total uranium assay and over 80 had submitted hair samples for isotope ratio ($^{238}U/^{235}U$) analysis. Test results had been negative for the presence of depleted uranium, but questions have arisen about the sensitivity of analytical methods employed. The purpose of the analytical exercise was to evaluate all available analytical techniques with the capability to measure uranium isotope ratios ($^{238}U/^{235}U$) at trace concentrations (subparts per billion) in biological fluids (e.g., urine). Synthetic urine was chosen in this study in order to negate any concern over biohazards. The techniques chosen for this study included sector field inductively coupled plasma mass spectrometry (ICP-SF-MS), quadrupole inductively coupled plasma mass spectrometry (ICP-Q-MS), thermal ionization mass spectrometry (TIMS), and neutron activation analysis (NAA).

A host laboratory was selected to prepared six identical sets (12 samples/sets) of synthetic urine samples (1 kg each) containing natural and depleted uranium and distributed these sample sets to the five participating laboratories [27]. The synthetic urine formulation (Table 9.4) is one that has been utilized by the U.S. Department of Energy (DOE) Laboratory Accreditation Program (DOELAP) to conduct performance assessment of all laboratories performing urine bioassay analyses on DOE personnel. Total uranium content, of the prepared solutions, was confirmed, by NIST, using inductively coupled plasma atomic emission spectrometry (ICP-AES).

TABLE 9.4
Synthetic Urine Composition as Proscribed by DOELAP

Component	g kg^{-1}
Urea (CH_4N_2O)	16.00
Sodium sulfate ($Na_2SO_4 \cdot H_2O$)	4.31
Potassium chloride (KCl)	3.43
Sodium chloride (NaCl)	2.32
Creatinine ($C_4H_7N_3O$)	1.10
Ammonium chloride (NH_4Cl)	1.06
Hippuric acid ($C_9H_9NO_3$)	0.63
Calcium chloride ($CaCl_2 \cdot 2H_2O$)	0.63
Citric acid ($C_6H_80_7$)	0.54
Glucose ($C_6H_{12}0_6$) {dextrose}	0.48
Magnesium sulfate ($MgSO_4$)	0.46
Sodium metasilicate ($Na_2SiO_3 \cdot 9H_2O$)	0.071
Pepsin	0.029
Oxalic acid ($C_2H_2O_4$)	0.02
Sodium phosphate, monobasic ($NaH_2PO_4 \cdot H_2O$)	2.73
Lactic acid ($C_3H_6O_3$)	0.094
2% v/v nitric acid	966

The participating laboratories had no prior knowledge as to the urine spiking levels (if any) and were asked to provide their results for determination of total uranium and its isotopic composition in urine for interlaboratory comparison. Each laboratory used in-house analytical methods (sample handling and analysis) for the detection and determination of total uranium and its isotopic composition in the provided urine samples.

The total uranium results, reported by four of five testing laboratories (a second ICP-Q-MS laboratory provided an incomplete set of results and was omitted from the analysis of the data), are presented in Table 9.5. Samples #2 and #11 were blank samples where the uranium present arose from uranium impurities in the synthetic urine samples. The blank values reported by the ICP-SF-MS lab and one of the two ICP-Q-MS labs, plus those reported in a previous U.S. Department of Energy (USDOE) interlaboratory comparison [28], were used to set the background uranium concentration (25.2 ± 2.5 ng U/l) in the synthetic urine. The concentration of uranium and the isotope ratio ($^{238}U/^{235}U$) in the 10 spiked synthetic urine samples were adjusted to take into account the background.

The test results (Table 9.5) are presented in Figure 9.2, where the values for total uranium as determined by the testing labs is plotted against the host lab values. Regression results for the ICP-SF-MS lab, the ICP-Q-MS lab, and the TIMS lab displayed excellent agreement between individual values and the fitted lines (correlation coefficients > 0.999), whereas the results for the neutron activation analysis (I/DNAA) laboratory exhibited greater scatter around the line of best fit (correlation coefficient ~0.95). The results from the ICP-SF-MS and ICP-Q-MS labs were in

TABLE 9.5
Total Uranium Determined by Participating Laboratories

Sample Number	Total Uranium (ng kg^{-1})[a] Host	ICP-SF-MS	ICP-Q-MS	TIMS	I/DNAA
1	250 ± 20	251 ± 6	260 ± 30	221.9	140 ± 20
2	25 ± 2	24 ± 1	25 ± 2	37.4	< DL[b]
3	420 ± 20	430 ± 10	440 ± 20	358.3	370 ± 60
4	570 ± 30	560 ± 40	580 ± 50	470.4	350 ± 60
5	320 ± 20	330 ± 20	330 ± 20	281.2	240 ± 50
6	130 ± 10	130 ± 20	136 ± 8	127.2	90 ± 40
7	80 ± 10	76 ± 4	74 ± 7	77.3	< DL[b]
8	55 ± 6	57 ± 5	54 ± 7	61.2	50 ± 30
9	100 ± 10	97 ± 6	95 ± 6	97.2	70 ± 40
10	770 ± 40	760 ± 30	770 ± 70	640.8	680 ± 80
11	25 ± 2	27 ± 1	20 ± 6	38.4	< DL[b]
12	200 ± 20	200 ± 10	210 ± 40	178.3	180 ± 40

[a] Mean ± 2 standard deviations.
[b] DL = detection limit.

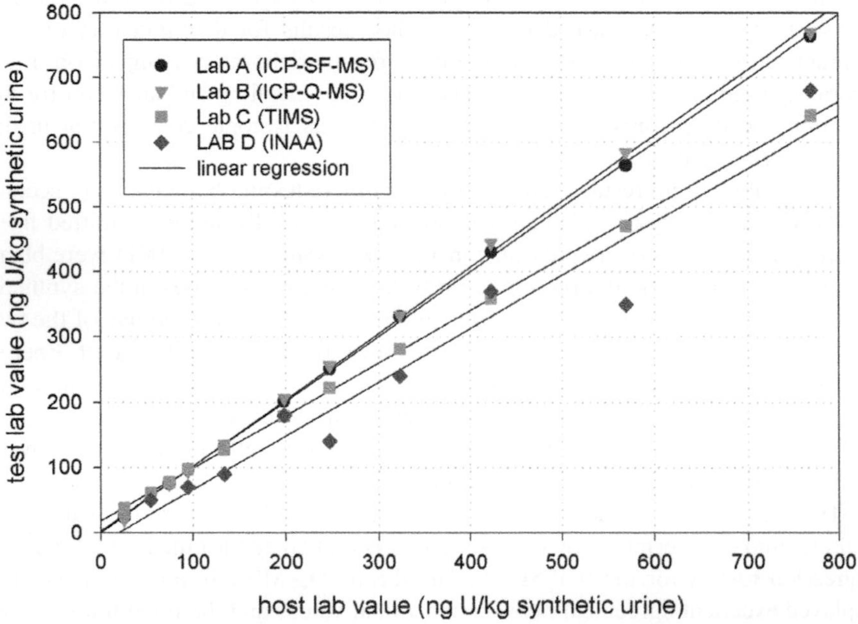

FIGURE 9.2 Total uranium (determined by participating laboratories) plotted against the known value (determined by the host laboratory).

TABLE 9.6
$^{238}U/^{235}U$ isotopic Ratio Determined by the Host and Participating Laboratories

Sample number	Total Uranium (ng kg^{-1})[a]	$^{238}U/^{235}U$ (± 2σ)				
		Host[a]	SF-ICP-MS	ICP-Q-MS	TIMS	I/DNAA
1	250 ± 20	168 ± 2	167 ± 1	170 ± 10	170.5 ± 0.1	200 ± 200
2	25 ± 2	140 ± 10	133 ± 8	138 ± 8	137.9 ± 0.1	
3	420 ± 20	220 ± 3	216 ± 5	220 ± 8	213.4 ± 0.2	200 ± 100
4	570 ± 30	186 ± 2	183 ± 2	187 ± 4	183.0 ± 0.2	90 ± 30
5	320 ± 20	199 ± 3	194 ± 3	198 ± 6	192.4 ± 0.4	200 ± 100
6	130 ± 10	206 ± 4	205 ± 1	199 ± 6	194.3 ± 0.2	100 ± 100
7	80 ± 10	159 ± 4	158 ± 5	160 ± 10	153.9 ± 0.5	
8	55 ± 6	215 ± 9	218 ± 8	200 ± 20	186.1 ± 0.3	
9	100 ± 10	174 ± 4	174 ± 2	170 ± 10	166.5 ± 0.3	
10	770 ± 40	158 ± 2	158 ± 2	157 ± 2	158.3 ± 0.3	130 ± 40
11	25 ± 2	140 ± 10	140 ± 10	130 ± 10	138.5 ± 0.1	
12	200 ± 20	158 ± 2	156 ± 5	160 ± 4	156.4 ± 0.1	200 ± 200

good agreement with the host lab values with the fitted lines overlapping the line of perfect agreement. The results from the TIMS and INAA laboratories deviated significantly from the line of perfect agreement.

The test results for the isotope ratio ($^{238}U/^{235}U$) measurements are presented in Table 9.6. (The neutron activation lab provided an incomplete set of isotopic ratios and was omitted from subsequent analysis of the isotope data.) It was expected that all three reporting laboratories would have little difficulty where the total uranium content was relatively high but would have increasing difficulty as total uranium concentration approached background levels. To visualize this effect on the precision of isotope measurements, the test lab isotope ratio ($^{238}U/^{235}U$) measurements vs. host lab isotope ratio ($^{238}U/^{235}U$) values have been plotted in Figure 9.3 to Figure 9.5 for total uranium concentrations between 25 ng kg^{-1} and 100 ng kg^{-1} (5 samples), 100 ng kg^{-1} and 350 ng kg^{-1} (4 samples), and > 350 ng kg^{-1} (3 samples), respectively. As expected, the best agreement between the host and the three MS laboratories was observed for sample 10 ($^{238}U/^{235}U = 158$), which had the highest concentration (770 ± 40 ng L^{-1}) of uranium (Figure 9.5) and the largest variance was observed for sample 8, which had the lowest concentration of uranium (55 ± 6 ng L^{-1}), in a spiked sample, and the highest percentages of depleted uranium (~49%).

The greatest deviation from host laboratory values was consistently seen in the TIMS results. Although the isotope ratios from the TIMS assay had the smallest standard deviations (< 0.4% RSD), the numbers were consistently outside the host laboratory values. Within experimental error, the ICP-SF-MS and ICP-Q-MS laboratories reported isotope ratios that were consistent with the host laboratory values.

The testing laboratories were also required to calculate method detection limits for total and $^{238}U/^{235}U$ isotope ratio assays (Table 9.7). The $^{238}U/^{235}U$ isotope ratio

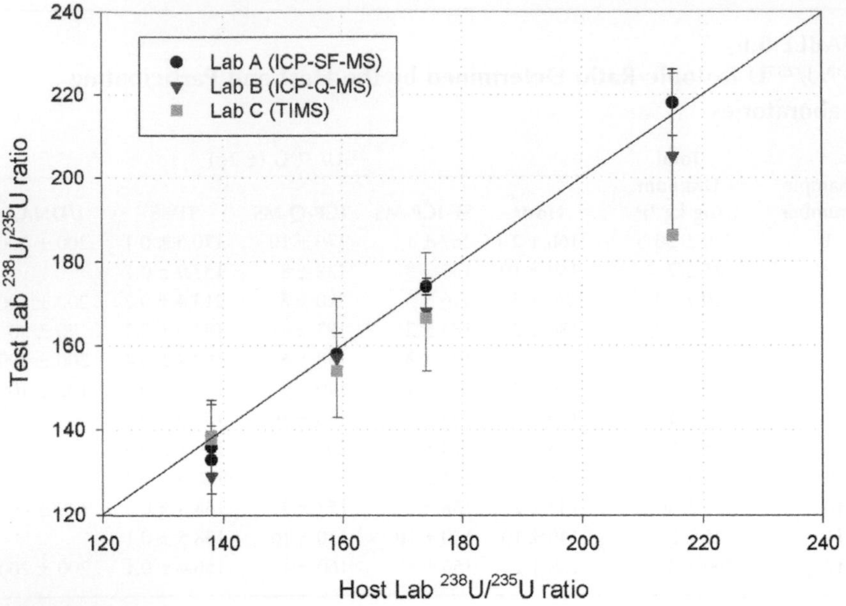

FIGURE 9.3 $^{238}U/^{235}U$ ratios determined by participating laboratories (for samples containing 25 ng kg^{-1} to 100 ng kg^{-1} of total uranium).

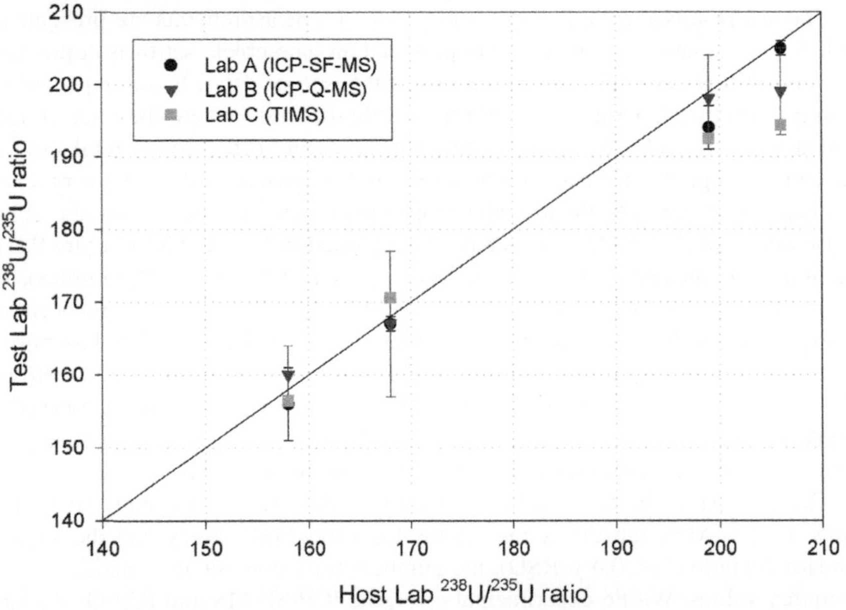

FIGURE 9.4 $^{238}U/^{235}U$ ratios determined by participating laboratories (for samples containing 100 ng kg^{-1} to 350 ng kg^{-1} of total uranium).

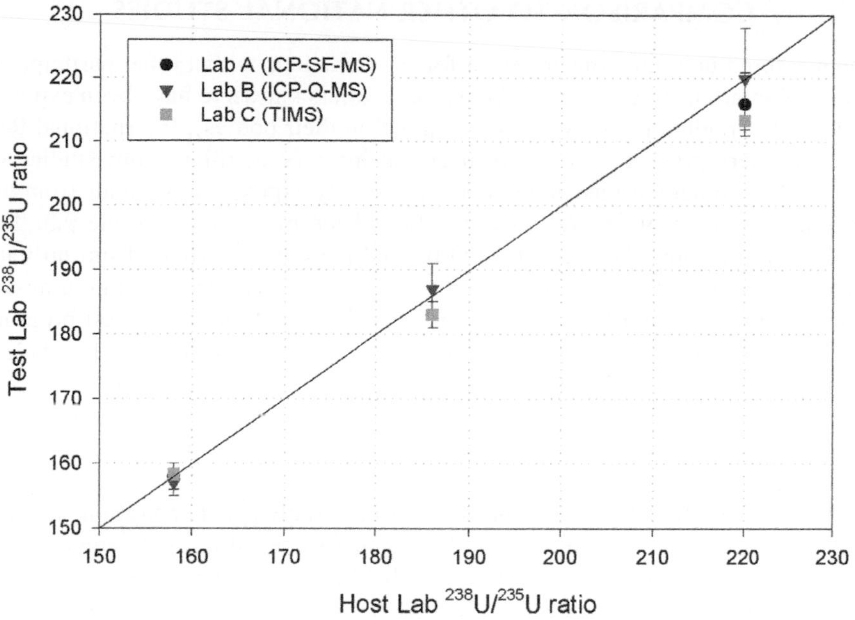

FIGURE 9.5 $^{238}U/^{235}U$ ratios determined by participating laboratories (for samples containing > 350 ng kg^{-1} of total uranium).

DL is based on the presence of 100% natural uranium and would be about three times higher if the sample were 100% depleted uranium. The reported limits are comparable to the limits that are generally accepted for the instruments. The ICP instruments method detection limits are approximately 10 times above the absolute instrument detection limits due to the need to reduce the percentage of total dissolved solids (TDS) through dilution of the urine sample prior to introduction into the ICP. The limits for I/DNAA could have been lowered if a stronger neutron source had been utilized during activation. Overall, the relative sensitivity of the four techniques (TIMS > ICP-SF-MS > ICP-Q-MS > NAA) was consistent with the available literature [29].

TABLE 9.7
Total Uranium and Isotopic ($^{238}U/^{235}U$) Ratio Detection Limits

Technique	U_{total}	$^{238}U/^{235}U$ isotopic ratio
ICP-SF-MS (lab A)	0.025 ng kg^{-1}	3.5 ng kg^{-1}
ICP-Q-MS (lab (B)	0.1 ng kg^{-1}	13.9 ng kg^{-1}
TIMS (lab C)	0.12 pg	16.5 pg
I/DNAA (lab D)	50 ng kg^{-1}	50 ng kg^{-1}

COMPARISON TO OTHER NATIONAL STUDIES

The measured uranium concentration for the Canadian soldiers that participated in the voluntary testing program, American soldiers known to have been exposed to depleted uranium (some with DU shrapnel in their bodies), International Red Cross workers present in Kosovo after the military conflict, from studies of soldiers from other nations, and nonoccupationally exposed individuals from the general population of several countries have been presented in Table 9.3. The results from testing of the general public and the control groups from military studies show (for the countries represented) a normal background uranium concentration in urine of 1 ng L^{-1} to 130 ng L^{-1}. The CF voluntary test program has observed urinary uranium concentrations ranging from 0.11 ng L^{-1}, below the range in the general populations, to 49.5 ng L^{-1}. This range is consistent with the normal background observed, and the assignment of the uranium as natural uranium has been validated by the $^{238}U/^{235}U$ isotope ratio measurements on hair samples [4].

Without additional verification, a low total uranium concentration in urine cannot be used as an unambiguous indicator that someone has not been exposed to uranium. Although there is a strong probability, when the concentrations are within the lower 50th percentile of normal background levels, that exposure is not due to DU exposure, there cannot be 100% confidence in the assignment. A recent American study [17] measured total uranium and $^{238}U/^{235}U$ isotope ratios for a group of nonexposed soldiers, a group of DU exposed soldier without embedded shrapnel, and a group of DU exposed soldiers with embedded shrapnel. The authors noted that there was significant overlap in the ranges of total uranium for the first two groups (nonexposed and exposed but without retained fragments) and that they were not statistically different from each other. Upon closer examination of the data it does appear that there is a greater than 99% probability that an individual with total uranium concentrations below this studies DU detection limit of 8.8 ng U L^{-1} urine, is negative for the presence of depleted uranium. This observation leads to the conclusion that any DU test on an individual, from the CF program, with total urinary uranium concentration in the lower 90th percentile can be considered, with 95% confidence, to be negative as supported by the hair bioassays. Another conclusion is that any individual with measured total uranium below the 50th percentile for the general population can also consider their test negative.

An interesting trend was observed in the results from the study of Swedish soldiers before (control) and after their deployment to Kosovo. The total uranium concentration in the urine of these solders actually dropped, from a mean of 38.3 ng L^{-1} in country to a mean of 9.9 mg L^{-1} when deployed. With all the concern over DU exposure, there is a tendency to ignore exposure to that uranium naturally present in the environment. These soldiers were actually exposed to larger amounts of uranium when at home, relative to when they were deployment abroad, and from a health point of view were at greater risk at home in Sweden than deployed to Kosovo.

FINAL COMMENTS

In its 5 years of operation, the CF uranium testing program has evolved in order to respond to problems encountered during testing and to take advantage of improvements in analytical procedures. The program started out employing two commercial laboratories (ICP-Q-MS and I/DNAA) to perform total and isotopic uranium bioassays on 24-h urine samples. Early into testing, it was recognized that I/DNAA was unable to perform accurate quantitative measurements of uranium in urine (at what would be considered natural background levels), and the lab was subsequently removed from the testing program. Due to low uranium content in 24-h urine samples (below the detection limit for $^{238}U/^{235}U$ isotope ratio measurements), a compromise was reached where hair samples replaced urine in $^{238}U/^{235}U$ isotope ratio bioassays. Because the completion of the interlaboratory study commissioned by DRDC [6], the remaining commercial laboratory has switched from a quadrupole to sector-field MS for analysis. Although the results from the study did show that ICP-Q-MS was a competent analytical procedure for DU testing, they also illustrated the fact that sector field instruments were superior to ICP-Q-MS instruments for both total uranium and isotope ratio measurements.

It should not be ignored that there is one improvement that could be made to the CF testing programs. As the CF testing laboratory is now equipped with the more sensitive ICP-SF-MS, it may be the time to return to urine for the $^{238}U/^{235}U$ isotope ratio measurement. The new instrument has an absolute uranium detection limit below 1 pg L^{-1} (part per quadrillion) [11], but this improvement may not be enough to overcome an interference problem recently reported [17]. Researchers observed higher than expected ^{235}U concentrations in spiked urine samples, which they attributed to interference from a poly-atomic. The results indicate that the traditional "shoot and dilute" method for low resolution ICP-MS (sector field or quadrupole) is not reliable for the accurate measurement of $^{238}U/^{235}U$ isotope ratios from urine samples. Three methods to overcome the interference are to use a collision cell to remove it, operate the instrument at higher resolution, and remove of the uranium from the urine matrix. A collision cell could be used to remove the interference, but a significant portion of the urine samples are expected, based on historical data, to have total uranium below the limit for straight "shoot and dilute." The use of a sector field instrument in medium resolution, as required to resolve the poly-atomic interference from ^{235}U, would increase the limit for $^{238}U/^{235}U$ isotope ratio measurements from 3.5 ng kg^{-1} to > 35 ng kg^{-1}, which is higher than the total uranium from > 95% of the submitted samples. Although the removal of the uranium from the urine matrix would require additional sample workup, which would increase the cost of analysis and lengthen analysis time, it would still be possible to acquire isotope ratio values. It is apparent that $^{238}U/^{235}U$ isotope ratio measurements are required for identification of the source of the uranium when an individual tests above the 50th percentile for the general population [17]. As the reliability of hair analysis will be questioned, with regard to exogenous vs. endogenous source, $^{238}U/^{235}U$ isotope ratio assays are ideally performed on urine samples where the uranium is first removed from the sample matrix prior to its introduction into an ICP-SF-MS.

ACKNOWLEDGMENTS

The author would like to acknowledge the financial support of the Directorate Medical Policy (National Defence Headquarters, Ottawa, ON, Canada).

REFERENCES

1. Birchard, K. Lancet: does Iraq's depleted uranium pose a health risk. *The Lancet* 351: 657, 1998.
2. United Nations Environment Program (UNEP). Depleted Uranium in Kosovo: Post-Conflict Environmental Assessment, UNEP scientific team mission to Kosovo (November 5–19, 2000). Geneva, March, 2001.
3. USCENTAF Assessment and Analysis Division. Operation IRAQI FREEDOM — By the Numbers (2003). Available at http://www.globalsecurity.org/ military/library/ report/ 2003/uscentaf_oif_report_30apr2003.pdf. Accessed 26 May 2004.
4. Ough, E.A., Lewis, B.J., Andrews, W.S., Bennett, L.G.I., Hancock, R.G.V., Scott, K. An examination of uranium levels in Canadian Forces personnel who served in the Gulf War and Kosovo. *Health Phys* 82(4): 527–532, 2002.
5. Ough, E.A., Lewis, B.J., Andrews, W.S., Bennett, L.G.I., Hancock, R.G.V., D'Agostino, P.A. "Determination of Natural and Depleted Uranium in Synthetic Urine Samples at Ultratrace Concetrations: An Interlaboratory Analytical Exercise, accepted for publication in Health Physics.
6. D'Agostino, P.A., Ough, E.A., Glover, S.A., Vallerand, A.L. Determination of Natural and Depleted Uranium in Urine at the ppt Level: An Interlaboratory Analytical Exercise. DRES TR 2002-024 Prepared for Defense Research and Development Canada, 2002.
7. National Council on Radiological Protection (NCRP). Uranium: Radiation Protection Guidelines, Draft Report No. SC 57-15, Bethesda, MD, August 1999.
8. Kramer, H.H., Molinski, V.J., Nass, H.W. Urinalysis for uranium-235 and uranium-238 by neutron activation analysis. *Health Phys* 13: 27–30, 1967.
9. Ide, H.M., Moss, W.D., Monor, M.M., Campbell, E.E. Analysis of uranium in urine by delayed neutrons. *Health Phys* 37: 405–408, 1979.
10. Bouvier-Capely, C., Baglan, N., Montegue, A., Ritt, J., Cossonnet, C., Validation of uranium determination in urine by ICP-MS. *Health Phys* 85: 216–219; 2003.
11. Thermoquest Finnegan MAT Element2 Brocure, Finnegan MAT GmbH, Barkhausenstr. 2, 28197 Bremen, Germany, 1998.
12. Horan, P., Dietz, L., Durakovic, A. the quantitative analysis of depleted uranium isotopes in British, Canadian, and U.S. Gulf War veterans. *Mil Med* 167: 620–627, 2002.
13. Roth, P., Werner, E., Paretzke, H.G. Research into Urinary Excretion of Uranium: Verification of Protective Measures in the German KFOR Army Contingent (English Translation). Neuherberg, Germany: Forschungszentrum für Umwelt und Gesundheit, Institut für Strahlenschutz; GSF-Bericht 3/01, 2001.
14. Hooper, F.J., Squibb, K.S., Siegel, E.L., McPhaul, K., Keogh, J.P. Elevated urine uranium excretion by soldiers with retained uranium shrapnel. *Health Phys* 77: 512–519, 1999.
15. McDiarmid, M.A., Hooper, F.J., Squibb, K., McPhaul, K. The utility of spot collection for urinary uranium determinations in depleted uranium exposed Gulf War veterans. *Health Phys* 77: 261–264, 1999.

16. Ejnik, J.W., Carmichael, A.J., Hamilton, M.M., McDiarmid, M., Squibb, K., Boyd, P., Tardiff, W. Determination of the isotopic composition of uranium in urine by inductively coupled mass spectrometry. *Health Phys* 78: 143–146; 2000.
17. Gwiazda, R.H., Squibb, K., McDiarmid, M., Smith, D. Detection of depleted uranium in urine of veterans from the 1991 Gulf War. *Health Phys* 86: 12–18, 2004.
18. Sandström, B. Levels of Uranium in Urine from Swedish Personnel that Have Been Serving or Will Serve in the Swedish KFOR Contingent: Part II. Follow-Up. Report No. FOI-R-0581-SE. Swedish Defense Research Agency, Umeå, 2001.
19. Meddings, D.R., Haldiman, M. Depleted uranium in Kosovo: an assessment of the potential exposure for aid workers. *Health Phys* 82: 467–472, 2002.
20. Private communication with Dr. Fred Leung, Trace Elements Laboratory, London Health Sciences Centre (LHSC), London, ON, Canada.
21. Lorber, A., Karpas, Z., Halicz, L. Flow injection method for determination of uranium in urine and serum by inductively coupled plasma mass spectrometry. *Anal Chim Acta* 334: 295–301; 1996.
22. Dang, H.S., Pullat, V.R., Pillai, K.C. Determining the normal concentration of uranium in urine and applications of the data to its biokinetics. *Health Phys* 62: 562–566, 1992.
23. Private communication with Dr. I. Rodushkin, SGAB Analytica, Luleå, Sweden.
24. Galletti, M., D'Annibale, L., Pinto, V., Cremisini, C. Uranium daily intake and urinary excretion: a preliminary study in Italy. *Health Phys* 85: 228–235, 2003.
25. Rodushkin, I., Axelsson, M.D. Application of double focusing sector field ICP-MS for multielemental characterization of human hair and nails. Part I. Analytical methodology. *Sci Total Environ* 250: 83–100; 2000.
26. Rodushkin, I., Axelsson, M.D. Application of double focusing sector field ICP-MS for multielemental characterization of human hair and nails. Part II. A study of the inhabitants of Northern Sweden. *Sci Total Environ* 262: 21–36, 2000.
27. Glover, S.E. Preparation of Uranium Spiked Urine Samples (draft report), in response to PWCSC 045SV.W7714-0-0462, Pullman, Washington State University; WA, 2002.
28. U.S. Department of Defense. Information Paper: Impact of Laboratory Performance of Urine Uranium Analysis on Exposure Evaluations for Gulf War Veterans. Available at http://deploymentlink.osd.mil/du_library/ lab_assessment/. Accessed April 4, 2005.
29. Ough, E.A., Stodilka, R.Z., Lewis, B.J., Andrews, W.S., Bennett, L.G.I., Hancock, R.G.V., Cousins, T., Haslip, D.S. Uranium: Detection of Contamination and Assessment of Biological Hazards — A Literature Survey (RMC-CCE-BJL01.001). Kingston, ON: Royal Military College of Canada, 2001.

10 Biokinetics of Uranium in the Human Body

R. W. Leggett

CONTENTS

INTRODUCTION

During the 1991 Gulf War, several U.S. fighting vehicles were hit by friendly-fire munitions containing DU. Soldiers in or near the vehicles were exposed to airborne DU, and some were left with embedded fragments of DU in their tissues. Evaluation of the radiological and chemical risks from these exposures to DU requires estimation of the rate of mobilization of U from the sites of entry into the body to blood and the time-dependent tissue concentrations of the mobilized U. The mobilization and distribution of U cannot be measured directly but must be inferred from the rate of urinary excretion of U, using biokinetic models.

Considerable information on the biokinetics of U in humans and laboratory animals has been published since the 1940s. Information available through the early 1990s is reflected in physiologically based biokinetic models published by the International Commission on Radiological Protection (ICRP, 1994a, 1994b, 1995a, 1995b). These models describe the respiratory and systemic behavior of forms of U commonly encountered in the workplace or environment. Their applicability to forms of DU arising from military uses has been investigated to some extent. For example, studies of the biokinetics and toxicology of DU in rats with implanted DU pellets simulating embedded fragments provide information on the behavior of DU migrating to blood from shrapnel. Also, the chemical and physical forms of DU resulting from impact or combustion of DU munitions have been evaluated for purposes of determining appropriate parameter values for the ICRP's respiratory model.

This chapter summarizes the biokinetic models for U currently recommended by the ICRP and their applications to different forms of U taken into the body, with emphasis on DU encountered in military operations. Derived limits on intake of U as a chemical toxin are compared with limits based on radiation protection guidance for a wide range of mixtures of ^{234}U, ^{235}U, and ^{238}U.

ICRP RESPIRATORY TRACT MODEL AS APPLIED TO U

The ICRP's current model of the behavior of inhaled radionuclides in the respiratory tract, called the Human Respiratory Tract Model or HRTM, was introduced in ICRP Publication 66 (1994a). Default parameter values describing retention, translocation, and absorption of material are provided for application to inhaled particles or gases, but material-specific parameter values solution rates have been developed for some forms of U and a few other elements (ICRP, 2002a).

The structure of the HRTM is shown in Figure 10.1. The HRTM divides the respiratory system into extrathoracic (ET) and thoracic tissues. The airways of the ET region are further divided into the anterior nasal passages in which deposits are removed by extrinsic means such as nose blowing and the posterior nasal passages (nasopharynx, oropharynx, and the larynx) from which deposits are swallowed or absorbed into the bloodstream. The airways of the thorax include the bronchi (BB), bronchioles (bb), and alveolar interstitium (AI). Uranium or other material deposited in the thoracic airways is cleared into blood by absorption — to the gastrointestinal tract by mechanical processes (i.e., transported upward and swallowed) and to the regional lymph nodes via lymphatic channels. In Figure 10.1, reference values for particle transport rate constants are shown beside the arrows and are in units of d^{-1}.

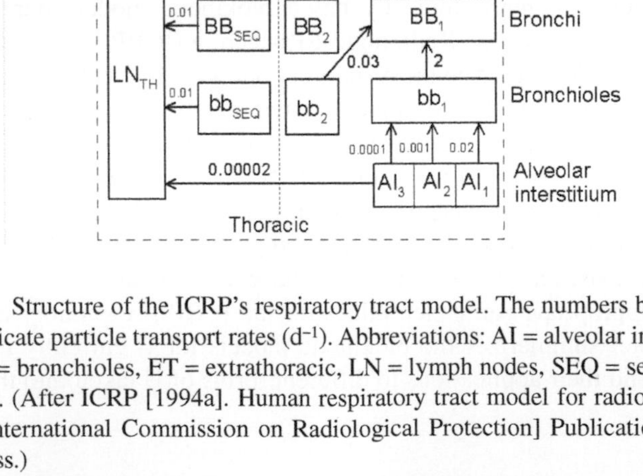

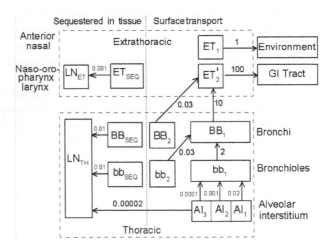

FIGURE 10.1 Structure of the ICRP's respiratory tract model. The numbers beside or above the arrows indicate particle transport rates (d^{-1}). Abbreviations: AI = alveolar interstitium, BB = bronchi, bb = bronchioles, ET = extrathoracic, LN = lymph nodes, SEQ = sequestered, and TH = thoracic. (After ICRP [1994a]. Human respiratory tract model for radiological protection. ICRP [International Commission on Radiological Protection] Publication 66. Oxford: Pergamon Press.)

For an inhaled compound, the mechanical clearances of particles indicated in Figure 10.1 are in competition with dissolution rates and absorption to blood, which depend on the chemical and physical form of the inhaled element. Material-specific dissolution rates have been developed for some forms of U, usually on the basis of *in vitro* dissolution studies or interpretation of measured clearance times from the lungs in laboratory animals (ICRP, 2002a). In most applications of the HRTM to U or other elements inhaled in particulate form, however, the material is assigned to one of three generic absorption types: Type F, representing fast dissolution and a high level of absorption to blood; Type M, representing a moderate rate of dissolution and an intermediate level of absorption to blood; and Type S, representing slow dissolution and a low level of absorption to blood. The user selects Type F, M, or S based either on ICRP recommendations or independent interpretation of site-specific data and information from the literature.

Extensive observations of the behavior of inhaled U in laboratory animals and in acutely or chronically exposed radiation workers provide the information needed to assign some commonly encountered U compounds to appropriate solubility classes or, in some cases, to assign material-specific parameter values (ICRP, 2002a; Leggett et al., 2005). For example, studies of laboratory animals and accidentally exposed workers indicate that U inhaled as UO_2F_2 is rapidly absorbed to blood, so that assignment of Type F is appropriate in the absence of case-specific information (ICRP, 1995b). Uranium inhaled as $UO_2(NO_3)_2$ typically is moderately soluble in the lungs, so that assignment of Type M is reasonable in the absence of specific information.

The three generic absorption types do not cover the spectrum of observed behavior of U in the human respiratory tract. For example, all three of these generic types predict a decreasing rate of dissolution of material in the respiratory tract, with a rapid phase of dissolution immediately after intake. By contrast, urinary data for a group of workers exposed to airborne uranium aluminide, a material used in fuels for research and test reactors, initially showed little dissolution in the lungs but began to dissolve rapidly after residing in the lungs for several weeks (Leggett et al., 2005). In effect, the material initially behaved in the lungs as a Type S (or even less soluble) material but gradually transformed into a Type M or Type F material.

The problem also arises that different studies may yield considerably different dissolution rates for apparently the same form of U. For example, considerable variation in the behavior of U_3O_8 has been observed, with some studies indicating Type M behavior and others Type S behavior. Similarly, inhaled UO_2 has exhibited Type M behavior in some studies and Type S behavior in others.

A report on effects of DU on Gulf War soldiers issued by the U.S. Department of Defense (2001) indicates that the DU oxides encountered during the Gulf War were generally mixtures of relatively soluble (Type F or M) and insoluble (Type S) materials. In dose estimates given in that report, DU oxides from impact of DU munitions were assumed to consist of 17% Type M and 83% Type S material as determined from test data.

The chemical form of particulate DU resulting from its use in munitions appears to depend on the conditions of formation (Etherington, 2004). Particles formed by impacts may consist largely of U_3O_8 mixed with smaller quantities of UO_2, while combustion produces an oxide that is expected to be almost entirely U_3O_8 (Royal Society, 2001). Nevetheless, the absorption behavior of DU formed as a result of its use in munitions remains uncertain due to limited information on factors such as particle size distribution.

Inhaled uranium may enter the gastrointestinal (GI) tract after being transported to the pharynx along with fluids escalating along the airways. Fractional absorption of the swallowed U to blood cannot be determined directly due to simultaneous absorption from the respiratory tract but can be estimated from studies of absorption of U after oral intake. These include controlled studies involving intake of U in fluids by human volunteers, a controlled balance study on adult male subjects in a metabolic research ward of a hospital, and a variety of environmental studies in which urinary U was related to total intake or total excretion of U (Leggett and Harrison, 1995; Harrison et al., 2001). Estimates of absorption as a percentage of the swallowed amount range from less than 0.1% to about 6% for individual subjects. Central values for different studies fall in the range 1–2.4%. Environmental studies yield central estimates in the range 0.3–3.2%. Expressed as a percentage of total intake of soluble U in food and fluids, average GI uptake of U in adult humans appears to be about 1–1.5%. Data from studies on laboratory animals indicate that fractional uptake of forms of U commonly encountered in the workplace depends strongly on the chemical form ingested. Absorption appears to be greatest for U ingested as $UO_2(NO_3)_2 6H_2O$, UO_2F_2, or $Na_2U_2O_7$, roughly half as great for UO_4 or UO_3, and 1–2 orders of magnitude lower for UCl_4, U_3O_8, UO_2, and UF_4 (Leggett and Harrison, 1995).

In current documents of the ICRP (1994b, 1995a, 1995b), fractional uptake from the alimentary tract to blood is assumed to be 0.02 for soluble forms of U including U in food or drink, and 0.002 for relatively insoluble forms of U. The lower value, 0.002, seems appropriate for DU arising from impact or combustion of DU munitions, because such DU may exist largely as U_3O_8 or UO_2.

BIOKINETICS OF U THAT REACHES BLOOD

The ICRP's current systemic biokinetic model for U was adopted in ICRP Publication 69 (1995). The model structure is shown in Figure 10.2. This is a generic model structure applied by the ICRP to several elements that tend to follow the movement of Ca in bone (Leggett, 1992). The analogy between U and Ca is weak in some respects, but U appears to follow the movement of Ca sufficiently closely in bone that the Ca model structure provides a useful framework for modeling the biokinetics of U (Leggett, 1994). There is evidence that UO_2^{++} exchanges with Ca^{++} at the surfaces of bone mineral crystals, although UO_2^{++} apparently does not participate in crystal formation or enter existing crystals. The early gross distribution of U in the skeleton is similar to that of Ca. Uranium is initially present on all bone surfaces but is most highly concentrated in areas of growth. Perhaps depending on the microscopic structure of the bone of each species, U on bone surfaces may gradually diffuse into bone volume. As is the case for Ca, a substantial portion of U deposited in bone apparently is lost to plasma by processes that occur more rapidly than bone resorption.

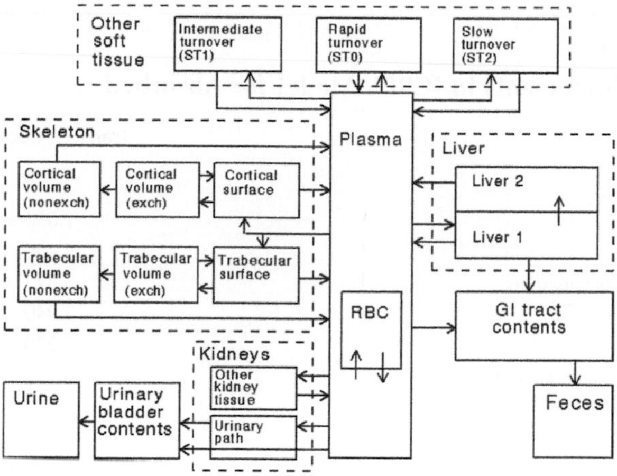

FIGURE 10.2 The ICRP's model structure for uranium. Exch = exchangeable; nonexch = nonexchangeable. (After ICRP [1995a]. Age-dependent doses to members of the public from intake of radionuclides: Part 3. International Commission on Radiological Protection Publication 69. Oxford: Pergamon Press.)

Parameter values for U in a reference adult are given in Table 10.1. Some of these values are generic for "calcium-like" elements. For example, values describing removal from nonexchangeable bone volume compartments are estimated in terms of bone remodeling rates and therefore are independent of the element. Most of the transfer coefficients in the model are element specific. Uranium-specific parameter values given in Table 10.1 were based mainly on the following sources of information: measurements of U in blood and excreta of several human subjects who were intravenously injected with U; postmortem measurements of U in tissues of some of those subjects; postmortem measurements of U in tissues of occupationally and environmentally exposed subjects; data on baboons, dogs, or smaller laboratory animals exposed to U for experimental purposes; and consideration of the physiological processes thought to determine retention and translocation of U in the body (Leggett, 1994; ICRP, 1995a). The methods of selection of the parameter values of the ICRP's systemic model for U are described by Leggett (1994) and will not be repeated here, but some of the data sets that figured most prominently in the selection process are summarized in the following paragraphs.

The most direct information on the biokinetics of U, particularly its rate of urinary excretion as a function of time after injection (Figure 10.3), comes from three controlled studies on human subjects. These subjects were intravenously injected with U isotopes and followed for periods varying from a few days to 1.5 y after injection. A study known as the Rochester study involved six ambulatory hospital patients, ages 24–61 y, suffering from diseases that were not immediately life threatening (Bassett et al., 1948). These subjects were injected with uranyl nitrate solutions enriched with ^{234}U and ^{235}U. A study known as the Boston study (Struxness et al., 1956; Bernard and Struxness, 1957; Luessenhop et al., 1958; Leggett 1994) involved at least 11 comatose patients, ages 26–63 y, in the terminal phases of

TABLE 10.1
Transfer Coefficients in ICRP's Model for Absorbed U

Path	Transfer Coefficient (d^{-1})
From plasma to:	
STO	1.050×10^{1}
RBC	2.450×10^{-1}
Urinary bladder contents	1.543×10^{1}
Kidney 1	2.940×10^{0}
Kidney 2	1.220×10^{-2}
Upper large intestine contents	1.220×10^{-1}
Liver 1	3.670×10^{-1}
ST1	1.630×10^{0}
ST2	7.350×10^{-2}
Trabecular bone surfaces	2.040×10^{0}
Cortical bone surfaces	1.630×10^{0}
To plasma from:	
STO	8.320×10^{0}
RBC	3.470×10^{-1}
Kidney 2	3.800×10^{-4}
Liver 1	9.200×10^{-2}
Liver 2	1.900×10^{-4}
ST1	3.470×10^{-2}
ST2	1.900×10^{-5}
Bone surfaces[a]	6.930×10^{-2}
Nonexch. trabecular bone volume	4.930×10^{-4}
Nonexch. cortical bone volume	8.210×10^{-5}
From Kidney 1 to urinary bladder contents	9.900×10^{-2}
From Liver 1 to Liver 2	6.930×10^{-3}
From bone surfaces to exchangeable bone volume[a]	6.930×10^{-2}
From exchangeable bone volume to bone surfaces[a]	1.730×10^{-2}
From exchangeable bone volume to nonexchangeable volume[a]	5.780×10^{-3}

[a]Applies both to trabecular and cortical bone compartments.
Source: ICRP (1995a). Age-dependent doses to members of the public from intake of radionuclides: Part 3. ICRP (International Commission on Radiological Protection) Publication 69. Oxford: Pergamon Press.

diseases of the central nervous system. The Boston subjects were injected with uranyl nitrate solutions enriched with ^{234}U and ^{235}U, or with UCl_4 (Leggett, 1994). A study by Terepka et al. (1966) (also see Hursh and Spoor, 1973) involved three control patients and seven patients with various bone disorders (Paget's disease, hyper- or hypoparathyroidism, osteomalacia, senile osteoporosis). These subjects received injections of natural U administered as the soluble uranyl ion.

The Boston study provided considerably more detailed information on the biokinetics of U than the other two human injection studies. For example, several bone biopsy samples were taken from the Boston subjects during the first day or

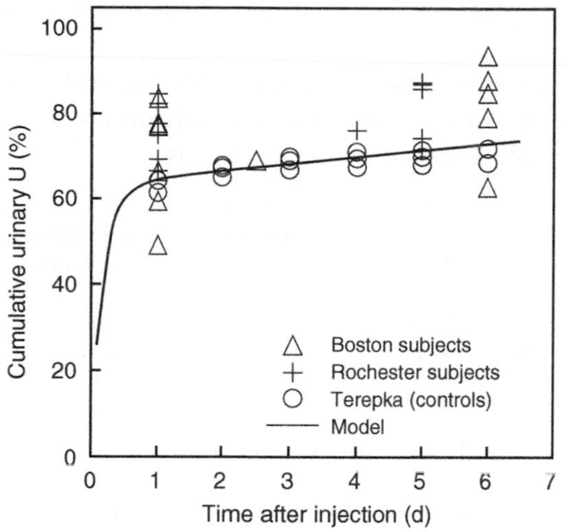

FIGURE 10.3 Observations and model predictions of cumulative urinary U in human subjects as a function of time after intravenous injection with U isotopes. (After Leggett, R.W. [1994]. Basis for the ICRP's age-specific biokinetic model for uranium. *Health Phys* 67: 589–610.) The three study groups indicated in the legend are described in the text.

two after injection. Also, autopsy samples were obtained from various bones and soft tissues of subjects dying at times from 2.5 d to 4 months after injection and from one subject dying 566 days after injection. The usefulness of these postmortem data for purposes of constructing a biokinetic model for U is limited, however, by the poor physical conditions of the subjects. For example, the bed-ridden condition of these subjects indicated a negative calcium balance, which might result in unusually rapid removal of U from the skeleton (Struxness et al., 1956). Another limitation of these data is that the subjects were administered relatively high masses of U, which may have altered the kinetics of U to some extent, particularly in the kidneys (Leggett, 1989). A third difficulty is that the postmortem data are not sufficiently detailed in some cases to allow a close determination of the total U content of some massive tissues such as the skeleton, muscle, fat, and skin.

Additional information on the biological fate of U in humans is provided by postmortem measurements of U in tissues of occupationally and environmentally exposed subjects (Table 10.2). These data provide the best available information on the long-term distribution of U in the human body. The value of the postmortem data is also limited by several factors: the small numbers of subjects examined; uncertainties in the exposure histories of those subjects; uncertainties in estimates of total-organ contents of the subjects based on small samples of tissue, particularly skeletal tissues; and technical difficulties in measuring the typically low concentrations of U in tissues of environmentally exposed subjects.

Various aspects of the biokinetics of U have been examined in baboons, dogs, rabbits, rats, mice, monkeys, sheep, and other animal species (Leggett, 1994). The animal studies yield much information not provided by the human studies. The data

TABLE 10.2
Relative Amounts of U in Total Liver, Total Kidneys, and Total Skeleton as Determined from Portmortem Measurements on Occupationally or Nonoccupationally Exposed Human Subjects

Subjects	Liver Content[a] (μg U)	Ratio of Kidney to Liver Content[b]	Ratio of Skeleton to Liver Content[b]
New York residents (Fisenne and Welford, 1986)	0.36	0.37	19 (V)[c]
Pennsylvania residents (Singh et al., 1986)	0.22	0.56	34 (R,V)
Utah residents (Singh et al., 1986)	0.59	0.47	25 (R,V)
Colorado residents (Singh et al., 1986)	0.54	0.57	36 (R,V)
Tokyo residents (Igarashi et al., 1985)	0.43	0.24	16 (F,R,Sk)
U worker, 26-y exposure (Kathren et al., 1989)	216	0.36	23 (R,V,S)
U worker, 10-y exposure (Donoghue et al., 1972)	35	0.86	29 (S)
U worker, 2-y exposure (Roberts et al., 1977)	6	3.2	74 (R,V,S)
U worker, 9-y exposure (Campbell, 1975)	6.6	3.8	45 (F,R,S,V)
U worker, 28-y exposure (Campbell, 1975)	8	0.18	50 (F,R,V)
U miller, 33-y exposure (Singh et al., 1987)	14	0.20	20 (V)
U miller, 4-y exposure ~30 y pre-death (Singh et al., 1987)	0.9	0.60	150 (V)
Uranium miner (Singh et al., 1987)	6	0.50	24 (V)
Median		0.50	29

[a]Based on measurement of ^{238}U, where available; otherwise, assumption made that isotopic ratios were same as for natural U.
[b]If organ contents and organ weights were not reported, ratio of kidney mass to liver mass was assumed to be 0.172 (ICRP, 1975). Reported concentrations in bones were converted to ash weight basis; percent ash weights of bones assumed to be 15% for vertebra and sternum, 25% for rib, and 35% for femur and skull. (*Source*: Fisenne, I.M., Welford, G.A. [1986]. Natural U concentrations in soft tissues and bone of New York City residents. *Health Phys* 50: 739–746.). Weight of total bone ash was assumed to be 1.556 times wet weight of liver (ICRP, 1975).
[c]Sampled bones indicated in parentheses: V = vertebra, R = rib, F = femur, Sk = skull, S = sternum.

for animals, particularly baboons and dogs, figured into the selection of several of the U-specific parameter values of the model.

APPLICABILITY OF THE SYSTEMIC MODEL TO U MIGRATING FROM EMBEDDED SHRAPNEL

It was not at first evident whether the ICRP's systemic model for U could be used to interpret urinary excretion data for the Desert Storm veterans with embedded DU shrapnel, due to the possibility that the form of U migrating from the fragments behaved much differently from that of forms of U used in the studies underlying the model. For example, the possibility was considered that a substantial portion of U migrating from the fragments could be released as relatively insoluble particulates that would be accumulated by the reticuloendothelial system, which is not addressed in the ICRP's model.

To assess the health risks associated with embedded DU fragments, the Armed Forces Radiobiology Research Institute (AFRRI) conducted an experimental study in which DU pellets were implanted in rats (Pellmar et al., 1999). As part of this study, the systemic distribution and rate of excretion of U were determined at 1 d and at 1, 6, 12, and 18 months after implantation of the pellets. The U concentration generally was measured in urine, serum, kidney, tibia, skull, liver, spleen, brain, and muscle, and limited measurements were made on testes, heart, lungs, and teeth.

Data from the AFRRI study were used to check whether the biokinetics of U migrating from embedded pellets is similar to that of other forms of U, at least in rats (Leggett and Pellmar, 2003). Biokinetic data on U-exposed rats were collected from the literature and used to develop a baseline biokinetic model for rats. The model structure is similar to that shown in Figure 10.2 but slightly simpler due to limitations in data for rats. The reader is referred to the paper by Leggett and Pellmar (2003) for a discussion of the basis of the model structure and individual parameter values and a listing of parameter values.

Predictions of the baseline model for rats were compared with observations for the DU-implanted rats. Because the rate of migration from the DU pellets to blood is not known precisely, model predictions and observations were compared only in a relative sense. Specifically, the time-dependent rate of uptake by the plasma compartment of the baseline model was set to produce the urinary excretion rates observed in the DU-implanted rats, and predicted time-dependent concentrations of U in various organs were then compared with the observed concentrations in the DU-implanted rats.

Three dosage levels, referred to as L (low dose), M (medium dose), and H (high dose), were used in the AFRRI study (Pellmar et al., 1999). Groups L, M, and H were implanted with 4, 10, and 20 cylindrical pellets, respectively. The pellets were of diameter 1 mm and height 2 mm and consisted of 99.25% DU and 0.75% titanium by weight. For purposes of comparison of the data with predictions of the baseline model for rats, the data for groups L, M, and H were reduced to a common basis by normalization to one pellet. That is, the measured concentration of U in tissues or fluids of rats in groups L, M, and H were divided by 4, 10, and 20, respectively, giving the U concentration per implanted DU pellet. This allowed simultaneous graphical comparison of data for all three groups with model predictions. Normalized values were labeled by dosage level, however, in view of the possibility of dosage dependence in the biokinetics of U in these animals.

Model predictions of time-dependent U in tissues and excreta were compared with median values for the DU-implanted rats. Reference organ masses or volumes at different ages were required to compare reported concentrations with data reported as organ contents. Reference organ masses were based on measurements on the animals used in the AFRRI study, supplemented where necessary with information from the literature.

For comparisons of model predictions with observations for the DU-implanted rats, model input (i.e., the time-dependent uptake rate to blood) was adjusted, as described below, so that model predictions of the urinary excretion rate agreed with the median time-dependent urinary excretion rates determined for the DU-implanted rats. Model predictions of the time-dependent concentrations of U in the skeleton (based on the average of values for tibia and skull), kidneys, liver, and other tissues and fluids of the rat were then compared with observations for the DU-implanted rats. For an organ such as the spleen that is not addressed explicitly in the model, predictions were derived by assuming a uniform distribution of U in other soft tissues and multiplying by the mass fraction of the organ.

Estimation of the time-dependent rate of uptake to blood required to reproduce an observed urinary excretion rate is a two-step process:

> *Step 1*: Suppose urinary excretion rates $u(t_i)$ have been determined at times t_i , i = 1, 2, …, n. The model is run, using these urinary excretion rates as uptake rates to blood at these times and basing uptake rates for times other than t_i , i = 1, 2, …, n, on linear interpolation between observed urinary excretion rates. For example, if the first three observation times are 10, 20, and 30 d and the observed urinary excretion rates at these times are 500, 600, and 620 ng d^{-1}, respectively, then the assumed rates at 4, 5, 15, 25, and 29 d are the linearly interpolated values, 440, 450, 550, 610, and 618 ng d^{-1}, respectively. The model run yields a new, generally lower, set of predicted urinary excretion rates $v(t_i)$, i = 1, 2, …, n. Because a substantial portion of U reaching plasma will be excreted in urine within a short time, the values $v(t_i)$ may not differ greatly from the urinary excretion rates $u(t_i)$. The important pieces of information revealed in this step are the ratios, $u(t_i)/v(t_i)$, which indicate the sizes of the errors in the assumed uptake rates.
>
> *Step 2*: Each observed urinary excretion rate $u(t_i)$ is multiplied by $u(t_i)/v(t_i)$. The product, $[u(t_i)]^2/v(t_i)$, is the final estimate of the uptake rate to blood. Again, uptake rates at times other than t_i , i = 1, 2, …, n, are estimated by linear interpolation. With the revised uptake curve, the urinary excretion rates predicted by the model are virtually identical to the observed urinary excretion rates, $u(t_i)$.

The same two-step process for estimating the rate of migration of U from embedded pieces of DU was later applied to the Desert Storm veterans (Squibb et al., 2005).

Comparisons of predictions of the baseline model for rats with observations of the median U concentrations in skeleton and kidneys of the DU-implanted rats are shown in Figure 10.4 and Figure 10.5, respectively. The high and low estimates indicated in the figures are uncertainty bounds on model predictions, associated with differences in reported distributions of U in rats as a function of time after introduction of common forms of U to blood. Normalized data for the three dosage groups L, M, and H in the AFRRI study turned out to be reasonably consistent in most instances, with data for individual animals typically showing considerable overlap. Although some degree of dosage dependence in the data was suggested in some cases (for example, Group L showed lower central skeletal concentrations than the other two groups at most times), differences in median values for the three groups may be attributable to the small number of measurement times, small sample sizes, and typically high variability of individual data points within dosage groups.

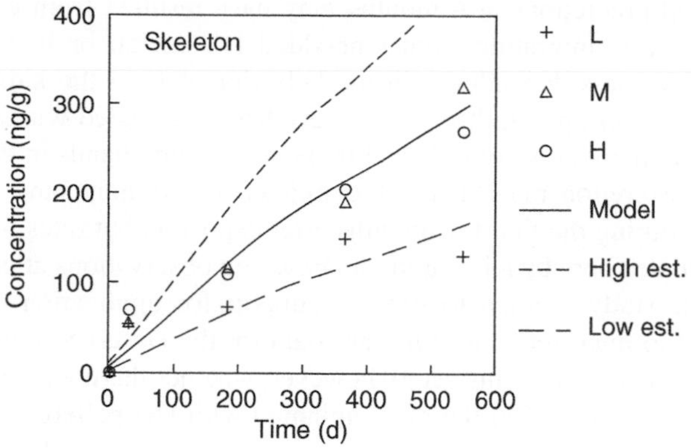

FIGURE 10.4 Comparison of predictions of accumulation of U in the rat skeleton, derived from a baseline model for rats, with data from the AFRRI study of rats implanted with DU pellets. (After Leggett, R.W., Pellmar, T.C. [2003]. The biokinetics of uranium migrating from embedded DU fragments. *J Environ Radioact* 64: 205–225.)

In most cases, the observations for the DU-implanted rats did not differ greatly from predictions of the baseline model for rats. Some of the more troublesome differences between predictions and observations were seen for the kidneys. For example, at 6 months after implant, the medians of observed values fell above the upper end of the uncertainty range for model predictions. Also, the model predicted that the kidney concentration at 6 months should be lower than that for 12 months, whereas the observed values fell sharply from six months to 12 months. At measurement times other than 6 months, model predictions of the U concentration in the kidneys agree reasonably well with observations. The discrepancy between

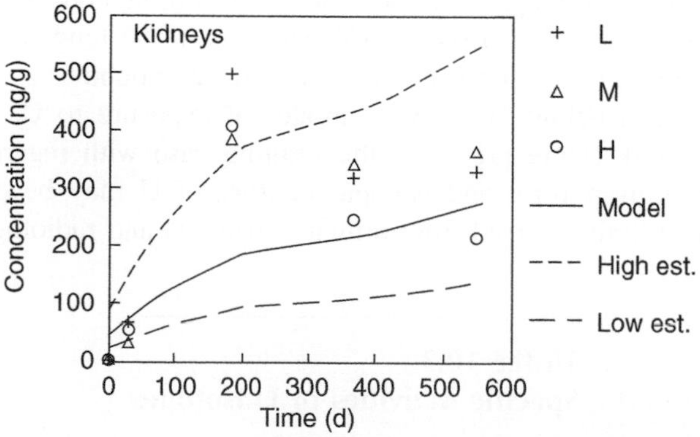

FIGURE 10.5 Comparison of predictions of accumulation of U in rat kidneys, derived from a baseline model for rats, with observations from the AFRRI study of rats implanted with DU pellets. (After Leggett, R.W., Pellmar, T.C. [2003]. The biokinetics of uranium migrating from embedded DU fragments. *J Environ Radioact* 64: 205–225.)

observations and predictions at 6 months may have resulted from a peculiarity in the biokinetics of U migrating from embedded DU metal, or it may have been associated with a mass dependence in the behavior of U in the kidneys (Leggett, 1989, 1994). As a third possibility, it may have been associated with a limitation of the model simulation that is not reflected in the uncertainty bands in the figure. That is, the urinary excretion rates in the DU-implanted rats may have actually been highly variable during the first few months after exposure, but rates for 0–6 months have been simulated here by a linear model based on observations at only two times.

Overall, the study seemed to provide support for application of the ICRP's systemic model to interpretation of urinary data for the Desert Storm veterans. The question was not completely answered, however, whether there were some peculiarities in the biokinetics of U in the rats implanted with DU pellets.

APPLICATIONS OF THE MODELS TO DU AND OTHER MIXTURES OF U ISOTOPES

Health risk associated with elevated intake of uranium risk may be divided into two categories: chemical toxicity (particularly interference with normal renal function) and radiogenic injury to lungs, bone, and other tissues. The relative significance of the chemical and radiological hazards depends on the mixture of ^{234}U, ^{235}U, and ^{238}U, and the chemical and physical form of U taken into the body. Nephrotoxicity generally has been considered the overriding hazard for intake of relatively soluble U compounds with nearly naturally occurring isotopic mixtures, based on studies with experimental animals (Wrenn et al., 1985). This would also apply to intake of relatively soluble forms of DU, because DU has an even lower specific activity (radioactive decays per second per gram of material) than natural U. The radiogenic risk increases with the level of ^{235}U-enrichment due mainly to an associated increase in the percentage of ^{234}U, which has a much higher specific activity than ^{235}U or ^{238}U (Table 10.3). Also, for the cases of inhalation of poorly soluble U compounds, the radiation dose to the lungs could become the prevailing consideration even for natural U or DU due to an increased residence time in the lungs and a decreased level of absorption to blood compared with soluble compounds.

As demonstrated below for selected modes of exposure to U, the prevailing consideration of risk (more precisely, the limiting case with regard to exposure guidelines) for a given form and isotopic mixture of U may be determined by reducing alternate limits derived from primary chemical and radiological guidance

TABLE 10.3
Specific Activities of U Isotopes

U isotope	Specific Activity (Bq/g)
^{234}U	2.32×10^8
^{235}U	8.01×10^4
^{238}U	1.25×10^4

to a common basis and taking the lower of the two limits. The analysis relies on the ICRP models described above and the following guidelines:

Primary guidance for U as a chemical hazard: The concentration of U in the kidneys should not exceed 3 μg U g^{-1} tissue. Since the early 1950s this value has served as a primary guidance level for avoidance of chemical toxicity in workers exposed to U (Voegtlin and Hodge, 1953; Spoor and Hursh, 1973). Although subtle effects on the kidneys may occur at much lower concentrations (Leggett, 1989), available evidence suggests that this may be a reasonable guidance level for avoidance of serious, irreversible effects. The chemically based annual limit on intake for a given mixture of U isotopes is calculated as (365 μg × 3 μg/g)/A, where 365 μg is the annual intake based on daily intake of 1 μg , and A is the steady-state concentration of U in the kidneys based on daily intake of 1 μg . In this analysis the mass of the kidneys is assumed to be 310 g, which is a reference value for an adult male (ICRP, 2002b)

Primary guidance for U as a radiation hazard: The effective dose (50-y integral) should not exceed 0.02 Sv from intakes of U over a 1-year period. This is the ICRP's current primary guidance on occupational exposures to radionuclides (ICRP, 1991). Although current ICRP guidance does not explicitly involve the concept of annual limit on intake, there is an implied annual limit on intake of 0.02 Sv/E, where E is the effective dose coefficient for the radionuclide of concern or, for purposes of this analysis, a given mixture of U isotopes.

The ICRP's systemic biokinetic model was used to predict the time-dependent concentration of U in the kidneys based on a mobilization rate of 1 μg/d (Figure 10.6). The predicted kidney concentration converged asymptotically to ~0.0105 μg/g. Therefore, a mobilization rate of (365 μg × 3 μg/g) / 0.0105 μg/g = 0.1 g U per year would eventually lead to a kidney concentration near the guideline value of 3 μg/g.

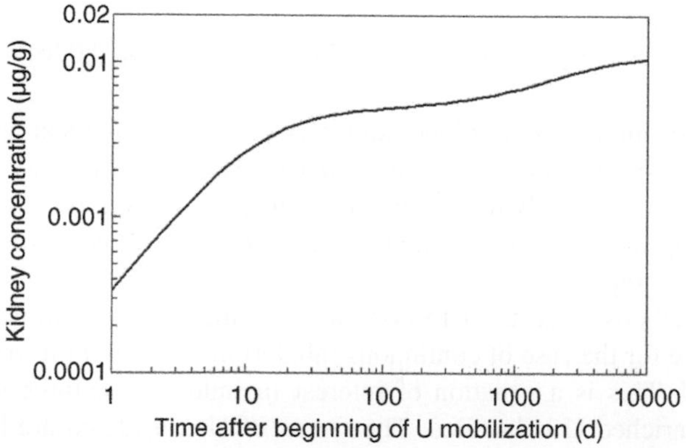

FIGURE 10.6 ICRP model predictions of the time-dependent concentration of U in the kidneys, assuming continuous inflow of U to blood at a rate of 1 μg/d. The kidney concentration converges asymptotically to a value of about 0.0105 μg/g.

TABLE 10.4
Effective Dose Coefficients for Intravenously Injected U Isotopes

	Effective Dose Coefficient	
U isotope	**Sv/Bq**	**Sv/g**
^{234}U	2.3×10^{-6}	5.3×10^{2}
^{235}U	2.1×10^{-6}	1.7×10^{-1}
^{238}U	2.1×10^{-6}	2.6×10^{-2}

For comparison of radiological guidance with the derived "chemical limit on annual intake" of 0.1 g, effective dose coefficients were derived for ^{234}U, ^{235}U, and ^{238}U for the case of direct input of radionuclides to blood, using the ICRP's systemic biokinetic model (Table 10.4). Values were derived in terms of a unit activity input to blood, i.e., in units of Sv/Bq, and were converted to units of Sv/g using the specific activities of ^{234}U, ^{235}U, and ^{238}U given in Table 10.3. Effective dose coefficients are similar for ^{234}U, ^{235}U, and ^{238}U when expressed in terms of activity (Bq), because these three radionuclides emit alpha particles with similar energies, and decay chain members contribute little to the total dose in each case. If intake is expressed in terms of mass, the dose coefficients depend strongly on the isotope, because the specific activity of ^{234}U is considerably greater than that of ^{235}U or ^{238}U. For the same reason, the effective dose coefficient for a mixture of these three isotopes expressed in terms of mass intake depends strongly on the percentage of ^{234}U present.

The effective dose coefficients given in units of Sv/g in Table 10.4 were used to derive an effective dose coefficient, E_{DU}, for the case of direct input of DU to blood, based on the assumption that DU consists of 0.25% ^{235}U, 0.0005% ^{234}U, and 99.7495% ^{238}U (Rich et al., 1988):

$$E_{DU} = (0.000005 \times 530) + (0.0025 \times 0.17) + (0.997495 \times 0.026) = 0.029 \text{ Sv g}^{-1}.$$

The level of input of U to blood equivalent to the radiological annual limit on intake (0.02 Sv from intakes over one year) is 0.02 Sv / 0.029 Sv g^{-1} = 0.69 g. Because this is considerably higher than the value derived on the basis of chemical guidance, it appears that nephrotoxicity would be the main concern in the case of embedded DU shrapnel.

Similar methods were used to compare chemically and radiologically based limits on intake for the case of continuous inhalation of different forms and isotopic mixtures of U. This is a situation of interest in industrial settings where natural, depleted, or enriched U is handled. The forms of U considered are highly soluble (Type F), moderately soluble (Type M), and relatively insoluble (Type S). The assumed particle size is 5 μm AMAD (activity median aerodynamic diameter), which is the ICRP's default value for occupational intakes (ICRP, 1994b). The isotopic mixtures considered are depleted, natural, and enriched U for all levels of ^{235}U enrichment.

Natural uranium is assumed to contain 0.0057% ^{234}U and 0.72% ^{235}U by mass. The ^{234}U content of natural or enriched U is related to the ^{235}U content by the equation

$$\%^{234}U = 0.0015 + 0.0058Y + 0.000054Y^2 \tag{10.1}$$

where Y = % ^{235}U by mass. Equation 10.1 is based on measurements of ^{234}U in material enriched by the gaseous diffusion technique.

The predicted change with time in the concentration of U in the kidneys based on the ICRP's respiratory and systemic biokinetic models is shown in Figure 10.7 for continuous inhalation of U (particle size 5 μm AMAD) as Type F, Type M, or Type S material at a rate of 1 μg d^{-1}. Model predictions asymptotically approaches 0.003 μg/g for Type F, 0.00075 μg/g for Type M, and 0.000078 μg/g for Type S. The annual limits on intake for Type F, for example, based on chemical guidance would be (365 μg × 3 μg/g)/0.003 μg/g = 0.37 g. Annual limits on intake of 1.5 g for Type M and 14 g for Type S are derived in a similar manner.

Effective dose coefficients for inhalation of a given mixture of U isotopes can be approximated as a linear combination of separate effective dose coefficients for ^{234}U, ^{235}U, and ^{238}U given in ICRP documents. The effective dose coefficients E (Sv/Bq) for ^{234}U, ^{235}U, and ^{238}U listed in the third column of Table 10.5 are from ICRP Publication 68 (1994b) and represent inhalation of the indicated absorption type by a worker, based on a particle size of 5 μm (AMAD). In the last column of Table 10.5, effective dose coefficients are given in terms of intake of a unit mass (1 g) of each of the U isotopes.

The calculation of effective dose per unit mass intake of U for a given level of ^{235}U enrichment or depletion is illustrated for the cases of DU, assumed to contain 0.25% ^{235}U and 0.0005% ^{234}U, and 10% enriched U, assumed to contain 0.065% ^{234}U based on Equation 10.1. Based on these percentages and the dose coefficients

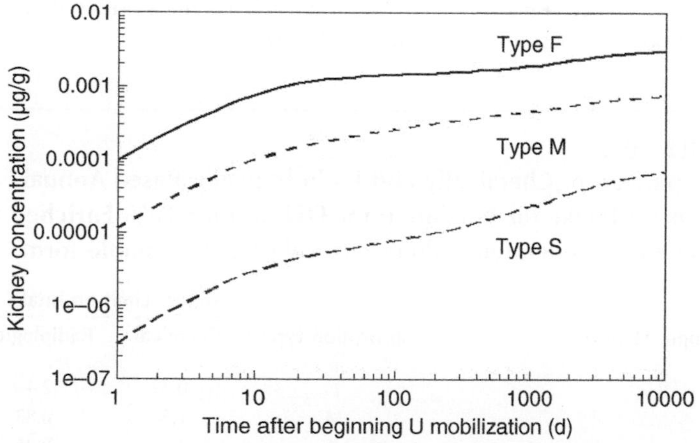

FIGURE 10.7 ICRP model predictions of the time-dependent concentration of U in the kidneys, assuming continuous inhalation of U of Type F, M, or S at a rate of 1 μg/d.

TABLE 10.5
Effective dose coefficients for inhalation of U isotopes of Type F, M, or S

Type	U isotope	E (Sv/Bq)	E (Sv/g)
F	^{234}U	6.4×10^{-7}	1.5×10^{2}
	^{235}U	6.0×10^{-7}	4.8×10^{-2}
	^{238}U	5.8×10^{-7}	7.3×10^{-3}
M	^{234}U	2.1×10^{-6}	4.9×10^{2}
	^{235}U	1.8×10^{-6}	1.4×10^{-1}
	^{238}U	1.6×10^{-6}	2.0×10^{-2}
S	^{234}U	6.8×10^{-6}	1.6×10^{3}
	^{235}U	6.1×10^{-6}	4.9×10^{-1}
	^{238}U	5.7×10^{-6}	7.1×10^{-2}

given in Table 10.5, the effective dose coefficient for inhalation of DU of Type S (particle size 5 μm) is approximated as

$$(0.000005 \times 1600) + (0.0025 \times 0.49) + (0.997495 \times 0.071) = 0.08 \text{ Sv g}^{-1}.$$

The effective dose coefficient for inhalation of 10% enriched U of Type S (AMAD 5 μm) would be

$$(0.00065 \times 1600) + (0.1 \times 0.49) + (0.89935 \times 0.071) = 1.15 \text{ Sv g}^{-1}.$$

The derived annual limit on intake based on radiological guidance is 0.02 Sv / 0.08 Sv g^{-1} = 0.25 g for DU of Type S and 0.02 Sv/1.15 Sv g^{-1} = 0.017 g for 10% enriched U of Type S. Thus, for intake of a given mass of U, the projected radiological hazard increases with the level of enrichment due mainly to the increased percentage of ^{234}U present, whereas the projected chemical hazard does not change because it depends only on the mass of U present (Table 10.6).

TABLE 10.6
Comparison of Chemically and Radiologically Based Annual Limits on Intake for Inhalation for DU and for 10% Enriched U in Soluble, Moderately Soluble, or Relatively Insoluble form

		Annual Limit on Intake (g)	
Isotopic Mixture	Absorption type	Chemical	Radiological
DU	F	0.37	2.4
	M	1.5	0.83
	S	14	0.25
Enriched U (10%)	F	0.37	0.18
	M	1.5	0.057
	S	14	0.017

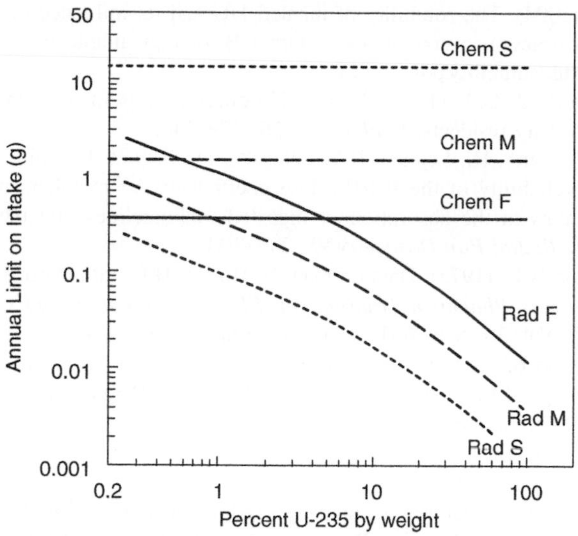

FIGURE 10.8 Comparison of chemically and radiologically based annual limits on intake for inhalation of soluble, moderately soluble, or relatively insoluble forms of depleted, natural, or enriched U. Relative masses of ^{234}U, ^{235}U, and ^{238}U for a given ^{235}U content are given in the text.

Results of the full analysis for inhalation of different mixtures of U are shown graphically in Figure 10.8. The labels Rad X and Chem X, where X = F, M, or S, refer to radiological and chemical limits, respectively, for inhalation of U of Type X. The annual limit on intake for a given isotopic composition is determined by the lower of the curves Chem X and Rad X directly above the ^{235}U content on the x-axis. Thus, for Type F, the limit on chronic inhalation of DU or natural U, or enriched U up to enrichment levels of about 5%, is determined by chemical guidance. For Type M and Type S, radiological guidance is dominant for all mixtures of U isotopes.

REFERENCES

Bassett, S.H., Frenkel, A., Cedars, N., VanAlstine, H., Waterhouse, C., Cusson, K. (1948). The Excretion of Hexavalent Uranium Following Intravenous Administration. II. Studies on Human Subjects. Rochester, New York: University of Rochester, pp. 1–57.

Bernard, S.R., Struxness, E.G. (1957). A study of the distribution and excretion of uranium in man. Oak Ridge National Laboratory, Oak Ridge, TN, ORNL-2304.

Campbell, E.E., McInroy, J.F., Schulte, H.F. (1975). Uranium in the tissue of occupationally exposed workers. In *Conference on Occupational Health: Experience with Uranium*. Arlington, VA, April 28–30, 1975, ERDA 93, 1975: 324–349.

Donoghue, J.K., Dyson, E.D., Hislop, J.S., Leach, A.M., Spoor, N.L. (1972). Human exposure to natural uranium: a case history and analytical results from some postmortem tissues. *Br J Ind Med* 29: 81–89.

Etherington, G. (2004). The solubility of inhaled DU and its influence on urine excretion. Depleted uranium oversight board, Great Britain. Available online at http://www.duob.org.uk/solubility.pdf.

Fisenne, I.M., Welford, G.A. (1986). Natural U concentrations in soft tissues and bone of New York City residents. *Health Phys* 50: 739–746.

Harrison, J.D., Metivier, H., Leggett, R.W., Nosske, D., Paquet, F., Phipps, A., Taylor, D. (2001). Reliability of the ICRP's dose coefficients for members of the public. II. Uncertainties in the absorption of ingested radionuclides and the effect on dose estimates. *Radiat Prot Dosimetry* 95: 295–308.

Hursh, J.B., Spoor, N.L. (1973). Data on man. In Hodge, H.C., Stannard, J.N., Hursh, J.B., Eds., *Uranium, Plutonium, Transplutonic Elements: Handbook of Experimental Pharmacology*, Vol. 36. New York: Springer-Verlag, pp. 197–240.

ICRP (1975). Report of the task group on reference man. ICRP (International Commission on Radiological Protection) Publication 23. Oxford: Pergamon Press.

ICRP (1991). 1990 recommendations of the International Commission on Radiological Protection. ICRP (International Commission on Radiological Protection) Publication 60. Oxford: Pergamon Press.

ICRP (1994a). Human respiratory tract model for radiological protection. ICRP (International Commission on Radiological Protection) Publication 66. Oxford: Pergamon Press.

ICRP (1994b). Dose coefficients for intakes of radionuclides by workers. ICRP (International Commission on Radiological Protection) Publication 68. Oxford: Pergamon Press.

ICRP (1995a). Age-dependent doses to members of the public from intake of radionuclides: Part 3. ICRP (International Commission on Radiological Protection) Publication 69. Oxford: Pergamon Press.

ICRP (1995b). Age-dependent doses to members of the public from intake of radionuclides: Part 4. ICRP (International Commission on Radiological Protection) Publication 71. Oxford: Pergamon Press.

ICRP (2002a). Supporting guidance 3. Guide for the practical application of the ICRP human respiratory tract model. International Commission on Radiological Protection. Oxford: Pergamon Press.

ICRP (2002b). Basic anatomic and physiological data for use in radiological protection: reference values. ICRP (International Commission on Radiological Protection) Publication 89. Oxford: Pergamon Press;

Igarashi, Y., Yamakawa, A., Seki, R., Ikeda, N. (1985). Determination of U in Japanese human tissues by the fission track method. *Health Phys* 49: 707–712.

Kathren, R.L., McInroy, J.F., Moore, R.H., Dietert, S.E. (1989). Uranium in the tissues of an occupationally exposed individual. *Health Phys* 57: 17–21.

Leggett, R.W. (1989). The behavior and chemical toxicity of uranium in the kidney: a reassessment. *Health Phys* 57: 365–383.

Leggett, R.W. (1992). A generic age-specific biokinetic model for calcium-like elements, *Radiat Prot Dosimetry* 41: 183–198.

Leggett, R.W. (1994). Basis for the ICRP's age-specific biokinetic model for uranium. *Health Phys* 67: 589–610.

Leggett, R.W., Eckerman, K.F., Boice, J.D. (2005). A respiratory model for uranium aluminide based on occupational data. *J Radiol Prot* 25: 1–12.

Leggett, R.W., Harrison, J.D. (1995). Fractional absorption of ingested U in humans. *Health Phys* 68: 484–498.

Leggett, R.W., Pellmar, T.C. (2003). The biokinetics of uranium migrating from embedded DU fragments. *J Environ Radioact* 64: 205–225.

Luessenhop, A.J., Gallimore, J.C., Sweet, W.H., Struxness, E.G., Robinson, J. (1958). The toxicity in man of hexavalent uranium following intravenous administration. *Am J Roentgenol* 79: 83–100.

Pellmar, T.C., Fuciarelli, A.F., Ejnik J.W., Hamilton, M., Hogan, J., Strocko, S., Emond, C., Mottaz, H.M., Landauer, M.R. (1999). Distribution of uranium in rats implanted with depleted uranium pellets. *Toxicol Sci* 49, 29–39.

Rich, B.L., Hinnefeld, S.L., Lagerquist, C.L., Mansfield, W.G., Munson, L.H., Wagner, E.R. (1988). Manual of Good Practices at Uranium Facilities — Draft. EGG-2530. Idaho Falls, Idaho: Idaho National Engineering Laboratory.

Roberts, A.M., Coulston, D.J., Bates, T.H. (1977). Confirmation of in vivo uranium-in-chest survey by analysis of autopsy specimens. *Health Phys.* 32: 435–437.

Singh, N.P., Bennett, D.B., Wrenn, M.E., Saccomanno, G. (1987). Concentrations of alpha-emitting isotopes of U and Th in uranium miners' and millers' tissues. *Health Phys* 53: 261–265.

Singh, N.P., Lewis, L.L., Wrenn, M.E. (1986). Uranium in human tissues of Colorado, Pennsylvania and Utah populations. In *Thirty-first annual meeting of the Health Physics Society*, Vol. 50, June 29–July 3. Pittsburgh, Pennsylvania. New York: Pergamon Press, S83 (abstract).

Spoor, N.L., Hursh, J.B. (1973). Protection criteria. In Hodge, H.C., Stannard, J.N., Hursh, J.B., Eds., *Uranium, Plutonium, Transplutonic Elements: Handbook of Experimental Pharmacology*, Vol. 36. New York: Springer-Verlag, pp. 241–270, chap. 5.

Squibb, K.S., Leggett, R.W., McDiarmid, M. (2005). Prediction of renal concentrations of depleted uranium and radiation dose in Gulf War veterans with embedded shrapnel. *Health Phys* 89: 267–273.

Struxness, E.G., Luessenhop, A.J., Bernard, S.R., Gallimore, J.C. (1956). The distribution and excretion of hexavalent uranium in man. In *Proceedings of international conference on the peaceful uses of atomic energy*, 1955. New York: United Nations, pp. 186–195.

Terepka, A.R., Toribara, T.Y., Neuman, W.F. (1966). Skeletal retention of uranium in man. Abstract 22, 46th meeting of the endocrine society, San Francisco, CA, 1964. *Nucl Sci Abstr* 20(3664), p. 439.

U.S. Department of Defense (2001). Environmental Exposure Report. Depleted Uranium in the Gulf (II) Bernard Rostker, Special Assistant for Gulf War Illnesses. Available online at http://www.deploymentlink.osd.mil/du_library/du_ii/du_ii_tabo1.htm.

Voegtlin, C. and Hodge, H. C., Eds. (1953). *Pharmacology and Toxicology of Uranium Compounds*. National Nuclear Energy Series, Division VI — Vol. I, Parts III and IV. New York: McGraw-Hill.

Wrenn, M.E., Durbin, P.W., Howard, B., Lipsztein, J., Rundo, J., Still, E.T., Willis, D.L. (1985). Metabolism of ingested U and Ra. *Health Phys* 48: 601–633.

11 Application of ICRP Biokinetic Models to Depleted Uranium

M.R. Bailey and A.W. Phipps

CONTENTS

INTRODUCTION

This chapter describes the methods currently in general use to assess uranium concentrations in tissues resulting from exposure to depleted uranium (DU) in environmental media, and the resulting radiation doses. Because of the importance of inhalation of airborne dust in such assessments, it includes an introduction to the mechanisms and terminology relating to aerosols. Much of this chapter is based on material prepared by the authors for ICRP publications relating to intakes of radionuclides, especially the ICRP Guide for the practical application of the ICRP human respiratory tract model (2002), and in support of the Royal Society (2001, 2002) assessments (Bailey and Phipps, 2002; Bailey, 2002a, 2002b), and reproduced here by kind permission of the ICRP and the Royal Society.

Current System of Radiation Protection for Intakes of Radionuclides

There is potential for people to be irradiated as a result of internal contamination by a wide range of radionuclides and in a variety of chemical forms. In only a few cases, however, such as exposure to radon, do epidemiological studies enable a direct assessment to be made of the risk to people. In all other cases, internal dosimetry provides a systematic basis for assessing the hazards from exposure to any radionuclide in any physicochemical form. The following description is based on the system of protection set out by the International Commission on Radiological Protection (ICRP) in its most recent Recommendations (ICRP, 1991), and applied in subsequent ICRP publications relating to intakes of radioactive materials (ICRP 1993, 1994a, 1994b, 1995a, 1995b, 1997, 2001, 2002, 2004). At the time of writing, ICRP is reviewing its Recommendations, and it is expected that new Recommendations might be published in 2007. Similarly, a new document on Occupational Intakes of Radionuclides is in preparation, which will replace those currently used (ICRP 1994b, 1997). This will take account of not only the new Recommendations, but also recent developments in the ICRP's internal dosimetry system.

Radiation Doses

Absorbed Dose

Internal dosimetry is based on the widely held premise that the risks of harm from irradiation of tissue by ionising radiation (such as alpha particles, beta particles, and gamma rays) are related to the mean absorbed dose to that tissue, i.e., the amount of energy deposited per unit mass of tissue. It is expressed in grays, symbol Gy. One Gy is equal to one joule per kilogram. The organ dose is taken to be the average absorbed dose in the organ or tissue of interest. For some organs, e.g., liver, kidneys, the organ dose calculated is the average dose to the whole organ, i.e., it is considered that the radiation sensitive cells are distributed throughout the whole organ. In other cases, notably in the parts of the gastrointestinal (GI) tract and respiratory tract, sensitive cells are identified, and the average dose to these is calculated, i.e., the average dose to the tissue at a specified range of depths from a specified organ surface.

Equivalent Dose

For the same absorbed dose to a tissue or organ, some types of ionizing radiation, notably alpha particles, can be more harmful than others. To take account of this, the absorbed dose is *weighted*, which means that it is multiplied by a factor (the radiation weighting factor, symbol w_R), to give the *equivalent dose*. This factor is taken to be equal to 1 for gamma rays, x-rays, and beta particles and 20 for alpha-particles. Equivalent dose is expressed in sieverts (symbol Sv), but as this is rather large, the millisievert (mSv), which is one thousandth of a sievert, and the micro-sievert (μSv) which is one thousandth of a millisievert are more commonly used. Equivalent dose provides an indication of the risk of harm resulting from irradiation of a particular tissue or organ.

Effective Dose

The risk of harm per mSv differs between organs: for example it is lower for the liver than for the lungs. To take account of this while obtaining an estimate of the overall risk of harm to a person, the equivalent dose to each important organ or tissue is multiplied by a factor (the tissue weighting factor, symbol w_T, Table 11.1). This factor can be thought of as representing the risk of harm per mSv to the tissue, compared to the risk of harm per mSv to the whole body. The sum of the weighted tissue equivalent doses is called the *effective dose*, and is also expressed in Sv. Effective dose provides a measure of the risk of harm resulting from irradiation of a person, taking account of different types of radiation and different doses to different organs. It is, therefore, a useful standardized measure of the risks from exposures to radiation, especially those resulting from intakes of radionuclides, which will often result in very different doses to different tissues. The primary standards of radiation protection (limits and constraints) are mainly expressed in terms of effective dose.

TABLE 11.1
Tissue Weighting Factors[a]

Organ or Tissue (T)	Tissue Weighting Factor w_T	Organ or Tissue (T)	Tissue Weighting Factor w_T
Gonads[b]	0.20	Liver	0.05
Bone marrow (red)	0.12	Esophagus	0.05
Colon[c]	0.12	Thyroid	0.05
Lungs[d]	0.12	Skin	0.01
Stomach	0.12	Bone surface	0.01
Bladder	0.05	Remainder[e,f]	0.05
Breast	0.05		

[a] Based on Table 11 of ICRP Publication 71, ICRP, 1995b; the values have been developed from a reference population of equal numbers of both sexes and a wide range of ages. In the definition of effective dose they apply to workers, to the whole population, and to either sex (ICRP 60, 1991).

[b] Dose coefficients are calculated for both the ovaries and testes, the higher of which is multiplied by the gonad tissue weighting factor for the calculation of effective dose.

[c] The colon w_T is applied to the mass average of the equivalent doses H_{ULI}, and H_{LLI}, in the walls of the upper large intestine (ULI) and lower large intestine (LLI) of the GI tract (ICRP 30, 1979).

[d] Thoracic airways.

[e] For purposes of calculation, the remainder is composed of the following 10 additional tissues and organs: adrenals, brain, extrathoracic airways, small intestine, kidneys, muscle, pancreas, spleen, thymus, and uterus. The extrathoracic airways described in the Human Respiratory Tract Model (ICRP 66) were added to the list of remainder tissues in ICRP 68. Their masses are given in Table 11 of ICRP 71.

[f] Whenever the most exposed remainder tissue or organ receives the highest committed equivalent dose of all organs, a weighting factor of 0.025 (half of remainder) is applied to that tissue or organ and 0.025 (half of remainder) to the mass-averaged committed equivalent dose in the rest of the remainder tissues and organs (ICRP, 1994b).

Doses from Intakes of Radionuclides

When a person is irradiated with gamma rays from a source outside the body (external), the radiation dose is received only while the person is in the presence of the source. However, after radionuclides are inhaled or ingested they can remain in the body for some time, and continue to irradiate tissues. For how long this continues depends upon the route of intake, the chemical form, the turnover rates in tissues, and the radioactive half-lives of the radionuclides. The radioactive half-life of uranium-238 is 4500 million years and, as a heavy metal, it is retained in tissues such as bone for many years. When uranium-238 enters the body it may continue to irradiate tissues, albeit at a decreasing rate, over the rest of the person's life. Thus, although the intake may occur over a short time — a few breaths or a meal — for long-lived radionuclides the person is committed to receiving a dose for a long time to come. The committed dose (equivalent or effective) is the dose that is expected to be received in a stated period after the intake, usually taken to be 50 years for workers or up to age 70 years for members of the public.

Dose Coefficients

A *dose coefficient* is the committed equivalent dose (in sieverts) to an organ (or the committed effective dose, also in sieverts) that results from an intake of unit activity of a radionuclide (1 becquerel, Bq, i.e., 1 nuclear transformation or radioactive decay per second). ICRP dose coefficients provide standard conversion factors to relate intakes of radionuclides to equivalent or effective doses, and so to approximate risks of harm which are internationally recognized (IAEA 1996, EC 1996). They are widely used for purposes such as planning, demonstrating compliance, and environmental impact assessment.

Biokinetic Models

In order to calculate committed doses and hence dose coefficients, we have to take account of how the radionuclide behaves after it has entered the body, and this is the role of biokinetic models. Once a radioactive material is inside the human body, it will irradiate the tissues surrounding the place where it is deposited (e.g., in the lungs), but generally it will, in time, move to other tissues or organs (e.g., the kidneys), and irradiate the surrounding tissues there. A biokinetic model describes how a radionuclide moves around a biological system such as a human body. A model summarizes our knowledge about the processes involved, but it describes what happens to the material in terms of mathematical equations. Using it, we can calculate how much of the material that entered the body at the time of the intake is present in each organ at any time afterwards. Generally, the model first describes how the atoms would behave if they were not radioactive, so it actually describes the behaviour of a stable isotope of the element. Radioactive decay is then taken into account to determine the amount of activity present in each organ at any time; the shorter the half-life, the more rapidly the activity decreases. A complicating factor, which also has to be taken into account, is that when some radionuclides decay, the atoms into which they are transformed are themselves radioactive but are atoms of a different element and so may behave differently in the body. This is the case for uranium, but the decay products generally only make small contributions to the effective dose compared to the uranium itself.

One major use of biokinetic models is in calculating committed doses. The biokinetic model is used to calculate how many nuclear transformations take place in each organ in a given time after intake. A dosimetric model is then applied to calculate how much of the energy released as each type of radiation (e.g., alpha- or beta-particles) is absorbed in each tissue or organ and, hence, the committed doses to the organs. The organ in which the radionuclide is located is known as the *source organ*, and that in which the energy is absorbed is known as the *target organ*. The dosimetric model takes account of energy absorbed in the source organ itself (*self-irradiation*), which predominates for nonpenetrating radiation such as alpha- and beta-particles, and also energy absorbed in other organs (*cross-fire*) which can be important for penetrating radiation such as gamma rays.

The other major use of biokinetic models is in the interpretation of measurements of activity present in the body or excreted in urine and/or feces (known as *individual monitoring* or *bioassay*). The biokinetic models are first used to calculate, for a given intake, how much activity is present in the organ measured or excreted at the specified time after intake. By comparing this with the measured value, the intake can be estimated.

Thus, the models may be used *prospectively*, i.e., in assessments of current, future or hypothetical exposures, or *retrospectively*, i.e., in assessments of past exposures based on measurements of activity in the body or excreted in urine or faeces. Rather than have a single biokinetic model for the whole human body, it has been found convenient to have separate respiratory tract, GI tract, and systemic models, which can be combined to provide an overall model for a particular intake.

Respiratory Tract Model

A respiratory tract model describes the processes involved in the entry of radionuclides into the body by inhalation, including:

- How much air is breathed in through the nose and mouth
- How much of the radioactive material in the inhaled air deposits in each part of the respiratory tract, (eg the nose or lungs)
- How quickly the radionuclides that have deposited are cleared, either by being carried in mucus to the throat where they are swallowed, or by being absorbed into the bloodstream

Radionuclides that are cleared from the respiratory tract are then handled by the GI tract or appropriate systemic model, according to whether they are swallowed or absorbed into the bloodstream.

GI Tract Models

A GI tract model describes the processes involved in the entry of radionuclides into the body by ingestion. The model accounts for the time it takes swallowed material to pass through each part of the GI tract, such as the stomach, and what fractions of the swallowed radionuclides are absorbed into the bloodstream, where they are handled by the appropriate systemic model. The rest is excreted in feces.

Systemic Models

Systemic models describe the behavior of radionuclides after they enter the bloodstream (or systemic circulation). Generally, a separate systemic model has been developed for each element, or group of similar elements. The systemic model describes how long the element remains in the circulation, how much goes to each important organ, or is excreted in urine or feces, and how long it remains in each organ. The behavior varies greatly between elements: some, like hydrogen, are uniformly distributed; many heavy metals are retained in liver and bone, while iodine concentrates in the thyroid.

Development of ICRP Biokinetic Models

Over the last half-century, the ICRP has, through its Task Groups and Committees, developed models to calculate radiation doses to organs from intakes of radionuclides. The ICRP models are used by most radiation protection professionals and organizations around the world.

Recent major assessments of the hazards resulting from intakes of DU (The Royal Society, 2001, 2002; Guilmette et al., 2004) have used the current ICRP models (see following). However, other reports concerned with DU and its potential hazards refer to earlier ICRP models, and therefore a brief history of their development follows.

The first ICRP biokinetic models for internal dose calculation were issued in Publication 2 (ICRP, 1960). At that time internal dose standards were based around the concept of a *critical organ*, the organ which received the highest dose. Exposures were to be controlled to prevent the dose to the critical organ exceeding the prescribed limit over a relevant, stated period.

A major advance in progress came with the publication of the "Task Group Lung Model" (TGLM) produced by the ICRP's Task Group on Lung Dynamics (TGLD, 1966). In this, the respiratory tract is divided into three regions; nasopharyngeal (N-P), tracheobronchial (T-B), and pulmonary (P). Deposition in each is dependent on the size distribution of the inhaled particles (see below). Clearance from each region to the blood, GI tract, and lymph nodes is related to the chemical form of the inhaled material. Clearance rates were recommended for three inhalation classes, and compounds were assigned to them on the basis of the long-term retention half-time in the pulmonary region: D (days) if this half-time was less than 10 d; W (weeks) if it was 10–100 d; and Y (years) if it was more than 100 d.

ICRP Publication 30

The TGLM (with minor changes) was adopted in ICRP Publication 30 (1979, 1980, 1981, 1988a) (generally known as "ICRP 30") which was based on the revised ICRP general recommendations issued in ICRP 26 (ICRP, 1977). These introduced major changes in approach. Rather than focusing on the "critical" organ, doses were to be calculated to all major organs and the effective (whole-body) dose derived (see subsection on "Biokinetic Models"). Intakes were to be limited on the basis of committed doses, rather than doses received in a given period. ICRP 30 provided "Annual Limits on Intake" for the several hundred radionuclides (isotopes of 91 elements) potentially relevant to occupational exposure. To do this, ICRP 30 also adopted a GI Tract model (see below), and provided a systemic biokinetic model for each element. These are generally simple compartment models, where retention in each organ in which significant deposition occurs is represented by one or more "compartments." In a compartment, the amount of material that leaves per unit time is proportional to the amount present at that time. Thus, if an amount A(0) is deposited in the compartment at time t = 0, the amount retained in the compartment A(t) at time t decreases with time according to a single exponential function:

$$A(t) = A(0)\exp(-\lambda t)$$

where λ is the fractional rate at which material leaves the compartment. It is often represented by the retention half-time $t_{1/2}$, where $t_{1/2} = \ln(2)/\lambda$. For radioactive materials, the amount retained will also decrease because of radioactive decay, at a rate determined by the radioactive half-life. A particular simplification of ICRP 30 models was that when activity left organs such as liver and bone it was taken to be immediately excreted, without returning to blood or the appropriate excretion pathway through the bladder or colon.

In the ICRP 30 systemic model for uranium (ICRP, 1979), of the uranium entering the transfer compartment (which represents the blood) from the GI or respiratory tract:

- Fractions 0.2 and 0.023 are assumed to go to mineral bone and be retained there with half-times of 20 and 5000 d, respectively
- Fractions 0.12 and 0.00052 are assumed to go to the kidneys and be retained there with half-times of 6 and 1500 d, respectively
- Fractions 0.12 and 0.00052 are assumed to go to "all other tissues of the body" and be retained there with half-times of 6 and 1500 d, respectively
- The remainder (about 54%) is assumed to go directly to excretion

ICRP 30 (Part 1, ICRP, 1979) assigned uranium hexafluoride (UF_6), uranyl difluoride (UO_2F_2), and uranyl nitrate ($UO_2(NO_3)_2$) to Class D on the basis of studies of Chalabreysse (1970). The trioxide (UO_3), tetrafluoride (UF_4) and chloride (UCl_4) were found to be less soluble in studies reported by Hursh and Spoor (1973), and Morrow et al. (1972), and were assigned to Class W. Uranium dioxide (UO_2) and uranium octoxide (U_3O_8) were assigned to Class Y on the basis of inhalation studies in dogs (Morrow et al., 1966).

ICRP 54 (1988b) gave guidance on the design of programs of bioassay measurements on workers and on their interpretation (assessment of intakes and doses), based on ICRP 30 models.

The recommendations in ICRP 26 and 30 formed the basis for radiation protection of workers from intakes of radionuclides adopted in most countries at the time of the Persian Gulf War and in some countries (including the U.S.) today.

Current ICRP Biokinetic Models

Following the Chernobyl accident, there was increased concern about exposure of the general public to intakes of radionuclides and, hence, a need for models that applied to a wider range of subjects and circumstances. (The ICRP 30 models are based on a reference worker who is a healthy young adult Caucasian male.) It was recognized that internationally-accepted models were needed for the public including children of various ages. It was also considered that there was a need for models that were more realistic (rather than conservative) and which made use of the additional information by then available.

ICRP has issued dose coefficients for workers and members of the public using the current models in several reports, as summarised in Table 11.2.

TABLE 11.2
Summary of ICRP Reports on Dose Coefficients for Workers and Members of the Public from Intakes of Radionuclides

ICRP Publication No. (year)	Application		Contents
56 (1989)[32]	Public[a]	Inhalation and ingestion	Age-dependent systemic models, and tissue dose coefficients for selected radioisotopes, for H, C, Sr, Zr, Nb, Ru, I, Cs, Ce, Pu, Am and Np. It was issued before ICRP 60[7], and hence gives dose equivalents using the tissue weighting factors from ICRP 26[26], rather than equivalent doses using the tissue weighting factors from ICRP 60[7]. It was also issued before ICRP 66[11] and hence used the ICRP 30 lung model[22]. The dose coefficients given in ICRP 56[32] were superseded by those in ICRP 67 and 71.
67 (1993)[10]	Public[a]	Ingestion	Age-dependent systemic models, and tissue dose coefficients for selected radioisotopes, for S, Co, Ni, Zn, Mo, Tc, Ag, Te, Ba, Pb, Po and Ra. Updated systemic models are given for Sr, Pu, Am and Np.
68 (1994)[12]	Workers	Inhalation and ingestion	Effective dose coefficients for workers, for about 800 radionuclides: selected radioisotopes of the 91 elements covered in ICRP 30, Parts 1-4[22–25]. The inhalation dose coefficients for workers exposed to ^{226}Ra given in ICRP 68 were revised in Annexe B of ICRP 72.
69 (1995)[13]	Public[a]	Ingestion	Age-dependent systemic models, and tissue dose coefficients, for selected radioisotopes, for Fe, Sb, Se, Th and U.
71 (1995)[14]	Public[a]	Inhalation	Tissue dose coefficients for selected radioisotopes of elements covered in ICRP 56, 67, and 69, plus Ca and Cm for which age-dependent systemic models are given.
72 (1996)[33]	Public[a]	Inhalation and ingestion	Effective dose coefficients for members of the public for radioisotopes of the 31 elements covered in ICRP 56, 67, 69, and 71, plus radioisotopes of the further 60 elements covered in ICRP 30 and 68.
CD-ROM 1 (1998)[34]	Public[a] and workers	Inhalation and ingestion	A database of effective and equivalent doses to individual tissues corresponding to the effective dose coefficients in ICRP 68 and 72. It also gives committed doses for a range of times after intake, and inhalation of aerosols (airborne particles) with a range of sizes.
88 (2001)[8]	Public and workers	Inhalation and ingestion	Fetal dose coefficients for intakes before and during pregnancy of the 31 elements covered in ICRP 56, 67, 69, and 71, including doses to the embryo and fetus and to the child from activity retained at birth.

(continued)

TABLE 11.2 (CONTINUED)
Summary of ICRP Reports on Dose Coefficients for Workers and Members of the Public from Intakes of Radionuclides

ICRP Publication No. (year)	Application		Contents
CD-ROM 2 (2001)[9]	Public and workers	Inhalation and ingestion	A database of effective and tissue equivalent fetal dose coefficients for 10 aerosol sizes and five times after birth, for all radionuclides covered in ICRP 88. Consistent with the dose coefficients in ICRP 88. Extensive help files also provided.
95 (2004)[16]	Public and workers	Inhalation and ingestion	Dose coefficients for the breast-feeding infant following intakes by the mother before and during pregnancy, and during breast-feeding. The 31 elements covered in ICRP 56, 67, 69, and 71 are addressed.

[a] Age-dependent dose coefficients (3 months, 1-, 5-, 10-, and 15-years and adult).

ICRP 69 (ICRP, 1995a) gave age-dependent systemic models, and ingestion dose coefficients for radioisotopes of five more elements, including uranium (Section 4).

ICRP 68 (ICRP 1994b) gave effective dose coefficients for workers, for about 800 radionuclides: selected radioisotopes of the 91 elements covered in ICRP 30, using the HRTM and, for those (31) elements for which new models were available, the new systemic models from ICRP 67, etc. For the other 60 elements the systemic models of ICRP 30 were applied. However, they were modified so that when activity left organs such as liver and bone it passed through and so irradiated the appropriate excretion pathway through the bladder or colon. As an interim measure, compounds that were assigned to inhalation Class D, W, or Y in ICRP 30 were assigned to the corresponding HRTM absorption Type (F, M, or S, see section heading "The ICRP Human Respiratory Tract Model (HRTM)" in ICRP 68. Equivalent dose coefficients to organs for workers are given in the ICRP Database of Dose Coefficients (ICRP, 1998).

ICRP 71 (ICRP, 1995b) gave age-dependent systemic models for two more elements. In addition, for each of the 31 elements for which age-dependent systemic models are given in ICRP 56, 67, 69, and 71, it gave inhalation dose coefficients for members of the public. For each element it also gave a brief review of information relating to inhalation of different chemical forms, updating the reviews in ICRP 30, but with emphasis on environmental exposure. Its treatment of uranium oxides is summarized in Section 5.3.4.

ICRP 72 (ICRP 1996) gave effective dose coefficients for members of the public for radioisotopes of the 31 elements covered in ICRP 71, plus radioisotopes of the other 60 elements covered in ICRP 30 and 68. The dose coefficients given in ICRP 68 and 72 have been adopted in the International Basic Safety Standards (IAEA, 1996) and in the Euratom Directive (EC, 1996). Thus the new generation of ICRP models are currently used in the European Union member states, and in many other countries.

ICRP 78 (1997) gave guidance on individual monitoring of workers for internal exposure based on the new generation of models, and so replaced ICRP 54.

THE ICRP HUMAN RESPIRATORY TRACT MODEL (HRTM)

As noted above (Section 1.3.1), a respiratory tract model describes the processes involved in the entry of radionuclides into the body by inhalation. For many years ICRP's respiratory tract models have taken account of the respiratory tract both as a route of entry to the body, and as a target in itself, the effect of the size distribution of the inhaled material on deposition in each part of the respiratory tract, and the effect of the chemical form of the inhaled material on its clearance and uptake to blood.

The current model, the Human Respiratory Tract Model for Radiological Protection (HRTM), is described in detail in ICRP 66 (ICRP, 1994a). Summaries are given in the ICRP Publications (68, 71, 72, 78) in which it is applied (ICRP 1994b, 1995b, 1996, 1997, 2002), and elsewhere (Bailey, 1994; Bailey et al., 1998, 2003). An outline is given here.

The main functions of the HRTM are to provide:

- A qualitative and quantitative description of the respiratory tract as a route for radionuclides to enter the body
- A method to calculate radiation doses to the respiratory tract for any exposure
- A method to calculate the transfer of radionuclides to other tissues

The HRTM is comprehensive. It applies to:

- Assessing doses from exposures, and assessing intakes from bioassay measurements.
- Radionuclides associated with particles (aerosols) of all sizes of practical interest (0.0006–100 μm) and to gases and vapors.
- All members of the population, giving reference values for children aged 3 months, 1, 5, 10, and 15 years, and adults. Guidance is provided for taking into account the effects of factors such as smoking, diseases, and pollutants.

MORPHOMETRY

In the HRTM the respiratory tract is represented by five regions, based on differences in radio-sensitivity, deposition and clearance (Figure 11.1). The extrathoracic (head and neck) airways (ET) are divided into ET_1, the anterior nasal passage (front of the nose), and ET_2, which consists of the posterior nasal and oral passages, the pharynx and larynx. The thoracic regions (the lungs) are bronchial (BB: trachea, generation 0, airway generations 1-8), bronchiolar (bb: airway generations 9–15), and alveolar-interstitial (AI, the gas exchange region). Lymph nodes are

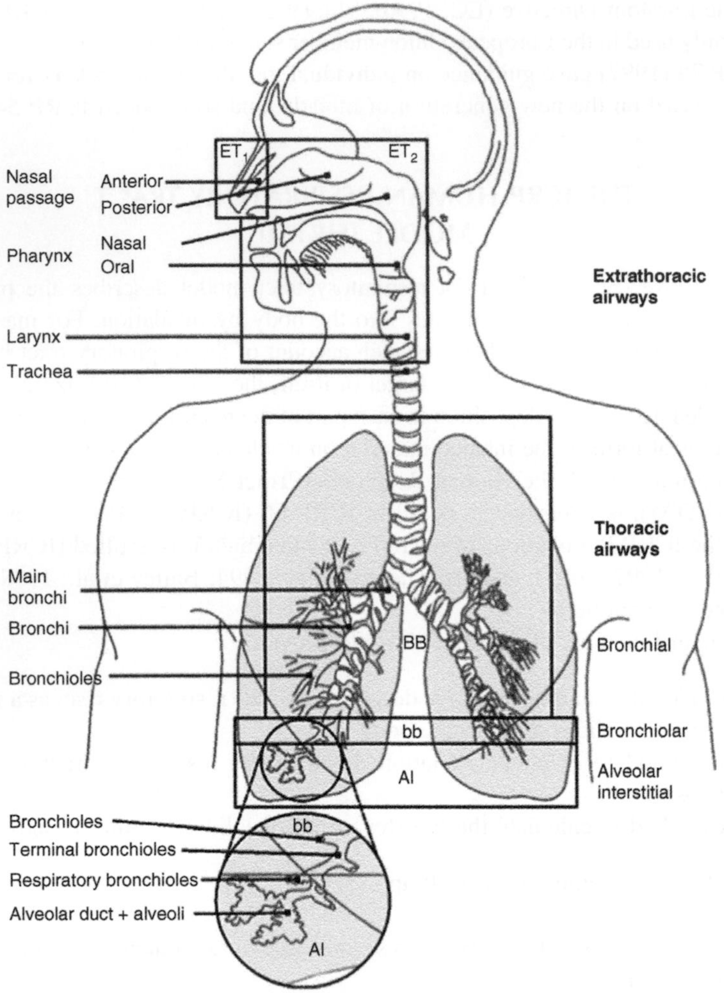

FIGURE 11.1 Respiratory tract regions defined in the HRTM.

associated with the extrathoracic and thoracic airways (LN_{ET} and LN_{TH}, respectively). Target cells are identified in each region: e.g., the basal cells of the epithelium in both ET regions and the basal and secretory cells in the bronchial epithelium. Reference values of dimensions that define the mass of tissue containing target cells in each region for dose calculations are given. They are assumed to be independent of age and sex.

Physiology

The breathing rate (frequency and volume) is the main factor in the model that depends on age and physical activity. Reference values of important parameters are recommended for the population groups noted above for four levels of exercise:

sleep, sitting, light and heavy exercise, and taking account of both nose and mouth breathing. These have been combined with habit survey data to give the reference volumes inhaled per working shift or per day. Thus, light work is a combination of light exercise and sitting. These parameters determine intakes per unit exposure (time-integrated air concentration) but are also used with the deposition model to determine regional deposition.

Deposition

Particle Sizes and Deposition Mechanisms

The behavior of an airborne particle depends on its size, shape, and density. If the particle is spherical, its size can be uniquely defined by its geometric diameter. If it is not spherical, its size is usually described in terms of an "equivalent diameter" — the diameter of a sphere (or circle) which gives the same result as the particle when measured in the same way. For example, the volume equivalent diameter, d_e, is the diameter of a sphere with the same volume as the particle.

There are three main mechanisms that affect the behavior of airborne particles in the respiratory tract and, indeed, their behavior in the ambient air and in many air sampling instruments: gravitational sedimentation, inertial impaction, and diffusion (For further information, see, e.g., IAEA, 1973; Hinds, 1982; ICRP 66, Annexe D, Deposition of Inhaled Particles.)

Gravitational Sedimentation

The gravitational force, F_g, on a particle of volume equivalent diameter d_e is given by:

$$F_g = g\rho\frac{\pi}{6}d_e^3 \tag{11.1}$$

where g is the acceleration due to gravity, and ρ is the particle density (strictly, the particle density minus the density of air).

As the particle falls, it experiences an opposing force due to viscous drag, which increases with particle velocity, u. For a spherical particle of diameter d, it is given approximately, by Stokes' law, as $3\pi u\mu d$, where μ is the viscosity of air. For particles with dimensions less than about 0.1 μm the molecular structure of the air becomes noticeable, and it acts less like a continuous fluid. As a result the drag is reduced, and this is taken into account using the Cunningham slip correction factor $C(d)$, which has a value of 1 for large particles and increases with decreasing size (ICRP 66, Section D.4.1.1). The drag on an irregular particle is usually greater than that on a sphere of the same volume, and this is taken into account using the "dynamic shape factor" χ, which has a value between 1 (for spheres) and about 2 (Hinds, 1982). (The HRTM uses default value of 1.5 for χ, typical of compact, irregular, i.e., nonspherical particles). Hence, the drag force F_D on an irregular particle is given by:

$$F_D = \frac{3\pi\mu u d_e \chi}{C(d_e)} \tag{11.2}$$

When the gravitational and drag forces are equal the particle falls at a constant rate u_g, known as the terminal or settling velocity. From Equation 11.1 and Equation 11.2, putting $u = u_g$ gives:

$$u_g = \frac{\rho\, d_e^2\, C(d_e)\, g}{18\,\mu\,\chi} \tag{11.3}$$

This is used to define the aerodynamic equivalent diameter, d_{ae}, which is widely used in occupational health, and in both the HRTM and the ICRP 30 lung model. It is the diameter of a unit density (1 g cm^{-3}) sphere with the same settling velocity as the particle:

$$u_g = \frac{\rho\, d_e^2\, C(d_e)\, g}{18\,\mu\,\chi} = \frac{d_{ae}^2\, C(d_{ae})\, g}{18\,\mu} \tag{11.4}$$

Hence d_{ae} is given by:

$$d_{ae} = d_e \sqrt{\frac{\rho}{\chi} \times \frac{C(d_e)}{C(d_{ae})}} \tag{11.5}$$

For larger particles, $C(d_e)$ and $C(d_{ae})$ are both approximately 1:

$$d_{ae} \approx d_e \sqrt{\frac{\rho}{\chi}} \tag{11.6}$$

Inertial Impaction

When an airstream changes direction, the inertia of the particles in it makes them tend to follow their original trajectories. It can be shown that if a particle with velocity u_0 enters still air it will come to rest in a distance L_s, known as the stop distance, which is given by:

$$L_s \approx \frac{u_0\, u_g}{g} \tag{11.7}$$

As u_g is a function of d_{ae} (Equation 11.4), aerodynamic diameter is also a useful indicator of deposition by impaction. The importance of inertia is also measured by the dimensionless Stokes' Number, $St = L_s / L$, where L is a "characteristic length" of the system, such as the diameter of an obstacle or aperture.

Diffusion

Collisions with gas molecules give rise to the random (Brownian) motion of a particle in a fluid, with average kinetic energy in each direction of $^1/_2 kT$, where k is Boltzmann's constant and T is the absolute temperature. The motion is opposed by viscous drag, i.e., subsequent collisions with air molecules. A useful measure of the effect of diffusion is the diffusion coefficient, D, which for a sphere of diameter d, is given by:

$$D = \frac{k\,T\,C(d)}{3\pi\,\mu\,d} \tag{11.8}$$

Note that D increases with decreasing particle size but not with particle density. So aerodynamic diameter is not appropriate when diffusion dominates. As diffusion is a thermodynamic process, the particle's behavior is described by means of its thermodynamic diameter, d_{th}, which is the diameter of a sphere with the same diffusion coefficient as the particle. In the HRTM, for simplicity, d_{th} is taken to be equal to the volume equivalent diameter, d_e, although in practice d_{th} would be determined by measuring the diffusion coefficient (ICRP 66, Paragraph D30). On this basis, d_{th} can be related to the aerodynamic diameter, d_{ae}, using Equation 5:

$$d_{th} = d_e = d_{ae}\sqrt{\frac{\chi}{\rho} \times \frac{C(d_{ae})}{C(d_e)}} \tag{11.9}$$

For larger particles, $C(d_{ae})$ and $C(d_e)$ are both approximately 1:

$$d_{th} \approx d_{ae}\sqrt{\frac{\chi}{\rho}} \tag{11.10}$$

Sedimentation and impaction are important above about 0.1 μm and increase with increasing size, whereas diffusion is important below about 1 μm and increases with decreasing size (Figure 11.2). In the range 0.1–1 μm all are important.

Particle Size Distributions

The particles produced by any source will generally have a wide range of sizes. A collection of airborne particles (solid or liquid) is known as an *aerosol*. In order to describe the size of the whole aerosol and its behavior, it is useful to represent it by a mathematical function. The one most frequently used for aerosols is the log-normal distribution. Its use was recommended by the ICRP Task Group on Lung Dynamics (TGLD, 1966), and it is applied in both the ICRP 30 lung model and the HRTM.

The Log-Normal Distribution

The log-normal distribution is often found suitable for describing the distribution of a parameter that shows a wide range of values. Moreover, although the function that represents it is complex, it easy to apply the distribution in practice using suitably

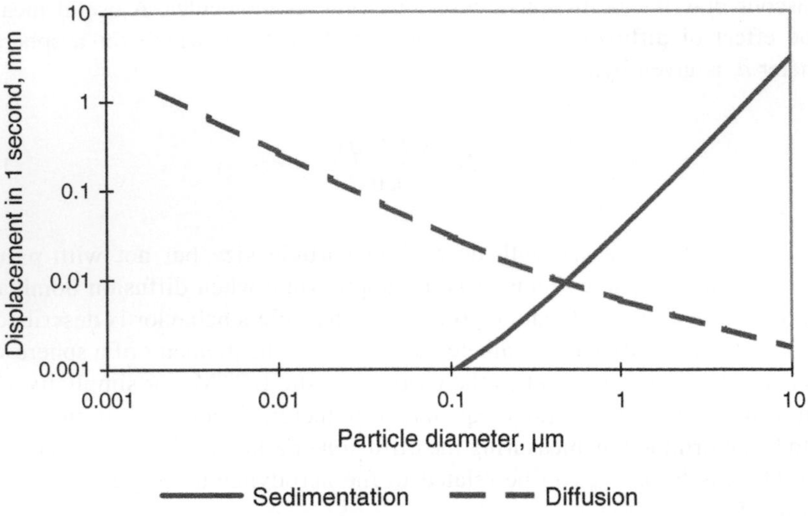

FIGURE 11.2 Relative importance of gravitational sedimentation and diffusion as a function of particle size. (For unit density spheres, data taken from Raabe, O.G., Characterisation of radioactive airborne particles, in *Internal Radiation Dosimetry*, Raabe, O.G., Ed., Medical Physics Publishing, Madison, WI, 1994, p. 111, chap. 7.) For gravitational sedimentation, the displacement is the vertical distance the particle falls in 1 sec. For diffusion the displacement is the root mean square distance the particle travels as a result of Brownian (random) motion in 1 sec.

formatted graph paper (see following). If a parameter y is normally distributed, then the probability of a value lying between y and y + *dy* is given by P(y)*dy*, where:

$$P(y)\,dy = \frac{1}{\sigma_y\sqrt{2\pi}} \exp\left[-\frac{(y-\bar{y})^2}{2\,\sigma_y^2}\right] dy \qquad (11.11)$$

where $\bar{y}$ is the (arithmetic) mean value of y, and σ_y is its standard deviation.

If a parameter x is such that the logarithm of x, ln x, is normally distributed, then x is said to be log-normally distributed. Substituting y = ln x and dy = dx/x in Equation 11.11 gives the probability, P(x)dx, of a value of the log-normally distributed parameter, x, lying between x and x + dx:

$$P(x)dx = \frac{1}{x(\ln\sigma_g)\sqrt{2\pi}} \exp\left[-\frac{(\ln x - \ln x_{50})^2}{2(\ln\sigma_g)^2}\right] dx \qquad (11.12)$$

where x_{50} is the median (50% of values lie below the median) and σ_g is the geometric standard deviation (GSD) of the distribution. A log-normal distribution is usually characterized by its median and GSD. Thus, $\ln\sigma_g = \sigma_y$ and $\ln x_{50} = \bar{y}$. However, whereas for a normal distribution the median, arithmetic mean and mode (most likely value) are all the same, for the log-normal distribution they are different, as the distribution is skewed. The mode $\hat{x}$ and arithmetic mean $\overline{x}$ are given by:

$$\hat{x} = x_{50}\exp[-(\ln\sigma_g)^2]$$
$$\overline{x} = x_{50}\exp[0.5(\ln\sigma_g)^2] \tag{11.13}$$

These quantities are shown in Figure 11.3 for a distribution with x_{50} = 1 µm and $\sigma_g = 2$.

For a radioactive aerosol, the amount of activity per unit size, rather than the number of particles, is usually considered. For particles of about 1 µm or larger, when sedimentation and impaction are important, and aerodynamic diameter, d_{ae}, is the appropriate measure of behavior, the aerosol would be characterized by the activity median aerodynamic diameter, AMAD: 50% of the activity is associated with particles larger than the AMAD. For smaller particles, for which diffusion dominates, and thermodynamic diameter, d_{th}, is the appropriate measure of behavior, the aerosol would be characterized by the activity median thermodynamic diameter, AMTD: 50% of the activity is associated with particles larger than the AMTD.

In practice, the parameters describing the distribution, the median and GSD, can easily be found graphically, using paper with a logarithmic scale on the x-axis, and with a probability scale on the y-axis. The (cumulative) percentage of activity associated with particles below a given diameter is plotted against the diameter. On these scales, a log-normal distribution is easily fitted as it gives a straight line.

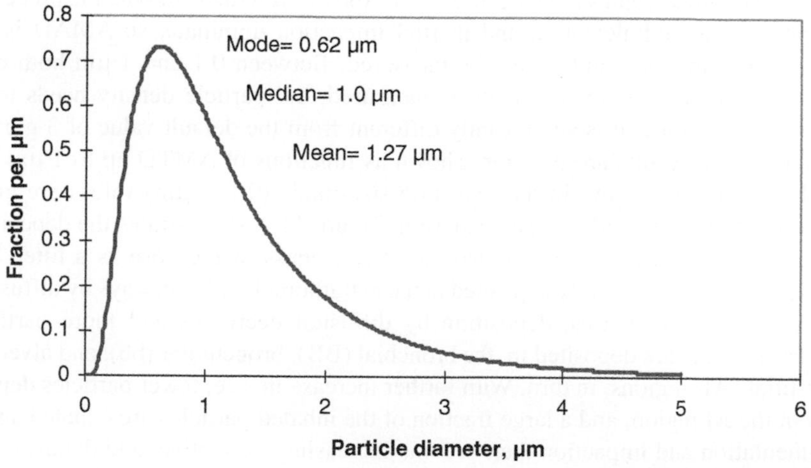

FIGURE 11.3 Log-normal distribution of particle sizes with median, x_{50} = 1 µm and GSD (σ_g) = 2.

The median is read from the x-axis, at the point corresponding to 50% on the y-axis. Similarly, the GSD can be found from the relationship:

$$\sigma_g \approx \frac{x_{84.13}}{x_{50}} \approx \frac{x_{50}}{x_{15.87}} \tag{11.14}$$

The diameter corresponding to 84.13% or 15.87% of the activity is read from the x-axis, and σ_g calculated from that and x_{50}.

In the HRTM it is assumed, by default, that σ_g is a function of AMTD:

$$\sigma_g = 1 + 1.5[1 - (100\ \text{AMTD}^{1.5} + 1)^{-1}] \tag{11.15}$$

Thus, the default value of σ_g increases from a value of 1.0 at 6 nm to a value of 2.5 above about 1 μm.

Respiratory Tract Deposition of Particles

The HRTM deposition model evaluates the fraction of activity in the inhaled air that is deposited in each region. Deposition in the ET regions was determined mainly from experimental data. For the lungs, a theoretical model was used to calculate particle deposition in each region, and to quantify the effects of the subject's lung size and breathing rate.

Figure 11.4(a) gives the fraction of inhaled activity deposited in each region as a function of aerosol size for a reference worker. (The HRTM default value of 3 g cm^{-3} is used for the particle density and the default value of 1.5 is used for the dynamic shape factor.) As noted above (Figure 11.2), for particle diameters below about 0.1 μm diffusion dominates, so AMTD is the appropriate parameter to characterize the aerosol, and should be measured. Above 1 μm the aerodynamic processes of gravitational sedimentation and inertial impaction dominate, so AMAD is the appropriate parameter and should be measured. Between 0.1 and 1 μm both contribute, and either parameter might be measured, but particle density needs to be taken into account if it is significantly different from the default value of 3 g cm^{-3}. In Figure 11.4, results are therefore shown as functions of AMTD up to 1 μm and of AMAD above 0.1 μm. In the transition size-range (0.1–1 μm) values are given for both AMAD and AMTD. The pattern in Figure 11.4(a) illustrates the deposition mechanisms outlined above and also the effectiveness of the nose as a filter. The smallest particles are mainly deposited in the extrathoracic (ET) airways by diffusion. As particle size increases, deposition by diffusion decreases and more particles penetrate to, and are deposited in, the bronchial (BB), bronchiolar (bb), and alveolar-interstitial (AI) regions, in turn. With further increase in size, fewer particles deposit even in the AI region, and a large fraction of the inhaled particles are exhaled again. Sedimentation and impaction then become increasingly effective, and deposition in the ET airways increases. The decrease in ET deposition at AMAD > 10 μm results from reduced "inhalability": because of their inertia, some particles in the inhaled air do not enter the nose and mouth.

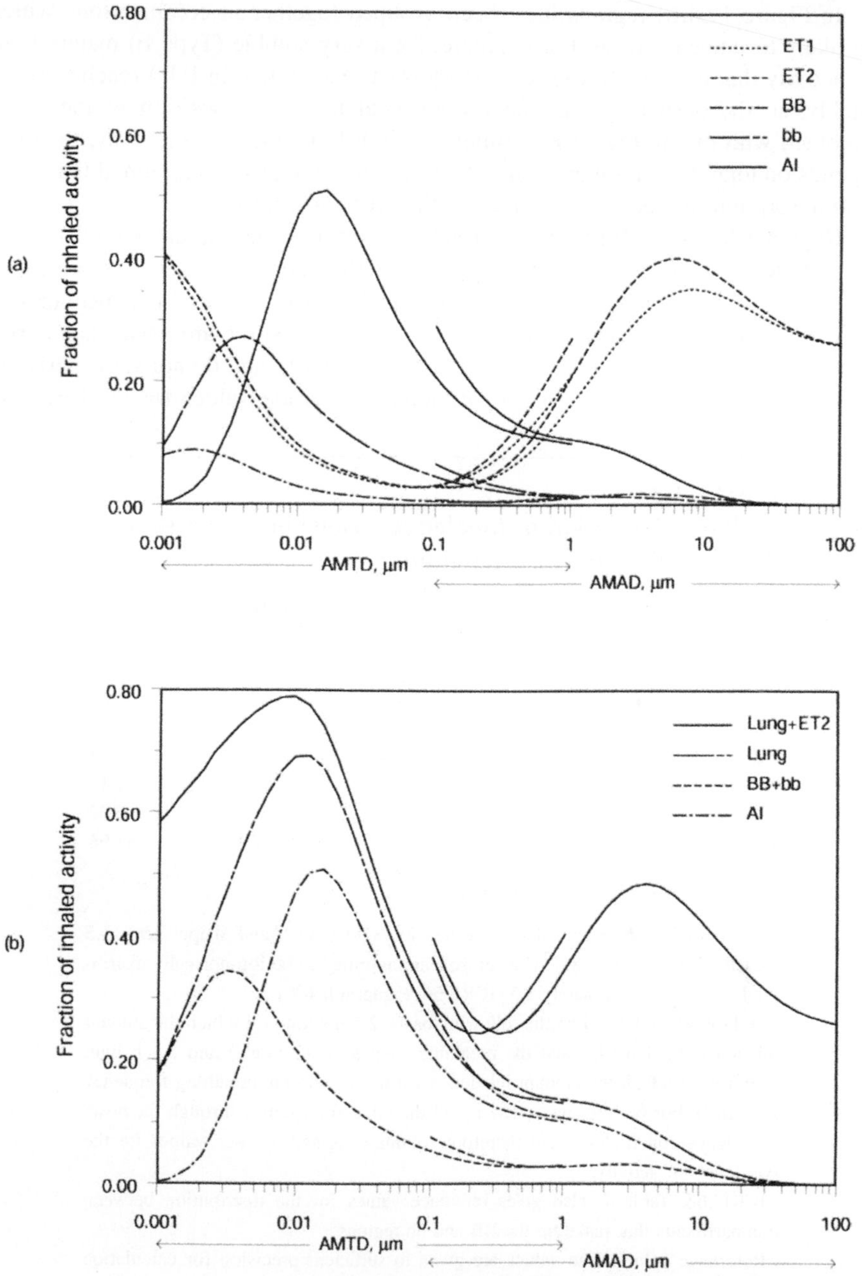

FIGURE 11.4 Deposition in (a) each respiratory tract region and (b) combinations of regions as predicted by the HRTM for a reference worker, as a function of aerosol median size. In the transition size range (0.1–1 μm) values are given for both AMAD and AMTD. (From ICRP, Guide for the practical application of the ICRP human respiratory tract model, Figure 4.3, Supporting Guidance 3, *Ann. ICRP* 32(1–2), Elsevier, Oxford, 2002.)

In Figure 11.4(b) regions have been grouped together in combinations which are likely to influence dose. For example, for a very soluble (Type F) material, all the activity that deposits in the respiratory tract (except that in ET_1) reaches blood quickly, so the dose may well be proportional to total deposition in the lungs combined with that in region ET_2 (lungs + ET_2). Expressed another way, the dose depends on total deposition excluding that in region ET_1, as it is assumed that there is no absorption to blood or clearance to the GI tract from ET_1.

Regional deposition for each age group and exercise level was calculated initially for particles of uniform size. The results were then applied to aerosols with log-normal particle size distributions, and tabulated as a function of the median size (AMAD or AMTD). In general, values of regional deposition are lower than corresponding values using the ICRP 30 model (ICRP, 1979), and do not vary markedly with age. Tables 11.3 and 11.4 give regional deposition values for workers, and

TABLE 11.3
Regional Deposition of Inhaled Aerosols[a] in a Reference Worker[b] (% of Inhaled Activity)

	AMAD	
Region[c]	1 μm	5 μm[d]
ET_1	16.52	33.85
ET_2	21.12	39.91
BB	1.24	1.78
bb	1.65	1.10
AI	10.66	5.32
Total[e]	51.19	81.96

Note: Based on ICRP 68, Table 11.1.

[a] The particles are assumed to have density 3.00 g cm^{-3}, and shape factor 1.5 (ICRP 66, Paragraph 181). The aerosols are assumed to be log-normally distributed with σ_g approximately 2.5 (ICRP 66, Paragraph 170).

[b] Light work is defined on the following basis: 2.5 h sitting (at which the amount inhaled is 0.54 m^3 h^{-1} and the breathing frequency 12 min^{-1}) and 5.5 h light exercise (at which the amount inhaled is 1.5 m^3 h^{-1} and the breathing frequency 20 min^{-1}). For both levels of activity all the inhaled air enters through the nose. The deposition fractions are therefore volume-weighted average values for the two levels of activity.

[c] ICRP 68, Table 7, also gives reference values for the distribution between compartments that make up the BB and bb regions.

[d] Reference values. The values are given to sufficient precision for calculation purposes and may be more precise than the biological data would support. Values for 1 μm AMAD are given for comparison with ICRP 30.

[e] The total depositions of 1 and 5 μm aerosols in the lung model of ICRP 30 were about 63% and 91%, respectively.

TABLE 11.4
Regional Deposition[a] of Inhaled Aerosols[b] in Reference Members of the Public[c] (% of Inhaled Activity)[d]

	Age					
Region	**3 Months**	**1 Year**	**5 Years**	**10 Years**	**15 Y (Male)**	**Adult (Male)**
ET_1	20.97	21.07	17.39	17.75	13.91	14.89
ET_2	27.20	27.30	22.32	22.86	18.13	18.97
BB	1.04	1.04	1.03	1.17	1.69	1.29
bb	2.05	1.71	1.85	1.70	2.00	1.95
AI	8.56	9.64	9.85	9.51	10.65	11.48
Total	59.82	60.76	52.44	52.99	46.38	48.58

[a] *Reference values*. The values are given to sufficient precision for calculation purposes and may be more precise than the biological data would support.
[b] The particles are assumed to have density 3.00 g cm^{-3}, and shape factor 1.5 (ICRP 66, Paragraph 181). The aerosols are assumed to be log-normally distributed with AMAD 1 μm and σ_g approximately 2.5 (ICRP 66, Paragraph 170).
[c] The distribution of time spent at each of the four reference exercise levels are as given in ICRP 71, Table 6. The deposition fractions are volume-weighted average values for deposition at the four exercise levels.
[d] Based on ICRP 71 Table 7, which also gives reference values for the distribution between compartments that make up each region.

members of the public, respectively. It is assumed by default for occupational exposure that the AMAD is 5 μm, and for environmental exposure of members of the public that the AMAD is 1 μm (ICRP, 1994a).

Clearance

The HRTM describes several routes of clearance from the respiratory tract (Figure 11.5), which involve three general processes. Material deposited in ET_1 is assumed to be removed by nose blowing and wiping. In other regions clearance results from a combination of movement of particles towards the GI tract and lymph nodes (*particle transport*), and movement of radionuclides from the respiratory tract into the blood and hence body fluids (*absorption*).

It is assumed that:

- All clearance rates are independent of age and sex.
- Particle transport rates are the same for all materials.
- Absorption into blood, which is material specific, occurs at the same rate in all regions except ET_1, where none occurs.

Fractional clearance rates vary with time but, to simplify, calculations are represented by combinations of compartments that clear at constant rates. As particle transport rates are the same for all materials, a single compartment model applies

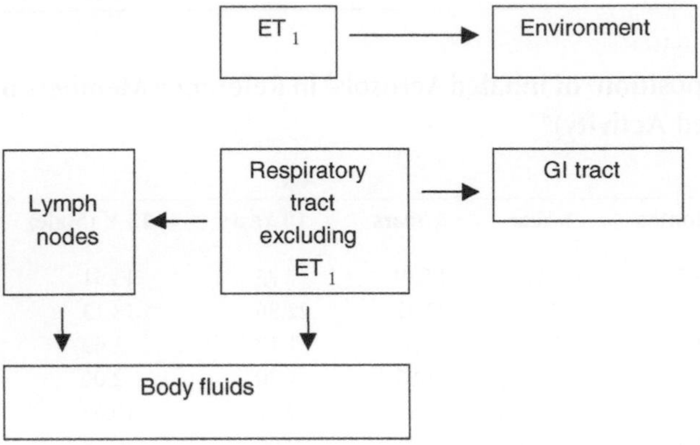

FIGURE 11.5 Routes of clearance from the respiratory tract.

to all (Figure 11.6). It was based, so far as possible, on human experimental data. It is assumed that:

- The AI deposit is divided between compartments AI_1, AI_2, and AI_3 in the ratio 0.3:0.6:0.1.
- The fraction of the deposit in BB and bb that is cleared slowly (BB_2 and bb_2) is 50% for particles of physical size < 2.5 μm and decreases with diameter > 2.5 μm.
- The fraction retained in the airway wall (BB_{seq} and bb_{seq}) is 0.7% at all sizes.
- Of material deposited in region ET_2, 0.05% is retained in its wall (ET_{seq}), and the rest in compartment ET_2' which clears rapidly to the GI tract.

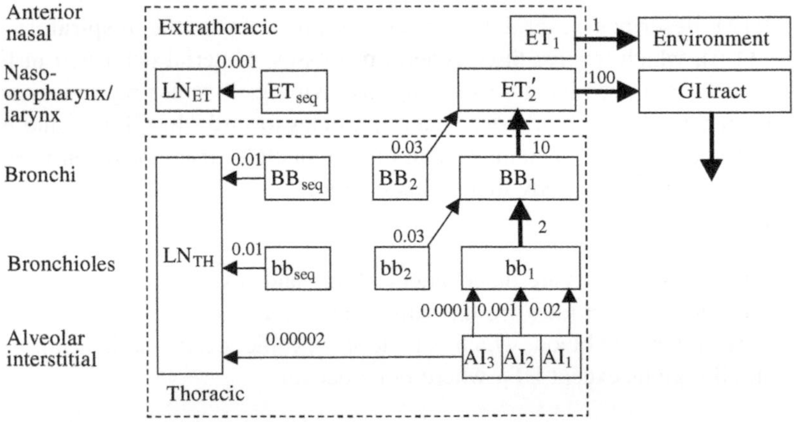

FIGURE 11.6 Compartment model representing time-dependent particle transport from each respiratory tract region. Rates shown alongside arrows are reference values in units of d^{-1}.

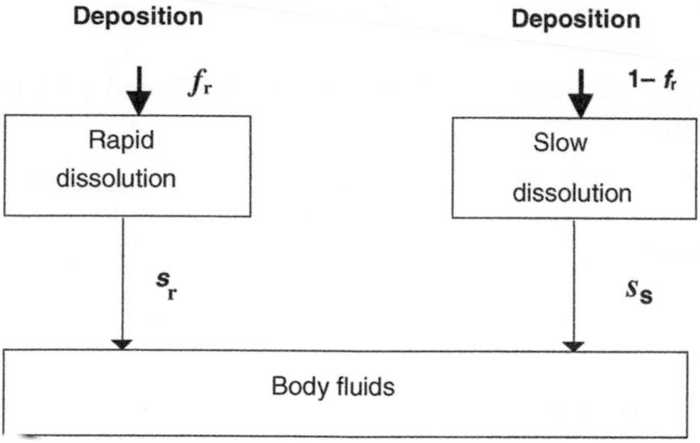

FIGURE 11.7 Compartment model representing time-dependent dissolution, followed by instantaneous uptake to body fluids. A fraction f_r of the deposit is initially assigned to the compartment labeled "rapid dissolution," and the rest $(1 - f_r)$ of the deposit is initially assigned to the compartment labelled "slow dissolution."

The model as shown in Figure 11.6 would describe the retention and clearance of a completely insoluble material. However, there is in general simultaneous absorption to body fluids of material from all the compartments except ET_1.

Absorption to blood is a two-stage process: dissociation of the particles into material that can be absorbed into blood (*dissolution*) and absorption into blood of soluble material and of material dissociated from particles (*uptake*). Both stages can be time dependent. The simplest representation of time-dependent dissolution is to assume that a fraction (f_r) dissolves relatively rapidly, at a rate s_r, and the remaining fraction $(1 - f_r)$ dissolves more slowly, at a rate s_s (Figure 11.7). Uptake to body fluids of dissolved material can usually be treated as instantaneous. In some situations, however, a significant fraction of the dissolved material is absorbed slowly. To enable this to be taken into account, the HRTM includes compartments in which activity is retained in each region in a bound state. However, it is assumed by default that uptake is instantaneous, and the bound state is not used.

It is recommended that material-specific rates of absorption should be used for compounds for which reliable experimental data exist. For other compounds, default values of parameters are recommended, according to whether the absorption is considered to be fast (Type F), moderate (M), or slow (S), corresponding broadly to inhalation Classes D, W, and Y in the ICRP 30 model (Table 11.5).

These absorption rates, expressed as approximate half-times, and the corresponding amounts of material deposited in each region that reach blood can be summarized as follows:

Type F: 100% absorbed with a half-time of 10 min. There is rapid absorption of almost all material deposited in the lungs (BB, bb, and AI), and 50% of material deposited in ET_2. The other 50% of material deposited in ET_2 is cleared to the GI tract by particle transport.

Type M: 10% absorbed with a half-time of 10 min and 90% with a half-time of 140 d. There is rapid absorption of about 10% of the deposit in BB and bb; and 5%

TABLE 11.5
Default Absorption Parameter Values for Type F, M, and S Materials (From ICRP, 1994a)[a]

Type		F(fast)	M (moderate)	S (slow)
Fraction dissolved rapidly	f_r	1	0.1	0.001
Dissolution rates:				
Rapid (d^{-1})	s_r	100	100	100
Slow (d^{-1})	s_s	—	0.005	0.0001

[a] The values of f_r, s_r, and s_s in this table are close approximations to the reference values.

of material deposited in ET_2. About 70% of the deposit in AI eventually reaches the blood.

Type S: 0.1% absorbed with a half-time of 10 minutes and 99.9% with a half-time of 7000 d. There is little absorption from ET, BB, or bb, and about 10% of the deposit in AI eventually reaches the blood.

For absorption Types F, M, and S, all the material deposited in ET_1 is removed by extrinsic means. Most of the deposited material that is not absorbed is cleared to the GI tract by particle transport. The small amounts transferred to lymph nodes continue to be absorbed into body fluids at the same rate as in the respiratory tract.

Dose Calculation

The deposition and clearance models enable the amounts of activity throughout the respiratory tract at any time after intake to be calculated. The dosimetric model enables the resulting doses to each part of the lungs to be calculated.

For dosimetric purposes, the respiratory tract is treated as two tissues: the thoracic airways (TH) and the extrathoracic airways (ET). It is considered that some regions in both TH and ET are more sensitive to radiation than others. To take account of these differences in sensitivity between tissues, the equivalent dose H_i to each region i is multiplied by a factor, A_i, representing the region's estimated radiosensitivity relative to that of the whole organ. The weighted sum is the equivalent dose to the thoracic airways (given as "lungs" in the tables of dose coefficients) or ET airways, respectively:

$$H_{TH} = H_{BB} A_{BB} + H_{bb} A_{bb} + H_{AI} A_{AI} + H_{LN_{TH}} A_{LN_{TH}} \tag{11.16}$$

$$H_{ET} = H_{ET_1} A_{ET_1} + H_{ET_2} A_{ET_2} + H_{LN_{ET}} A_{LN_{ET}} \tag{11.17}$$

The tissue weighting factor, w_T, of 0.12 specified for lungs in ICRP 60 is applied to the equivalent dose to the thoracic airways, H_{TH}, to calculate its contribution to

TABLE 11.6
Target Tissues of the Respiratory Tract (Based on ICRP 66, Table 31)

	Region	Target Cells	Assigned Fractions[a] A_i, of w_T
Extrathoracic airways	ET_1 (anterior nose)	Basal	0.001
	ET_2 (posterior nose, mouth, pharynx, larynx)	Basal	0.998
	LN_{ET} (lymphatics)	c	0.001
Thoracic airways (lungs)	BB (bronchial)	secretory (BBsec) Basal (BBbas)	0.333b
	bb (bronchiolar)	Secretory	0.333
	AI (alveolar-interstitial)	c	0.333
	LN_{TH} (lymphatics)	c	0.001

[a] Reference values, i.e., the recommended default values for use in the model. Independent of age and sex.

[b] The dose to BB, H_{BB}, is calculated as the arithmetic mean of the doses to BB_{sec} and BB_{bas}.

[c] Average dose to region calculated.

effective dose. The extrathoracic airways were included in the list of remainder tissues and organs in ICRP 68. Hence when the equivalent dose to the extrathoracic airways, H_{ET}, is higher than any other organ dose (as it often is), a w_T of 0.025 is applied to H_{ET} to calculate its contribution to effective dose.

The recommended values of A_i (the fraction of w_T assigned to each region) are given in ICRP 66, Table 31 (Table 11.6 this chapter) and are assumed to be independent of age and sex.

The dose to each respiratory tract region is taken to be the dose to the cells at risk (target cells) in that region, and is calculated as the average dose to the target tissue in that region. In the alveolar region (AI) and lymph nodes (LN_{TH} and LN_{ET}), the cells at risk are taken to be distributed throughout the region, and the average dose to the whole region is calculated. For the regions making up the conducting airways (ET_1, ET_2, BB, and bb) the target cells are considered to lie in a layer of tissue at a certain range of depths from the airway surface.

In each respiratory tract region there are also several possible sources of radiation. In bb, for example, particles in the fast phase of clearance (bb_1, Figure 11.6) are taken to be in a layer of mucus above the cilia; particles in the slow phase of clearance (bb_2) are taken to be in the fluid between the cilia; particles retained in the airway wall (bb_{seq}) are taken to be in macrophages which lie in a layer which is further from the surface than the target cells; activity bound to the epithelium is uniformly distributed in it. Account is also taken of irradiation from activity present in the alveolar region.

For each source/target combination, only a fraction of the energy emitted in the source is absorbed in the target: the absorbed fraction. ICRP 66, Annexe G provides photon absorbed fractions as a function of energy for the thoracic (TH) and extrathoracic (ET) airways as sources (and all other organs as targets) and for

TH and ET as targets (and all other organs as sources). For each respiratory tract region, and each source/target combination (see earlier description), ICRP 66, Annexe H provides absorbed fractions for nonpenetrating radiations: alpha particles, beta particles, and electrons in each case, as a function of energy. To obtain these absorbed fractions, a single cylindrical geometry was used to represent each region of the conducting airways. The representative bronchus for BB is 5 mm in diameter and the representative bronchiole for bb is 1 mm diameter (ICRP 66, paragraphs 48 and 54).

THE ICRP GI TRACT MODEL

A gastrointestinal (GI) tract model describes the processes involved in the entry of radionuclides into the body by ingestion. The GI tract model defined in ICRP 30 Part 1 (ICRP, 1979) was based on the models developed by Eve (1966) and Dolphin and Eve (1966). It is also used in ICRP 67, 68, 69, 71, 72, 78, and 88 to describe the behavior of radionuclides in the GI tract, and to calculate doses from radionuclides in the contents of the GI tract. It is used for intakes by ingestion and also for intakes by inhalation, when allowance has to be made for material passing through the GI tract after it has been cleared from the respiratory tract by particle transport.

Briefly, the GI tract is represented by four compartments, each of which clears to the next with a constant fractional rate (Figure 11.8). Material from the mouth or ET_2 enters the stomach and passes in turn to the small intestine (SI), upper large intestine (ULI), and lower large intestine (LLI), from which it is excreted in feces. The rates of transfer of material are taken to be independent of the material, and of the age and sex of the subject.

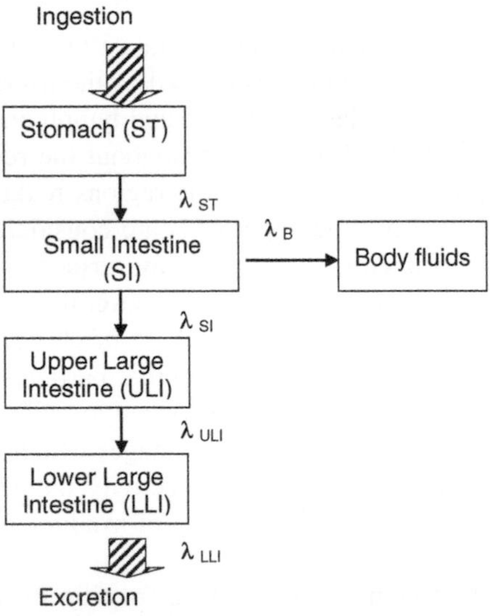

FIGURE 11.8 Mathematical model used to describe the kinetics of radionuclides in the GI tract.

Stomach

It is assumed that no absorption takes place from the stomach and that material moves to the next compartment, the small intestine, with a mean residence time of 1 h.

Small Intestine

The mean residence time is taken to be 4 h. This is the compartment from which absorption is assumed to take place. It is normal to quantify absorption by using the 'f_1 value' which is the fraction of material reaching body fluids following ingestion.

$$f_1 = \frac{\lambda_B}{\lambda_B + \lambda_{SI}} \tag{11.18}$$

λ_B = rate constant for transfer from the small intestine to body fluids.
λ_{SI} = rate constant for transfer from small intestine to upper large intestine.

Upper Large Intestine

The mean residence time is taken to be 13 h. In practice, water is absorbed from the gut content in the upper large intestine.

Lower Large Intestine

The mean residence time is taken to be 24 h.

The f_1 values adopted by ICRP for uranium (Table 11.7) depend on age and, for inhalation, on the absorption Type. The f_1 values for ingestion of uranium compounds by workers were given in Table 4 of ICRP 68 and were carried over from ICRP 30. The f_1 values for ingestion of uranium compounds by members of the public were given in ICRP 69, based on a review of the literature.

For occupational exposure, the f_1 values for inhaled materials passing through the GI tract after clearance from the respiratory tract used in ICRP 68 were mainly based on those in ICRP 30. Exceptions were made in cases where it was considered

TABLE 11.7
ICRP f_1 Values for Uranium

Age		f_1 Value 3 Months	f_1 Value 1 Year – 15 Years	f_1 Value Adult
Inhalation	Type F	0.04	0.02	0.02
	Type M	0.04	0.02	0.02
	Type S	0.02	0.002	0.002
Ingestion	Worker: UO_2, U_3O_8, most tetravalent compounds			0.002
	Worker: All other compounds			0.02
	Public	0.04	0.02	0.02

that more appropriate f_1 values for ingestion were given in ICRP 56, 67, or 69: these values were applied to both ingested and inhaled materials.

The f_1 values applied to materials inhaled by adult members of the public were assigned in ICRP 71 for 31 elements, including uranium. In assigning these values, it was considered that for environmental exposure the radionuclides might typically be present as minor constituents of the inhaled particles, and that, therefore, the absorption rate into body fluids would depend on dissolution of the particle matrix, as well as on the elemental form of the radionuclide. In ICRP 71, higher f_1 values were adopted for 3-month-old infants using the same approach as that applied to ingestion in ICRP 56, 67, and 69: for adult f_1 values of 0.002 or less, an increase by a factor of 10; for adult f_1 values between 0.01 and 0.5, an increase by a factor of 2; and for adult f_1 values greater than 0.5, complete absorption ($f_1 = 1$). For most elements (including uranium), adult values are applied to 1-, 5-, 10-, and 15-year-old children.

A new, age-dependent, model to describe the behaviour of ingested material, known as the Human Alimentary Tract Model (HATM) has been developed, but has not yet been published.

ICRP PUBLICATION 69 SYSTEMIC MODEL FOR URANIUM

The model used for uranium that enters the bloodstream and systemic tissues is described in ICRP 69 (1995a). This model describes the deposition of material from blood into various organs or regions, the transfer from region to region, the return of material to blood, and the eventual excretion of the material. In keeping with ICRP's move toward physiological realism in its models, the uranium model includes recycling, i.e., the possibility for material to pass from region to region via blood (Leggett and Eckerman, 1994). The current uranium model is thus a marked improvement on earlier models, which were simple catenary, or straight-chain models.

The model is based on a number of sources, including animal experiments (using baboons, dogs, and rats) and studies on humans. Clearly, human data are to be preferred, and in the case of uranium ICRP was fortunate in being able to draw on a number of good studies; this is not the case for many other elements. In particular, there are data from the so-called Boston Subjects, a group of terminally ill patients who were injected with uranium in the 1950s. A brief overview of the human data that support the ICRP model is given in ICRP 69. Other reviews are provided by Leggett and Harrison (1995) and Leggett (1989, 1992).

The ICRP 69 uranium model is shown in Figure 11.9. Parameter values are given in Table 11.8. The principal sites of deposition in the body of uranium are the kidneys, the liver, and bone. In addition, some material is deposited in various other tissues, generally at lower concentrations than the main sites of deposition; these are usually referred to as *soft tissues*. Of the activity in blood the model assigns 30% to soft tissues (rapid turnover, ST0); this represents a pool of activity distributed throughout the body which exchanges rapidly with blood. The remaining activity is apportioned as follows, kidneys 12%, liver 2%, bone 15%, red blood cells 1%, soft tissue (intermediate turnover, ST1) 6.7%, and soft tissue (slow turnover ST2) 0.3%, with 63% being excreted in urine via the bladder.

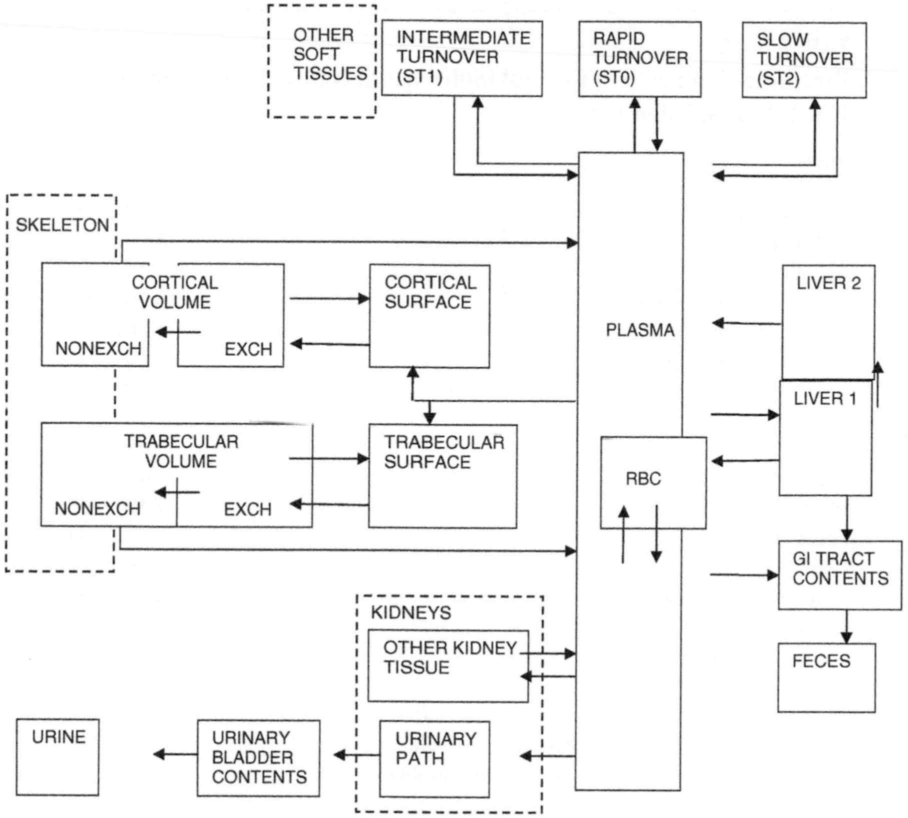

FIGURE 11.9 Diagram of the biokinetic model for uranium. (From ICRP, Age-dependent doses to members of the public from intake of radionuclides: Part 3 Ingestion dose coefficients, ICRP Publication 69. *Ann. ICRP* 25[1]. Elsevier, Oxford, 1995a.)

Some of the material initially deposited in these regions can be returned to blood whereas some is transferred to other regions of tissues. For example, material in the soft tissue compartments is returned only to blood, but material in the liver can be exchanged with blood or transferred to other regions of the liver (Liver 2 in Figure 11.9). The bone warrants additional comment. Material is initially deposited on the bone surface (either trabecular or cortical), from where it can be transferred to bone volume (exchangeable) or returned to blood. Material which does reach the exchangeable bone volume can be buried deeper in the bone volume (nonexchangeable) or returned to the surface. Material in nonexchangeable volume is transferred slowly to blood. All the pathways used in the model are illustrated in Figure 11.9. In time, most of the systemic uranium is excreted in urine via the bladder, but a small fraction is also excreted in feces.

The length of time that material remains in these regions is partly governed by a removal half-time, i.e., the time that it takes to remove half of the material present. This time varies from organ to organ; for example, the removal half-time for ST0

TABLE 11.8
Parameter Values (d^{-1}) for Biokinetic Models of Uranium (based on ICRP, 1995a, Table 5.1

Route	Transfer Rate
ST0* to plasma	8.32E + 00
ST1[a] to plasma	3.47E − 02
ST2[a] to plasma	1.90E − 05
Cortical surface to plasma	6.93E − 02
Nonexchangeable cortical volume to plasma	8.21E − 05
Trabecular surface to plasma	6.93E − 02
Nonexchangeable trabecular volume to plasma	4.93E − 04
Liver1 to plasma	9.20E − 02
Plasma to ST0	1.05E + 01
Plasma to ST1	1.63E + 00
Plasma to ST2	7.35E − 02
Plasma to cortical surface	1.63E + 00
Cortical volume to cortical surface	1.73E − 02
Cortical surf to exchangeable cortical volume	6.93E − 02
Exchangeable cortical volume to nonexchangeable cortical volume	5.78E − 03
Plasma to trabecular surface	2.04E + 00
Exchangeable trabecular volume to trabecular surface	1.73E − 02
Trabecular surface to exchangeable trabecular volume	6.93E − 02
Exchangeable trabecular volume to nonexchangeable trabecular volume	5.78E − 03
Plasma to liver1	3.67E − 01
Plasma to urinary bladder	1.543E + 01
Plasma to ULI[b]	1.22E − 01
Kidneys (other tissue) to plasma	3.80E − 04
Liver2 to plasma	1.90E − 04
Plasma to kidneys (other tissue)	1.22E − 02
Liver1 to liver2	6.93E − 03
Kidneys (urinary path) to bladder	9.90E − 02
Plasma to RBC[c]	2.45E − 01
Plasma to kidneys (urinary path)	2.94E + 00
RBC to plasma	3.47E − 01

Note: Values are given to sufficient precision for calculation purposes and may be more precise than the biological data would support.

[a] ST: soft tissue.
[b] ULI: upper large intestine.
[c] RBC: red blood corpuscles.

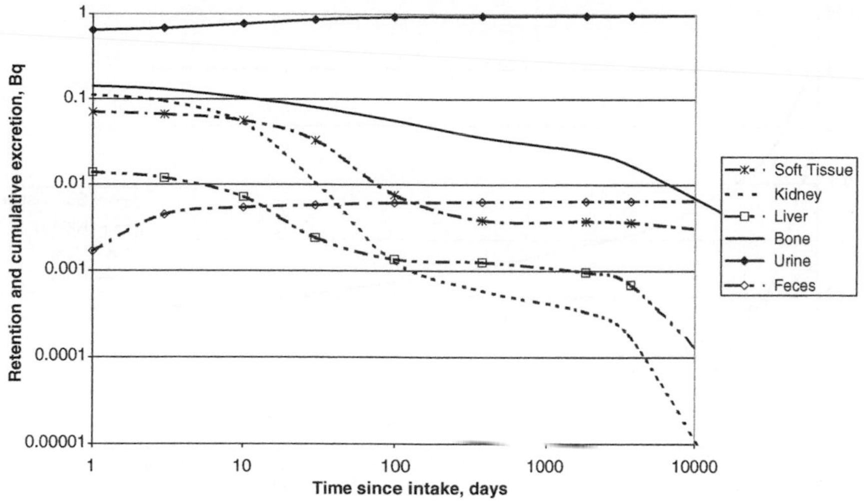

FIGURE 11.10 Predictions of the ICRP 69 systemic model for uranium (long term). The Royal Society, 2001, Figure A9.

is as little as 2 h, while for ST2 it is 100 years. The net or apparent time that is takes to halve the amount of material in an organ, however, can be very different from the removal half-time, as material is continually being redeposited by the recycling nature of the model. The net half-time thus results from a combination of removal of existing material and deposition of new material from blood. It is difficult to state values simply for net half-times. Figure 11.10 shows net retention in a number of important sites following a single input of material directly into blood. Clearly, retention in bone is more tenacious than that in kidney or liver. Table 11.9 complements Figure 11.10. It gives retention in liver, kidneys, bone

TABLE 11.9
Retention and Daily Excretion Following an Acute Uptake of 1 Bq of DU Directly into Blood (The Royal Society, 2001, Table A8)

Time (Days)	Feces (Bq per Day)	Urine (Bq per Day)	Liver	Kidneys	Bone	Soft Tissues	Whole Body
1	1.68E − 03	6.45E − 01	1.40E − 02	1.12E − 01	1.43E − 01	7.06E − 02	3.53E − 01
3	9.47E − 04	1.80E − 02	1.20E − 02	9.48E − 02	1.31E − 01	6.65E − 02	3.10E − 01
10	3.73E − 05	9.43E − 03	7.21E − 03	5.27E − 02	1.04E − 01	5.63E − 02	2.22E − 01
30	1.11E − 05	2.39E − 03	2.41E − 03	1.08E − 02	8.08E − 02	3.30E − 02	1.27E − 01
100	2.30E − 06	3.51E − 04	1.36E − 03	1.26E − 03	5.61E − 02	7.60E − 03	6.63E − 02
1000	5.37E − 08	8.09E − 06	1.12E − 03	4.51E − 04	2.95E − 02	3.79E − 03	3.49E − 02
10,000	5.41E − 09	8.15E − 07	2.13E − 04	1.80E − 05	8.03E − 03	3.26E − 03	1.15E − 02

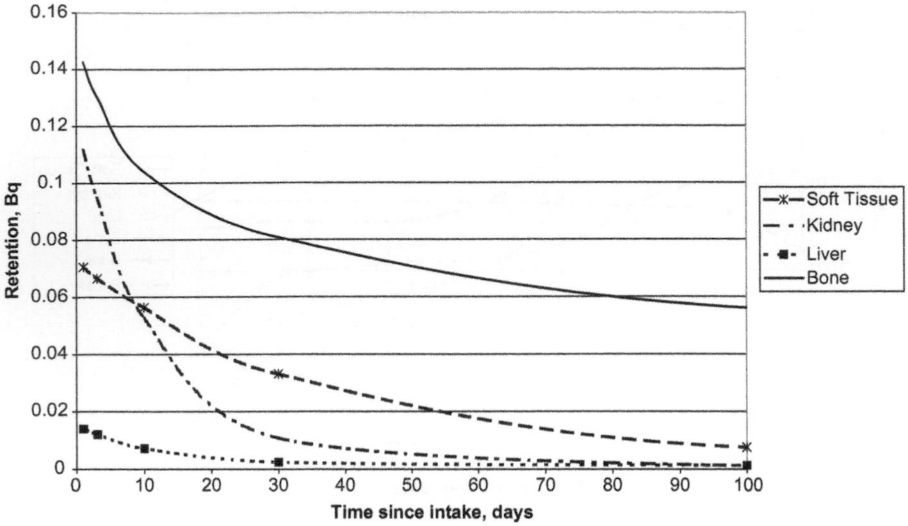

FIGURE 11.11 Predictions of the ICRP 69 systemic model for uranium (short term). The Royal Society, 2001, Figure A10.

(comprising the six skeleton compartments of Figure 11.9) and the whole body at a number of times after an acute intake directly into blood. It also gives the amount of activity excreted in urine and feces per day.

Approximate net retention half-times can be estimated directly from Figure 11.11 which shows the early section of Figure 11.10 on an expanded scale. For bone, the time taken for the activity after 1 d to fall to half of its original value is about 50 d, whereas for kidney it is 10 d.

APPLICATION OF THE HRTM TO DEPLETED URANIUM

Although the ICRP biokinetic models were developed primarily for assessment of doses to workers in industrial or laboratory situations, and of members of the public following releases of radioactive materials to the environment, they can in principle be applied to other situations, such as battlefield exposures to depleted uranium (DU).

As noted in Section 2 above, although ICRP provides reference values for all parameters used in its biokinetic models, it recognizes the desirability of using material-specific values when appropriate. This applies both to the inhaled (or ingested) material and to the exposed subject. Derivation of specific parameter values is considered in detail in both recent major assessments of DU hazards (The Royal Society, 2001, 2002; Guilmette et al., 2004). With regard to the material, the important parameter values are those that define the size distribution (for inhaled particles) and those that define absorption to blood in the HRTM and GI Tract model, as these clearly depend on the physical and chemical form of the material entering the body. Clearance by particle transport from the respiratory tract and transit through the GI tract are considered to be independent of the material, and the systemic model for uranium is taken to be independent of the chemical form of the uranium that entered the body.

Subject-Related Parameter Values

Parameter values relating to the subject are determined by the exposure scenario under consideration. The ICRP reference parameter values for a worker assume a healthy young adult male at light work, which is a defined combination of sitting and light exercise (ICRP, 1994a, 1994b). ICRP reference parameter values for members of the public assume defined combinations of sleep, sitting, and light and heavy exercise (ICRP, 1995b). It is relatively straightforward to adopt a more realistic combination, if known (ICRP, 2002), that will affect the breathing rate and, hence, both the intake, and the division of total deposition between regions. Thus, assessments of exposures immediately following impact of a penetrator (The Royal Society, 2001, 2002; Guilmette et al., 2004) assume more vigorous (heavy) exercise for those in the struck vehicle and first responders, at least in the first few minutes after impact.

Aerosol Size Distributions

Measurements of DU aerosol size distributions made following the impact of DU penetrators on armor plate are summarized in Table 11.10. In all cases cascade impactors (CI) were used, and results were expressed as mass (or activity) median aerodynamic diameter, MMAD or AMAD, and geometric standard deviation, GSD. Table 11.10 is based on a review carried out in support of the Royal Society assessment, but updated in the light of later French and U.S. studies (as are Table 11.11 to Table 11.14, which relate to dissolution rates and chemical forms). The Royal Society assessment concluded that for initial exposure near a target it is reasonable to take MMAD ~2 μm, with a large GSD (~10) but, at later times or further away, to take lower values, i.e., MMAD 1 μm, with GSD ~2.5, the HRTM defaults for environmental exposure. The AMADs measured by Chazel et al. (2003) are consistent with that, although the GSDs are smaller close to the impact. The far more comprehensive results from the Capstone Study (Parkhurst et al., 2004a, 2004b) are also broadly consistent, but provide more information about the change in size distribution with time, and about variability. They confirm the most important overall observation, namely, that most of the aerosol, at least after the first minute or so, is readily respirable, with an AMAD of about 1 μm.

Absorption from the Respiratory Tract to Blood

There are two routes by which information can be used to assess absorption parameter values for relevant forms of DU:

- Direct measurements on DU formed from penetrator impacts or in fires
- Determination of the chemical forms of uranium produced, combined with information relating to those forms.

TABLE 11.10
Aerosol Size Distributions from DU Penetrator Impacts[a]

Report	MMAD, μm	GSD	"Respirable Fraction" (%)[b]
	Reports Obtained		
Hanson et al., 1974	2.1 – 3.3 (Entrance chamber)	1.8 – 3.3	42 – 64
	2.4 – 4.2 (Exit chamber)	1.8 – 3.1	
Glissmeyer and Mishima, 1979	0.8 – 3.1	1.6 – 18	51 – 70
Patrick and Cornette, 1978	c	c	
Chambers et al., 1982	1.6 (1.4 – 2.0)	13 (12 – 17)	~70
Brown, 2000	3.7 (1.1 – 7.5) inside	3.5 (2.8 – 4.2)	
	1.8 (1.3 – 2.7) outside	4.1 (3.9 – 4.5)	
Chazel et al., 2003	1.05 (glacis shot) outside	3.7	
	2 (turret shot) outside	2.5	
Capstone: Parkhurst et al., 2004b[d]	BFV: first 10 sec: 0.6 – 4		
	BFV: after 10 sec: 0.4 – 4		
	Abrams: first 10 sec: 0.2 – 8		
	Abrams: after 10 sec: 0.3 – 7		
	Abrams DU armor: first 10 sec, 0.8 – 8; after 10 sec, 0.1 – 5		
	Reports (Restricted Distribution) Not Obtained[e]		
Gilchrist et al., 1979	2.1 (High volume, preferred)		
	5.8 (Low volume)		

[a] Partly based on Table G2 of The Royal Society, 2001.

[b] Here the term *respirable fraction* is used to mean the fraction of the airborne material that is small enough to be readily resuspended and inhaled, i.e., less than about 10 μm d_{ae}, and not, as usually defined for occupational health purposes, to mean the fraction of the aerosol that if inhaled could reach the alveolar region, i.e., the deep lungs. However, different definitions are used in different reports.

[c] Size distribution not measured, but a qualitative statement is made that a very wide range size was observed: from fragments >50 μm to submicron.

[d] No concise summary was found in the Capstone Report. These ranges are based on Attachment 1, Table 6.40, which gives results individually for each type of shot and time. BFV refers to Bradley Fighting Vehicle.

[e] A number of reports relating to DU hazards are cited in documents published by U.S. Government related sources, but are restricted in distribution, and so were not available to the authors. Summaries of several important restricted documents are given in OSAGWI, 2000 Tab L and provided the information given here.

TABLE 11.11
Dissolution Characteristics of Material Formed from DU Penetrator Impacts[a]

Report	Fraction Dissolved Rapidly (%)	Dissolution Rate of the Rapid Fraction, d^{-1}	Dissolution Rate of the Slow Fraction, d^{-1}	Duration of Measurements, d
		Reports Obtained		
Glissmeyer and Mishima, 1979	43 (34 – 49) respirable	—	<0.01	28
	15 (11 – 18) total		<0.01	
Scripsick et al., 1985a,b	25 (air filter) respirable	1.7	0.0014	~30
	4 (core sample) respirable	4.7	0.004	
Chazel et al., 2003	47 (glacis)	0.06	0.00018	30
	57 (turret)	0.07	0.00034	
Mitchell and Sunder, 2004	~5	~1	—	7
Capstone: Parkhurst et al., 2004b	1 – 28	0.1 – 30	0.0004 – 0.0095	46
		Reports (Restricted Distribution) Not Obtained[a]		
Jette et al., 1990	24 – 43 "Class D"	—	—	?
Parkhurst et al., 1990	17 "soluble"	—	—	?

[a] Partly based on Table G3 of The Royal Society, 2001.
[b] OSAGWI 2000, Table L ; see also footnote e to Table 11.10.

For use with the HRTM, it is necessary to estimate values of three parameters:

- The fraction that dissolves rapidly, f_r
- The dissolution rate of the rapid fraction, s_r d^{-1}
- The dissolution rate of the slow fraction, s_s d^{-1}

These are easily obtained from *in vitro* tests where the undissolved fraction is expressed as a two-component exponential function. However, determination from the results of *in vivo* studies requires application of an appropriate model to assess values from the results, which will typically be of excretion rates and organ contents at a limited number of times after administration.

TABLE 11.12
Dissolution Characteristics of Material Formed from Combustion of Uranium[a]

Report	Fraction Dissolved Rapidly (%)	Dissolution Rate of the Rapid Fraction, d^{-1}	Dissolution Rate of the Slow Fraction, d^{-1}	Duration of Measurements, d
		Reports Obtained		
Mishima et al., 1985	0.5 (<10 μm d_{ae})	—	0.0005	60
Scripsick et al., 1985a	6 – 10 (respirable)	0.4 – 10	0.0017 – 0.0034	~30
		Reports (Restricted Distribution) Not Obtained (OSAGWI 2000, Tab L)*		
Haggard et al., 1986	4 ("within 10 days")	—	—	?
Parkhurst et al., 1990	6.8 ("slightly soluble")	—	—	?

[a] Partly based on Table H4 of The Royal Society, 2001.

[b] See footnote e to Table 11.10.

TABLE 11.13
Chemical Composition of Material Formed from DU Penetrator Impacts[a]

Report	Amorphous (%)	UO_2 (%)	U_3O_8 (%)
Glissmeyer and Mishima, 1979		25	75
Patrick and Cornette, 1978	b	b	b
Scripsick et al., 1985(a,b)	—	60 (air filter, total)	40
	20	18 (air filter, respirable)	62
	—	97 (core sample, total)	3
	—	54 (core sample, respirable)	46
Chazel et al., 2003			30–40[c]
Mitchell and Sunder, 2004		9	44[d]

[a] Partly based on Table G4 of The Royal Society, 2001.

[b] Qualitative: air samples mainly U, Fe. Soil also Si, Al, and W.

[c] Also 25–40% U_4O_9 and 20% UO_3.

[d] Also 47% U_3O_7.

TABLE 11.14
Chemical Composition of Material Formed from Combustion of Uranium

Report	UO_2 (%)	U_3O_8 (%)
	Reports Obtained	
Elder and Tinkle, 1980	0.2 – 4	96 – 99.8
Scripsick et al., 1985a	<0.02 – 1.3 (respirable)	98.7 – >99.98
	Reports Not Obtained (OSAGWI 2000, Tab L)[b]	
Haggard et al., 1986	"Predominantly"	
Parkhurst et al., 1999	~100%	

[a] Partly based on Table H5 of the Royal Society, 2001.
[b] See footnote e to Table 11.10.

Direct Measurements of Dissolution of DU Formed from Penetrator Impacts

In several studies, measurements have been made of the rate of dissolution of particles formed from the impact of a DU penetrator on armour plate, in a medium designed to be a simulant of the fluid present in the lungs, with which the particle might be in contact after inhalation. Such *in vitro* dissolution tests have a number of advantages over *in vivo* studies, in which particles are deposited in the lungs of laboratory animals, including cost and ease of interpretation, and so are more frequently used. However, dissolution rates, like chemical reaction rates in general, are potentially very sensitive to conditions, and therefore great care is needed to be reasonably sure that the results are representative of dissolution in the human lungs. The advantages and disadvantages of *in vitro* and *in vivo* methods to determine dissolution *in vivo* are discussed in ICRP (2002).

Results are summarized in Table 11.11. In a recent study by Mitchell and Sunder (2004) material was administered to rats. However, only rough estimates of f_r and s_s can be made from the results. Although it has the merit of being an *in vivo* study, it is of limited value for risk assessment because of factors including lack of information about the material, its large particle size, and the short duration of measurements.

All other measurements of DU penetrator impact aerosols were made *in vitro* and used broadly similar procedures. Again, by far the most comprehensive results come from the Capstone Study (Parkhurst et al., 2004b). There, dissolution in simulated lung fluid was measured for 46 d on 27 samples. Most of these were obtained using a cascade cyclone, or *cyclone train*: a series of cyclones, which collected progressively smaller aerodynamically sized fractions. Time-dependent retention of undissolved DU was fitted by two- and/or three-component exponential functions. Based on the two-component fits, values of the rapid fraction, f_r, ranged from 1% to 28%, broadly similar, but somewhat lower in range than in the previous studies (4–57%, Table 11.10). Values of the slow dissolution rate, s_s, ranged from

0.0004 to 0.0095 d^{-1}, again broadly similar, but somewhat higher in range than the previous studies (0.0002 to 0.004 d^{-1}, Table 11.11). Although comprehensive, the results have some limitations. The dissolution tests ran for 46 days, at which time most of the material remained undissolved, and so the results have to be extrapolated in time, as well as extrapolated from the *in vitro* system used to the human lungs.

Direct Measurements of Dissolution of DU Formed in Fires

There have been a number of studies relating to the effects of fire on DU munitions and penetrators in order to address concerns about fires during storage or transport. In several, measurements were made of dissolution *in vitro* which, as noted above, should be treated with caution. Results are summarized in Table 11.12. The rapid fraction is typically a few percent, and the slow rate of the order of 0.001 d^{-1}, both broadly similar to the results for penetrator impact aerosols.

Chemical Composition

X-ray analysis was used in several studies to identify the oxides present and to attempt to quantify the proportions. Results are summarized in Table 11.13 and Table 11.14, respectively, for DU produced from penetrator impacts, and from combustion. Studies of penetrator impacts generally indicate that most of the crystalline uranium oxide is present as UO_2 or U_3O_8 or intermediates (U_3O_7 and U_4O_9). However, there is variation in the oxides chosen (Table 11.11), perhaps reflecting the difficulty in distinguishing between some, as noted in the Capstone Report. Chazel et al., like the Capstone Study, report U_4O_9 to be an important constituent. Glissmeyer and Mishima (1979) noted that the proportion of U_3O_8 may increase with decreasing size. This is consistent with the Capstone report observation that the proportion of U_3O_8/UO_3 (which could not be distinguished) increased with decreasing size. Following combustion most of the uranium is reported to be present as U_3O_8, with a small amount of UO_2 (Table 11.14).

Dissolution of Uranium Oxides

As described in the previous section, following the impact of a DU penetrator, or combustion of DU objects in a fire, any DU inhaled is likely to be in oxide form, predominantly UO_2, U_3O_8, or intermediates (U_3O_7 and U_4O_9). The processing of uranium into reactor fuel elements may involve one or more oxide forms and, as a result, there have been many *in vivo* studies of the biokinetics of uranium following the deposition of the various oxides in the lungs (usually of rats) and *in vitro* studies of the dissolution of uranium oxides under conditions intended to simulate lung deposition.

As noted above, ICRP 71 (ICRP, 1995b) included a brief review of information relating to inhalation of different chemical forms, updating the reviews in ICRP 30 but with emphasis on environmental exposure. ICRP 71 also introduced criteria for assigning compounds to the three HRTM default absorption types (F, M, or S) on the basis of experimental data. In discussing the behavior of each compound, consideration was given to the assignment to the appropriate absorption type. With respect to uranium oxides it made the following observations:

> Uranium trioxide, ammonium diuranate (ADU) and uranium octoxide are found in various hydration states alone or more often mixed in various proportions in industrial processes. The human data from accidental intakes (West et al., 1979; Eidson, 1990), and from monitoring data in workers from processing facilities (Barber and Forrest, 1995), the many animal studies in rats, dogs and monkeys (Morrow et al., 1972; Eidson and Damon, 1985a,b; Stradling et al., 1985; Métivier et al., 1992), and extensive *in vitro* studies (Mansur, 1988; Hengé-Napoli et al., 1989) show that the behavior depends on particular processes but, in most cases, is consistent with assignment to Type M, although pure UO_3 would be assigned to Type F. Considerable variation in the behavior of U_3O_8 was observed, with some studies indicating Type M behavior, and others, Type S. Human studies have shown that UO_2 can be very insoluble (Pomroy and Noel, 1981; Price, 1989). Experiments in rats, dogs, monkeys and baboons (Leach et al., 1973; Stradling et al., 1989; Métivier et al., 1992) also support the assignment of UO_2 to Type S.

In more recent studies, efforts have been made to derive specific values of HRTM absorption parameters. In doing so consideration has to be given to the extent of binding of dissolved uranium to lung tissues (Section 2.4), but experimental evidence suggests that this is unimportant. Cooper et al. (1982) and Ellender (1987) followed the behavior of ^{233}U after instillation of uranyl nitrate and bicarbonate into the pulmonary (AI) region of the lungs of rats. Cooper et al. (1982) found that less than 2% of the initial lung deposit (ILD) remained at 7 d. Ellender (1987) found that about 3% remained at 30 d. Detailed analysis, however, indicated that clearance over this period was mainly by particle transport, and that the results did not provide evidence for binding of uranium (Hodgson et al., 2000).

Hodgson et al. (2000) derived HRTM absorption parameter values for a number of uranium compounds produced during the manufacture of nuclear fuel, using the results of previously published experiments. Values for uranium oxides are given in Table 11.15. Ansoborlo et al. (2002) compiled HRTM parameter values for uranium compounds handled during nuclear fuel fabrication in France. *In vivo* results for pure oxides are also given in Table 11.15. For each compound there is considerable variation, reflecting differences in methodology, and in the physicochemical properties of the materials. However, there is a marked distinction between the relatively soluble UO_4 and UO_3, for which f_r, the rapidly-dissolving fraction is more than 50%, and the relatively insoluble U_3O_8 and UO_2, for which f_r is < 10% and for UO_2 about 1%.

Choice of Absorption Parameter Values for Assessments of DU Exposures

Different approaches were taken in the two major recent assessments to estimating appropriate HRTM parameter values for assessing the behavior of DU deposited in the respiratory tract and, hence, organ doses and maximum kidney uranium concentrations.

The Royal Society (2001) assessment considered a wide range of battlefield exposures to DU aerosols originating from the impact of DU penetrators on armor and from DU involved in fires. It aimed at obtaining "central" estimates of doses

TABLE 11.15
Summary of Absorption Parameter Values for Uranium Oxides

	Absorption Parameters			
Compound	f_r	s_r, d^{-1}	s_s, d^{-1}	**Reference**
UO_4 (n = 4)	0.87	0.93	0.024	Ansoborlo et al., 2002
UO_3	0.75	14	0.02	Bailey et al., 1998
UO_3	0.92	1.4	0.0036	Hodgson et al., 2000
UO_3	0.71	0.28	0.0011	Ansoborlo et al., 2002
U_3O_8	0.044	0.49	0.00035	Hodgson et al., 2000
U_3O_8	0.046	2.3	0.0012	Ansoborlo et al., 2002
U_3O_8	0.03	2.1	0.00038	Ansoborlo et al., 2002
UO_2 — nonceramic	0.011	0.95	0.00061	Hodgson et al., 2000
UO_2 — ceramic	0.008	1.3	0.00026	Hodgson et al., 2000
UO_2	0.03	1.3	0.0015	Ansoborlo et al., 2002
UO_2	0.01	nd[a]	0.00049	Ansoborlo et al., 2002
UO_2	0.01	nd[a]	0.00058	Ansoborlo et al., 2002
Defaults (ICRP 68)				
Type F	1	100	—	
Type M	0.1	100	5.0×10^3	
Type S	0.001	100	1.0×10^4	

[a] Nd = not determined.

and kidney concentrations, which would be typical for those exposed under a given set of conditions, and "worst cases," which it was unlikely that any individual would exceed. However, the worst case for radiation arises when absorption is slowest, giving longest retention in the lungs, whereas the worst case for maximum kidney concentration arises when absorption is most rapid, giving high uptake to blood and subsequent deposition in the kidneys. Hence, two worst-case sets of parameter values were derived for each exposure scenario considered. In each case, the values of f_r were based on the information then available from *in vitro* studies (Table 11.11 and Table 11.12). These gave a central value of 0.3 (range 0.1–0.5) for aerosols formed from impacts and a central value of 0.05 (range 0.05–0.1) for aerosols formed by combustion. Results of *in vivo* experiments on U_3O_8 and UO_2 (Table 11.15) were used to assess the central values of the rapid dissolution rate s_r (1 d^{-1}) and the slow dissolution rate s_s (0.001 d^{-1}) for both types of aerosol. The range of values taken for s_r (0.4–14 d^{-1}) was based on the range observed for uranium oxides (Table 11.15). The range of values taken for s_s was from 0.0001 d^{-1} (as for default Type S) to 0.0015 d^{-1} (the highest value for U_3O_8 or UO_2 in Table 11.15). However, a number of limitations of the data in Table 11.15 were recognized:

- It is difficult to determine dissolution rates less than about 0.001 d^{-1} in such experiments, and so there will be considerable errors on the values of s_s.
- The studies were conducted on rats: it is assumed that the same rates apply to man.

- As noted above, s_s depends on the physical and chemical form of the material, its mode of formation, and history before inhalation. There are some obvious differences between the DU aerosols formed in penetrator impacts and/or fires and the industrial uranium oxides studied *in vivo*:
  - The DU is not pure uranium but typically contains 0.75% titanium.
  - In impacts, in particular, the oxide may be a mixture of uranium with other metals from the target, notably iron.

The Capstone Human Health Risk Assessment (HHRA, Guilmette et al., 2004) considered only aerosols formed within a vehicle struck by a large caliber DU penetrator and only used information derived from the Capstone Aerosol Study (Parkhurst et al., 2004a, 2004b). As noted above, dissolution in simulated lung fluid was measured for 46 d on 27 samples, many of which were fractionated by size. Time-dependent retention of undissolved DU was fit by two- and/or three-component exponential functions. Based on the two-component fits, values of the rapid fraction, f_r, were in the range 0.01–0.28, broadly similar to, but somewhat lower than, the range adopted by the Royal Society assessment. Values of the slow dissolution rate, s_s, were in the range 0.0004–0.0095 d^{-1}, again broadly similar to, but in this case somewhat higher than, the range adopted by the Royal Society assessment. The HHRA incorporated the results (taking the more detailed three-component fits) into a probabilistic assessment that derived distributions of doses for each exposure scenario, using the results of the cascade impactor measurements to give distributions of DU air concentrations as functions of particle size and time after impact. Appropriate sets of absorption parameter values (according to the scenario and size) were assigned to each size fraction.

SUMMARY

The ICRP internal dosimetry models provide a sophisticated, internationally recognized, system for prospective and retrospective assessment of organ concentrations and doses from intakes of radionuclides. Because of the extensive use of uranium for the production of nuclear reactor fuel over the last 60 years, many studies have been carried out of its behavior in the body following intake, especially by inhalation. Thus, there is a large amount of information available on which appropriate parameter values for the models can be based. The current ICRP models have been applied in recent major assessments of the health hazards arising from the use of DU munitions. A number of factors specific to such exposures, and which differ from the workplace exposures mainly considered in the past, have been identified and addressed. The impact of a penetrator can result in generation of an aerosol with a very high concentration and a very broad size distribution, because several different mechanisms are involved. These mechanisms, together with the presence of titanium in the DU, and other metals derived from the target vehicle, result in aerosol particles consisting of a complex mixture of physical and chemical forms, although predominantly uranium oxides. Studies on aerosols formed following DU penetrator impacts, notably the recent comprehensive Capstone Study, provide a great deal of directly relevant information, although even the Capstone Study had limitations in scope.

The ICRP models are themselves subject to revision, and a major review is currently under way to produce a series of documents titled "Occupational Intakes of Radionuclides," which will replace those currently used (ICRP Publications 68 and 78). Although changes to the models that would have a large impact on recent major DU assessments are not anticipated, there will inevitably be some changes within the next few years.

REFERENCES

Ansoborlo, E., Chazel, V., Hengé Napoli, M.H., Pihet, P., Rannou, A., Bailey, M.R. and Stradling, N., Determination of the physical and chemical properties, biokinetics and dose coefficients of uranium compounds handled during nuclear fuel fabrication in France, *Health Phys.* 82, 279, 2002.

Bailey, M.R., The new ICRP model for the respiratory tract, *Radiat. Prot. Dosimetry* 53, 107, 1994.

Bailey, M.R., Summaries of Source Documents Relating to Depleted Uranium Penetrator Impacts, Annexe G to the Health Hazards of Depleted Uranium Munitions Part I. Policy Document 6/01, London, U.K., Online report available at www.royalsoc.ac.uk/du 2002a.

Bailey, M.R., Summaries of Source Documents Relating to Combustion of Depleted Uranium in Fires, Annexe H to the Health Hazards of Depleted Uranium Munitions Part I. Policy Document 6/01, London, U.K., online report available at www.royalsoc.ac.uk/du 2002b.

Bailey, M.R. and Phipps, A.W., Current ICRP Models Used to Assess Intakes of Uranium. Annexe A to the Health Hazards of Depleted Uranium Munitions Part I. Policy Document 6/ 01, London, U.K., online report available at www.royalsoc.ac.uk/du 2002.

Bailey, M.R., Guilmette, R.A., Jarvis, N.S., and Roy M., Practical application of the new ICRP human respiratory tract model, *Radiat. Prot. Dosimetry* 79, 17, 1998.

Bailey, M.R., Ansoborlo, E., Guilmette, R.A., and Paquet F., Practical application of the ICRP human respiratory tract model, *Radiat. Prot Dosimetry* 105, 71, 2003.

Barber, J.M. and Forrest, R.D., A study of uranium lung clearance at a uranium processing plant, *Health Phys.* 68, 661, 1995.

Brown, R., Depleted Uranium Munitions and Assessment of the Potential Hazards, Notes of a presentation to the Royal Society Working Group on January 19, 2000.

Chalabreysse, J., Etude et résultats d'examens effectués à la suite d'une inhalation de composés dits solubles d'uranium naturels, *Radioprotection* 5, 1, and 305, 1970.

Chambers, D.R., Markland, R.A., Clary, M.K., and Bowman R.I., Aerosolization Characteristics of Hard Impact Testing of Depleted Uranium Penetrators, Technical Report ARBRL-TR-023435, Aberdeen Proving Ground, MD: Ballistic Research Laboratory, October 1982.

Chazel, V., Gerasimo, P., Dabouis, V., Laroche, P., and Paquet F., Characterisation and dissolution of depleted uranium aerosols produced during impacts of kinetic energy penetrators against a tank, *Radiat. Prot. Dosimetry,* 105(1–4), 163, 2003.

Cooper, J.R., Stradling, G.N., Smith, H., and Ham, S.E., The behavior of uranium-233 oxide and uranyl-233 nitrate in rats, *Int. J. Radiat. Biol.* 41(4), 421, 1982.

Dolphin, G.W. and Eve, I.S., Dosimetry of the gastrointestinal tract, *Health Phys.* 12, 163, 1966.

EC (European Commission) Council Directive 96/29/EURATOM of May 13, 1996, Laying Down the Basic Safety Standards for the Protection of the Health of Workers and the General Public against the Dangers Arising from Ionizing Radiation.

Eidson, A.F., Biological Characterisation of Radiation Exposure and Dose Estimates for Inhaled Uranium Milling Effluents. NUREG/CR–5489 TI90 012914, U.S. Nuclear Regulatory Commission, 1990.

Eidson, A.F. and Damon, E.G., Biologically significant properties of refined uranium ore, in *Int. Conf. on Occupational Radiation Safety in Mining,* Stocker, H., Ed., Canadian Nuclear Association, Ontario, Canada, 1, 248, 1985a.

Eidson, A.F. and Damon, E.G., Comparison of uranium retention in dogs exposed by inhalation to two forms of yellowcake, in *Int. Conf. on Occupational Radiation Safety in Mining,* Stocker, H., Ed., Canadian Nuclear Association, Ontario, Canada, 1, 261, 1985b.

Ellender, M., The clearance of uranium after deposition of the nitrate and bicarbonate in different regions of the rat lung. *Human Toxicol.* 6, 479, 1987.

Elder, J.C. and Tinkle, M.C., Oxidation of Depleted Uranium Penetrators and Aerosol Dispersal at High Temperatures, Report LA-8610-MS, Los Alamos National Laboratory. http://lib-www.lanl.gov/la-pubs/00313603.pdf, 1980.

Eve, I.S., A review of the physiology of the gastrointestinal tract in relation to radiation doses radioactive materials. *Health Phys.* 12, 131, 1966.

Gilchrist, R.L., Nickola P.W., Glissmeyer, J.A., and Mishima, J., Characterisation of Airborne Depleted Uranium from April 1978 Test Firings of the 105 mm, APFSDS-T, M735E1 Cartridge, Report PNL-2881, Battelle Pacific Northwest Laboratory, Richland, WA, 1979 (initial release) June 1999 (publication date). (Restricted circulation; summary #11 in OSAGWI, 2000, Tab L).

Glissmeyer, J.A. and Mishima, J., Characterization of Airborne Uranium from Test Firings of XM774 Ammunition, Report PNL-2944, Pacific Northwest Laboratory, Richland, WA, 1979.

Guilmette, R.A., Parkhurst, M.A., Miller, G., Hahn, F.F., Roszell, L.E., Daxon, E.G., Little, T.T., Whicker, J.J., Cheng, Y.S., Traub, R.J., Lodde, G.M., Szrom, F., Bihl, D.E., Creek, K.L., and McKee, C.B., Human Health Risk Assessment of Capstone Depleted Uranium Aerosols. Attachment 3 of Depleted Uranium Aerosol Doses and Risks: Summary of U.S. Assessments, PNWD-3442. Prepared for the U.S. Army by Battelle under Chemical and Biological Defense Information Analysis Center Task 241, DO 0189, Aberdeen, MD, 2004.

Haggard, D.L., Hooker, C.D., Parkhurst, M.A., Sigalla, L.A., Herrington, W.M., Mishima, J., Scherpelz, R.I., and Hadlock, D., Hazard Classification Test of 120 mm APFSDS-T, M829 Cartridge: Metal Shipping Container, Report PNL-5928, Pacific Northwest Laboratory, Richland, WA, 1986. (Restricted circulation; summary #21 in OSAUWI, 2000, Tab L)

Hanson, W.C., Elder, J.C., Ettinger, H.J., Hantel, L.W., and Owens J.W., Particle Size Distribution of Fragments from Depleted Uranium Penetrators Fired against Armor Plate Targets, Report LA-5654, Los Alamos Scientific Laboratory, Los Alamos, 1974.

Hengé-Napoli, M.H., Rongier, E., Ansoborlo, E., Chalabreysse, J., Comparison of the in vitro and in vivo dissolution rates of two diuranates and research on an early urinary indicator of renal failure in humans and animals poisoned with uranium. *Radiat. Prot. Dosimetry* 26(1–4), 113, 1989.

Hinds, W.C., *Aerosol Technology: Properties, Behavior and Measurement of Airborne Particles,* John Wiley & Sons, New York, 1982.

Hodgson, A., Moody, J.C., Stradling, G.N., Bailey, M.R., and Birchall A., Application of the ICRP Respiratory Tract Model to Uranium Compounds Produced during the Manufacture of Nuclear Fuel, Report NRPB-M1156, Chilton, 2000.

Hursh, J.B. and Spoor, N.L., Data on man, in *Uranium, Plutonium, Transplutonic Elements,* Hodge, H.C., Stannard, J.N., and Hursh, J.B., Eds., Springer-Verlag, Berlin, 1973, p. 197.

IAEA, Inhalation risks from radioactive contaminants, Technical Report Series No. 142, International Atomic Energy Agency, Vienna, 1973.

IAEA, International Basic Safety Standards for Protection against Ionising Radiation and for the Safety of Radioactive Sources, Jointly sponsored by FAO, IAEA, ILO, OECD/NEA, PAHO, WHO, IAEA Safety Series No. 115, International Atomic Energy Agency, Vienna, 1996.

ICRP, Recommendations of the International Commission on Radiological Protection, Report of Committee II on Permissible Dose for Internal Radiation (1959), ICRP Publication 2, Pergamon Press, Oxford, 1960.

ICRP, Recommendations of the International Commission on Radiological Protection, ICRP Publication 26, *Annals of the ICRP* 1(3), Pergamon Press, Oxford, 1977. (Reprinted with additions in 1987.)

ICRP, Limits for intakes of radionuclides by workers, ICRP Publication 30, Part 1, *Annals of the ICRP* 2(3–4), Pergamon Press, Oxford, 1979.

ICRP, Limits for intakes of radionuclides by workers, ICRP Publication 30, Part 2, *Annals of the ICRP* 4(3–4), Pergamon Press, Oxford, 1980.

ICRP, Limits for intakes of radionuclides by workers, ICRP Publication 30, Part 3 (including addendum to Parts 1 and 2), *Annals of the ICRP* 6(2–3), Pergamon Press, Oxford, 1981.

ICRP, Radionuclide transformations: energy and intensity of emissions, ICRP Publication 38, *Annals of the ICRP* 11–13, Pergamon Press, Oxford, 1986.

ICRP, Limits for intakes of radionuclides by workers: an Addendum, ICRP Publication 30, Part 4, *Annals of the ICRP* 19, 4, Pergamon Press, Oxford, 1988a.

ICRP, Individual monitoring for intakes of radionuclides by workers: design and interpretation, ICRP Publication 54, *Annals of the ICRP* 19(1–3), Pergamon Press, Oxford, 1988b.

ICRP, Age-dependent doses to members of the public from intake of radionuclides: Part 1, ICRP Publication 56, *Annals of the ICRP* 20(2), Pergamon Press, Oxford, 1989.

ICRP, 1990 Recommendations of the International Commission on Radiological Protection, ICRP Publication 60, *Annals of the ICRP* 21(1–3), Pergamon Press, Oxford, 1991.

ICRP, Age-dependent doses to members of the public from intake of radionuclides: Part 2, Ingestion dose coefficients, ICRP Publication 67, *Annals of the ICRP* 23(3–4), Elsevier Science Ltd., Oxford, 1993.

ICRP, Human respiratory tract model for radiological protection, ICRP Publication 66, *Annals of the ICRP* 24(1-3), Elsevier Science Ltd., Oxford, 1994a.

ICRP, Dose coefficients for intakes of radionuclides by workers, ICRP Publication 68, *Annals of the ICRP* 24(4), Elsevier Science Ltd., Oxford, 1994b.

ICRP, Age-dependent doses to members of the public from intake of radionuclides: Part 3 Ingestion dose coefficients, ICRP Publication 69. *Annals of the ICRP* 25(1). Elsevier Science Ltd., Oxford, 1995a.

ICRP, Age-dependent doses to members of the public from intake of radionuclides: Part 4 Inhalation dose coefficients, ICRP Publication 71, *Annals of the ICRP* 25(3–4), Elsevier Science Ltd., Oxford, 1995b.

ICRP, Age-dependent doses to members of the public from intake of radionuclides: Part 5 Compilation of ingestion and inhalation dose coefficients, ICRP Publication 72, *Annals of the ICRP* 26(1), Elsevier Science Ltd., Oxford, 1996.

ICRP, Individual monitoring for internal exposure of workers, replacement of ICRP Publication 54, ICRP Publication 78, *Annals of the ICRP* 27(3–4), Elsevier Science Ltd., Oxford, 1997.

ICRP, ICRP Database of Dose Coefficients: Workers and Members of the Public, Version 1.0, ISBN 0 08 042 7510, CD-ROM distributed by Elsevier Science Ltd., Oxford, 1998.

ICRP, Doses to the embryo and fetus from intake of radionuclides by the mother, ICRP Publication 88, *Annals of the ICRP* 31(1–3), Elsevier Science Ltd., Oxford, 2001.

ICRP, ICRP database of dose coefficients: Embryo and Fetus, CD-ROM distributed by Elsevier Science Ltd., Oxford, 2001b.

ICRP, Guide for the practical application of the ICRP human respiratory tract model, Supporting Guidance 3, *Annals of the ICRP* 32(1–2), Elsevier Science Ltd., Oxford, 2002.

ICRP, Doses to infants from ingestion of radionuclides in mother's milk, ICRP Publication 95, Annals of the ICRP 34(3–4), Elsevier Science Ltd., Oxford, 2004.

Jette, S.J., Mishima, J., and Haddock, D.E., Aerosolization of M829A1 and XM900E1 Rounds Fired against Hard Targets, Report PNL-7452, Richland, WA: Battelle Pacific Northwest Laboratory, August 1990. (Reference in AEPI, 1995; OSAGWI, 1998, Tab M). [Restricted circulation, summary #31 in OSAGWI, 2000, Tab L.]

Leach, L.J., Yuile, C.L., Hodge, H.C., Sylvester, G.E., and Wilson, H.B., A five-year inhalation study with natural uranium dioxide (UO2) dust. II. Postexposure retention and biologic effects in the monkey, dog, and rat. *Health Phys.* 25, 239, 1973.

Leggett, R.W., The behavior and chemical toxicity of uranium in the kidney: a reassessment. *Health Phys.* 57, 365, 1989.

Leggett, R.W., A generic age-specific biokinetic model for calcium-like elements, *Radiat. Protect. Dosimetry* 41, 183 1992.

Leggett, R.W. and Eckerman, K.F., Evolution of the ICRP's biokinetic models, *Radiat. Prot. Dosimetry* 53, 147, 1994.

Leggett, R.W. and Harrison, J.D., Fractional absorption of ingested uranium in humans, *Health Phys.* 68, 484, 1995.

Mansur, E.S. and Carvalho, S.M., Solubility Classification of Yellowcake Produced by a Brazilian Uranium Mill, Report IRPA 1988, Vol. III, Pergamon Press, Sydney.

Métivier, H., Poncy, J.L., Rateau, G., Stradling, G.N., Moody, J.C., and Gray, S.A., Uranium behavior in the baboon after the deposition of a ceramic form of uranium dioxide and uranium octoxide in the lungs: implications for human exposure, *Radioprotection* 27, 3, 263, 1992.

Mishima, J., Parkhurst M.A., Scherpels R.L., and Hadlock D.E., Potential Behavior of Depleted Uranium Penetrators under Shipping and Bulk Storage Accident Conditions, Report PNL-5415, Pacific Northwest Laboratory, Richland, WA, 1985.

Mitchell, R.E.J. and Sunder, S., Depleted uranium dust from fired munitions: physical, chemical and biological properties, *Health Phys.* 87(1), 57, 2004.

Morrow, P.E., Gibb, F.R., and Leach, L.J., The clearance of uranium dioxide dust from the lungs following single and multiple inhalation exposures, *Health Phys.* 12, 1217, 1966.

Morrow, P.E., Gibb, F.R., and Beiter, H.D., Inhalation studies of uranium trioxide, *Health Phys.* 23, 273, 1972.

OSAGWI (Office of the Special Assistant for Gulf War Illnesses), Depleted Uranium in the Gulf (II), Environmental (Second Interim) Exposure Report, Falls Church, VA, online report available at www.gulflink.osd.mil in the Environmental Exposure Reports Section, 2000.

Parkhurst, M.A., Mishima, J., Hadlock, D.E., and Jette, S.J., Hazard Classification and Airborne Dispersion Characteristics of the 25-MM, APFSDS-T XM919 Cartridge, PNL-7232, Battelle Pacific Northwest Laboratory, Richland, WA, April 1990. [Restricted circulation, summary #29 in OSAGWI, 2000, Tab L.]

Parkhurst, M.A., EG Lodde, G.M., Szrom F., Guilmette R.A., Roszell, L.M., and Falo, G.A., Depleted Uranium Aerosol Doses and Risks: Summary of U.S. Assessments Report PNWD-3476, prepared for the U.S. Army by Battelle, 2004a.

Parkhurst, M.A., Szrom, F., Guilmette, R.A., Holmes, T.D., Cheng, Y.S., Kenoyer, J.L., Collins, J.W., Sanderson, T.E., Fliszar, R.W., Gold, K, Beckman, J.C., and Long, J.A., Capstone Depleted Uranium Aerosols: Generation and Characterization, Vol. 1. Main Text. Attachment 1 of Depleted Uranium Aerosol Doses and Risks: Summary of U.S. Assessments. Report PNNL-14168, prepared for the U.S. Army by Pacific Northwest National Laboratory, Richland, WA, 2004b.

Patrick, M.A. and Cornette, J.C., Morphological Characteristics of Particulate Material Formed from High Velocity Impact of Depleted Uranium Projectiles with Armor Targets, Report AFATL-TR-78-117, Air Force Armament Laboratory, November 1978.

Pomroy, C. and Noel, L., Retention of uranium thorax burdens in fuel fabricators, *Health Phys.* 41, 393, 1981.

Price, A., Review of methods for assessment of intake of uranium by workers at BNFL Springfields, *Radiat. Prot. Dosimetry* 26, 35, 1989.

Raabe, O.G., Characterisation of radioactive airborne particles, in *Internal Radiation Dosimetry,* Raabe, O.G., Ed., Medical Physics Publishing, Madison, WI, 1994, p. 111, chap 7.

Scripsick, R.C., Crist, K.C., Tillery, M.I., Soderholm, S.C., Rothenberg, S.J., Preliminary Study of Uranium Oxide Dissolution in Simulated Lung Fluid, Report LA-10268-MS, Los Alamos National Laboratory, 1985a. http://lib-www.lanl.gov/la-pubs/00318819.pdf.

Scripsick, R.C., Crist, K.C., Tillery, M.I., Soderholm, S.C., Differences in *in vitro* dissolution properties of settled and airborne uranium material, in *Occupational Radiation Safety in Mining,* Stocker, H., Ed., Canadian Nuclear Association 255, 1985b. http://lib-www.lanl.gov/la-pubs/00374828.pdf.

Stradling, G.N., Stather, J.W., Strong, J.C., Sumner, S.A., Moody, J.C., Towndrow, C.G., Hodgson, A., Sedgwick, D., and Cooke, N., Metabolism of an industrial UO_3 dust after deposition in the rat lung, *Human. Toxicol.* 4, 563, 1985.

Stradling, G.N., Stather, J.W., Price, A., and Cooke, N., Limits on intake and the interpretation of monitoring data for workers exposed to industrial uranium bearing dusts, *Radiat. Prot. Dosimetry* 26(1-4), 83, 1989.

TGLD (Task Group on Lung Dynamics), Deposition and retention models for internal dosimetry of the human respiratory tract, *Health Phys.* 12, 173, 1966.

The Royal Society, The Health Hazards of Depleted Uranium Munitions Part I. Policy Document 6/01, London, U. K., online report available at www.royalsoc.ac.uk/du, 2001. in the Science Policy Section, 2001.

The Royal Society, The Health Hazards of Depleted Uranium Munitions Part II. Policy Document 5/02, London, U.K., online report available at www.royalsoc.ac.uk/du, 2002.

West, C.M., Scott, L., and Schultz, N.B., Sixteen years of uranium personnel monitoring experience — in retrospect, *Health Phys.* 36, 665, 1979.

Yeh, H.C., Phalen, R.F., and Raabe, O.G., Factors influencing the deposition of inhaled particles, *Environ. Health Perspect.* 15, 147, 1976.

12 Depleted Uranium and Radiological Hazard during Operations in Kosovo

Patrik Gerasimo, Pierre Laroche, and Gérard Romet

CONTENTS

INTRODUCTION

A radiological assessment was conducted in Kosovo in December 1999. It showed the importance of carrying out a radiological survey as soon as possible when deploying troops in territories under reconstruction, so that appropriate protective measures can be taken. The results show that it is not possible to detect depleted uranium in the environment. This radionuclide, however, could be identified in some smear samples collected in the shell hole and in those collected in shell casings. The risks turned out to be negligible.

The North Atlantic Treaty Organization (NATO) used depleted uranium ammunition during the conflict in Kosovo to destroy the tanks of the Serbian Army and made it known in a letter sent to the United Nations Organization by Lord Robertson (NATO 2000).

A potential radiological hazard from depleted uranium was thus identified. That is why, when French troops were deployed in Kosovo in 1999, the authorities requested monitoring of military personnel and their working environment in order to know the radiological conditions of this deployment and evaluate the actual risks.

HISTORICAL BACKGROUND

The first operational use of uranium-based ammunition dates back to the Gulf War in 1990. This use then was a surprise, although this type of ammunition was already well known in France. At that time, the doctrine of protecting and monitoring personnel participating in military operations was basically the result of the posture adopted during the Cold War on the assumption of massive use of nuclear, bacteriological, or chemical weapons. This doctrine proved effective but quite heavy; also, the history of the hazard was available only several years later and was incomplete.

Since the Gulf War in 1990, the use of depleted uranium munitions, therefore, has been considered as likely. When such use is established, it is necessary to monitor personnel and their operational environment. It is also necessary to arrange follow-up after exposure.

During the peacekeeping operations in Kosovo, the intelligence services reported that a concentration of Serbian tanks and armoured vehicles had been attacked by American A10 aircraft in the area of Likovac, 35 km south of Mitrovica, where the French troops were stationed. The A10 aircraft are capable of firing depleted uranium munitions. Alerted in July 1999 to conduct an assessment mission on site, the French Defence Radiological Protection Service (Service de Protection Radiologique des Armées [SPRA]) could not dispatch a team until December, for safety reasons, after the bomb disposal experts had identified the only tank remnants left in the field.

ASSESSMENT FACILITIES

To conduct their assessment, the SPRA used a documentary database, a field laboratory, and a fixed laboratory.

Documentary Database

From a scientific standpoint in regard to uranium, the database contains numerous studies and publications. It contains data on health risk, on the biological evolution of incorporated uranium, and on the laboratory techniques used in searching for uranium and identifying the isotopes of uranium. All these studies and documents have been compiled into a dossier that subsequently formed the subject of a scientific publication (Laroche et al., 2003).

The health risk associated with uranium incorporation has been studied for a long time. In the case of natural uranium, and *a fortiori* in that of depleted uranium, it is established that chemical toxicity, comparable to that of the other heavy metals, significantly predominates over radiological toxicity.

The modeling of the biological evolution of uranium is welldocumented, also. The "recycling model" proposed by publication 78 of the International Commission of Radiological Protection (ICRP 78) allows the assertion that, 10 years after significant contamination, urinary excretion of uranium can still be detected if the analytical tools available are adequate.

The radiotoxicological data show that the investigation, after past or chronic incorporation, should be performed in urine. Two major analytical techniques are proposed for uranium: ICP-MS (inductively coupled plasma mass spectrometry) or classical alpha spectrometry (Baglan et al., 1997).

Field Laboratory

To ensure the radiological protection of military personnel in all places, the SPRA has built a field laboratory capable of performing assessments in a theater of operations. The equipment of this laboratory can be packed in crates for speedy transportation by air. It can also be installed in air-transportable vans. The modular design of this laboratory provides high flexibility of use. It is possible to perform both a radiotoxicological analysis of personnel and a radiological analysis of the site.

The operational aspect of the information is of paramount importance in the conduct of the assessment. It is on the basis of information provided by NATO and confirmed in the field that the decision to carry out an assessment was made.

Fixed Laboratory

The radiotoxicological analysis method used by the SPRA to look for depleted uranium is alpha spectrometry. The advantage of this method is that it is possible to obtain a spectrum of all alpha emissions in a sample and thereby to identify all alpha emitting radioelements. Depleted uranium is identified by a $^{234}U/^{238}U$ activity ratio lower than 50% (it is around 100% for natural uranium).

This analysis by alpha spectrometry can be supplemented by gamma spectrometry. Depleted uranium is recognized from:

1. The emitting energies of 766 keV and 1001 keV of protactinium 234, a metastable daughter of uranium 238.
2. A $^{235}U/^{238}U$ activity ratio lower than 3%. (Natural uranium is 4.66%.)

FIELD LABORATORY OPERATION

The tank remnants identified by the bomb disposal experts are those of a T55 in very poor condition. The assessment conducted by the SPRA concerns:

- The personnel most likely to have been exposed: i.e., the nine bomb disposal experts who conducted the initial surveys
- The tank itself
- The ground around the tank
- Two 30 mm shell casings from A10 aircraft

To supplement the measurements made near the tank and its direct surroundings, the SPRA's team was asked to assess two sites:

- A zinc ore processing plant located on a height overhanging Mitrovica. This plant had been disused shortly before and stripped of anything merchantable, but a suspicious chest had been discovered in it.
- An old Serbian ammunition depot located south of Mitrovica, near the railway connecting Belgrade to Pristina. This depot, very close to the future French camp, had been destroyed by aerial bombardment.

Personnel Examinations

It appeared very quickly that the facilities available on site could reveal only obvious contamination of the personnel. They could not give meaningful results in the case of low-level contamination, the uranium detection limits of the mobile laboratory being in the order of 50 to 100 mBq/24-h urine.

The need for a superior detection limit of the fixed laboratory (0.5 mBq/24-h urine and 3 mBq/feces) caused the analyses to be transferred to this laboratory. The analyses were mainly aimed at looking for depleted uranium but also for artificial radioelements and gamma emitters, as well as strontium.

Destroyed Tank Examinations

Dose rate measurements and investigations for contamination have been carried out on the destroyed T55 tank. (See Figure 12.1.)

The dose equivalents are measured using an EASYSPEC analyzer. In addition to evaluating dose rates, this device can identify by spectrometry the radioelements that may be present.

To investigate for contamination, smear samples were collected on the outside of the armor, in the shell hole through the armor, and inside the vehicle. Part of these smear samples were analyzed on site, using the mobile laboratory.

The technique used in the field is based on the extraction of the soluble fraction of uranium, followed by scintillation (photon electronic rejecting alpha liquid scintillation [PERALS]). Then the results are confirmed by alpha spectrometry in the fixed laboratory.

FIGURE 12.1 Destroyed tank examination: dose equivalent measurement.

Environmental Examinations

The environment examinations consisted of looking for uranium in the soil, in surface water, and in the water of a well below the position. The area was divided into sectors of about 10 m^2 each. The geographical coordinates were measured using a GPS device.

As for the smear samples, a first analysis was performed locally by the PERALS technique, and a finer analysis by alpha spectrometry was carried out in the SPRA laboratory.

Measurements on Two Shell Casings Found on Site

As for the other samples, these casings were investigated for uranium by the PERALS technique and, subsequently, in the fixed laboratory in Clamart.

Mitrovica Zinc Plant

The assessment in the zinc plant essentially concerned a chest containing suspicious metal bars. It was carried out using the EASYSPEC analyzer.

Old Serbian Ammunition Depot

The measurements were made in an area of about 1 km^2, covered with the remnants of the old Serbian depot. An extemporaneous analysis was performed on soil samples and was supplemented by finer analysis in the SPRA laboratory. Also, a stainless steel disk 5 cm in diameter which puzzled the bomb disposal experts was analyzed on site (spectrometry and dose rate); a thousand identical disks in leather bags had been destroyed by incineration shortly before.

RESULTS

Personnel Examinations

A complete radiotoxicological analysis of the bomb disposal experts' urines and feces for uranium and artificial radioelements was carried out in the fixed laboratory to take advantage of the lower detection limits. The examinations revealed no contamination of this personnel. The detection limits are indicated in Table 12.1.

Destroyed Tank Examinations

The dose equivalent was very low, about 0.03 to 0.05µSv/h inside the tank examined and about 0.07 to 0.08 µSv/h around it. The difference is due to the fact that the armor acts as a shield against telluric and cosmic radiation.

The EASYSPEC device (CANBERRA) detected no radioactivity in the subject tank, apart from potassium 40, which was of natural origin.

Fourteen smear samples were collected. Uranium was found only in the three smear samples collected from a shell hole through the tank. The analysis in the SPRA laboratory of the $^{238}U/^{234}U$ isotopic ratio allows asserting that the uranium found in the dust collected in the hole was indeed depleted uranium (Table 12.2). The spectrometric analyses performed in the field and in the fixed laboratory are shown in Figure 12.2.

Soil and Water

The analyses performed on soil samples revealed the presence of uranium (1.4 ± 0.5 µg/g). The $^{238}U/^{234}U$ isotopic ratio points to natural uranium (Table 12.2).

Dust Inside the Shell Casing

The analyses confirm the presence of depleted uranium (Table 12.2).

TABLE 12.1
Radionuclides Tested for in Military Personnel and Their Detection Limits

URINE	Alpha emitters	Uranium isotope ^{238}U/ ^{234}U, $^{235}U + ^{236}U$)	Detection limit: 10 mBq/l
	Gamma emitters	^{137}Cs, ^{60}Co, ^{256}Ra	Detection limit: 10 Bq/l
	Beta emitters	^{90}Sr	Detection limit: 10 Bq/l
FECES	Alpha emitters	Uranium isotope^{238}U, ^{234}U, $^{235}U + ^{236}U$)	Detection limit: 10 mBq/24-h feces
	Alpha emitters	Plutonium isotope ^{238}Pu, $^{239}Pu + ^{240}Pu$	Detection limit: 10 mBq/24-h feces

TABLE 12.2
Presence of Depleted or Natural Uranium in the Samples

Type of Sample	Location	Uranium Detection
	Tank front armor	No detection
6 Smear samples	Inside the tank	No detection
2 Smear samples	Inside the shell hole	Depleted uranium Alpha spectrometry: $^{234}U/^{238}U$ (%) = 12 ± 4 Gamma spectrometry: $^{235}U/^{238}U$ (%) < 3 Presence of the spectrum lines of $^{234}Pa_m$ (766 and 1001 keV)
Dust	Inside the shell casing	Depleted uranium Alpha spectrometry: $^{234}U/^{238}U$ (%) = 10 ± 5 Gamma spectrometry: ^{235}U / ^{238}U (%) < 3 Presence of the spectrum lines of $^{234}Pa_m$ (766 and 1001 keV)
5 Soil samples	Around the tank	Natural uranium: $^{234}U/^{238}U$ (%) = 97 ± 5
4 Surface water samples	Below the site	No detection
2 Groundwater samples	Well	No detection

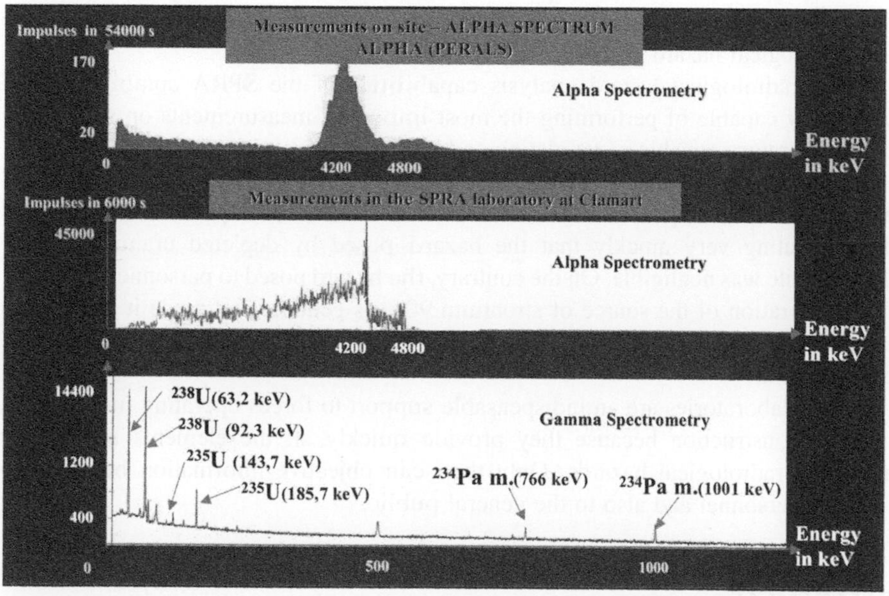

FIGURE 12.2 Analyses of dust samples from the orifice of a shell hole.

Zinc Plant of Mitrovica

The assessment of the chest discovered in this factory quickly identified, using the EASYSPEC spectrometer, bars of americium 241. These bars came from dismantled lightning conductors. The activity in these bars was 10 to a 100 times higher that in the lightning conductors used in France. As to the dose rate near the chest, it was such that staying one hour close to this chest was sufficient to give an exposure equal to the annual limit for the general public.

Old Serbian Ammunition Depot

The analyses carried out in the field revealed very localized contamination by radium. Decontamination was easy because of the small area to be treated.

The stainless steel disk that puzzled the bomb disposal experts was quite difficult to analyze. It was actually a calibration source consisting of strontium 90 (37 kBq). This radioelement has the distinctive feature of emitting only beta radiation.

The recent incineration of a thousand such sources, although their radioactivity had not been identified, led to the screening for strontium 90 of all persons having participated in this incineration. The debris and ashes containing strontium 90 have been gathered, put into crates, and stored in such way as to make them inaccessible.

LESSONS LEARNED

This example highlights the importance of making available in the field efficient analytical capabilities as early as possible. It may lead to the elaboration of a doctrine for radiological hazard analysis and management.

The radiological hazard analysis capabilities of the SPRA combine a field laboratory capable of performing the most important measurements on site with a fixed laboratory capable of validating and refining the results obtained in the field.

The importance of early analysis is obvious: a hazard identified and quantified early is better managed. The assessment of the tank and its environment allowed demonstrating very quickly that the hazard posed by depleted uranium on the Likovac site was negligible. On the contrary, the hazard posed to personnel following the incineration of the source of strontium 90 was genuine and made it possible to take appropriate measures for the personnel likely to be contaminated and for the environment.

Field laboratories are an indispensable support to forces operating in territories under reconstruction because they provide quickly all the elements needed for assessing radiological hazards. Only then can objective information be given to military personnel and also to the general public.

REFERENCES

Baglan, N., Amaral, A., Cossonet, C., Franck, D., Ritt, J. (in French). Complémentarité des mesures nucléaires et de spectrométrie de masse, dans le cadre de l'analyse radiotoxicologique. Application aux urines et aux selles. *Radioprotection* 32, 1997, 673–683.

ICRP 1997. International Commission on Radiological Protection. Individual Monitoring for internal Exposure of Workers. ICRP publication 78. Oxford, Pergamon Press.

Laroche, P., Gérasimo, P., Tréguier, J.Y. (in French). Risques liés à l'uranium appauvri en isotope 235. Encycl Méd Chir Elsevier SAS, Paris, tous droit réservés. Toxicologie-Pathologie professionnelle, 16-008-U-20, 2003, 8 p.

NATO 2000. Letter SG(2000)0108 sent on February 7 by Lord Robertson to the UN Secretary General, Kofi Annan. This letter was published on March 21, 2000.

13 United Nations Environment Programme Results Based on the Three DU Assessments in the Balkans and the Joint IAEA/UNEP Mission to Kuwait

Mario Burger and Henri Slotte

CONTENTS

INTRODUCTION

The United Nations Environment Programme (UNEP) work on depleted uranium (DU) started in the summer of 1999 when UNEP carried out an assessment of the impacts of the Kosovo conflict on the environment and human settlements. As part of that review, UNEP conducted a desk assessment study of the potential effects of the possible use of DU during the conflict. In 2000, the North Atlantic Treaty Organization (NATO) provided UNEP with new information concerning the use of DU during the Kosovo conflict. This information included maps, the amount of DU ammunition used, and coordinates of the targeted areas. It enabled UNEP to carry out the first-ever international assessment on the environmental behavior of DU, following its use in a real-conflict situation.

After the publication of the report, the assessment of DU in the Balkans was, however, not closed. During the Kosovo conflict, a small number of sites outside Kosovo, in Serbia and Montenegro, had also been targeted with ordnance containing DU. To reduce uncertainties about the environmental impacts of DU, it was evident that a second phase of scientific appraisal would be needed. This second phase has been carried out in Serbia and Montenegro, starting with the field mission in October 2001 to collect samples and followed by laboratory work during the winter and the early spring. The report finally has been published in spring 2002.

When, in the summer of 2002, the Council of Ministers of Bosnia and Herzegovina (BiH) requested UNEP to conduct a similar assessment in BiH related to the use of DU ordnance during the war 1994–1995, UNEP was naturally ready to initiate action. UNEP was choosing the same procedure as already successfully introduced in the earlier missions to the Balkans. The report "Depleted Uranium in Bosnia and Herzegovina — Post Conflict Environmental Assessment" was published in spring 2003. In addition, IAEA was responsible for specific DU field work in Kuwait in 2002 on the DU risk from the 1991 Gulf War, which led to the report "Radiological Conditions in Areas of Kuwait with Residues of Depleted Uranium." UNEP contributed both to the mission and to this report.

Based on the work in the Balkans and in Kuwait, UNEP has become a reference in the scientific community regarding the impacts of DU when used in a conflict situation. It is necessary to point out that in the Balkans and during the IAEA Kuwait mission, mainly air-to-ground ammunition of the type 20/25mm penetrators (Kuwait) and 30mm penetrators (Balkans) was available for study. In general, there exists only very poor data on the arrow-type penetrators used in the ground-to-ground battles in a real-postconflict situation. It is not expected that the overall situation in the battle fields may differ dramatically.

Based on UNEP's work to date, the residue of DU munitions does not present a significiant risk to human health at the national level. On a site-specific basis, the main risks are toxological, based on exposure to a heavy metal. The radiological risks are insignificiant and less than or equal to background natural radiation. However, based on the precautionary principle and on existing scientific uncertainty regarding the environmental behavior of DU, risk reduction measures such as access restrictions and cleanup should be adopted. In addition, long-term monitoring of groundwater should be employed.

BACKGROUND

Natural Uranium and Depleted Uranium (DU)

Depleted uranium (DU) is the main by-product of uranium enrichment and, like any other uranium compound, has both chemical and radiological toxicity; it is mildly radioactive, having about 60% of the activity of natural uranium.

The total specific activity of natural uranium (i.e., the activity per unit mass of natural uranium metal) is 25.4 [Bq/mg]. However, in nature uranium isotopes are in radioactive equilibrium with the other isotopes, such as ^{234}Th, ^{231}Th, ^{226}Ra, ^{223}Ra, ^{222}Rn, ^{210}Pb, and ^{210}Po, created as a result of radioactive decay.

In DU only traces of decay products beyond ^{234}Th and ^{231}Th are present, as these decay products have not had time to form since the DU was produced.

The specific activity of DU including the decay products is shown in Table 13.2.

DU is a strong emitter of alpha radiation (from ^{238}U and ^{234}U), and beta radiation (from ^{234m}Pa). The gamma radiation is weak (from ^{234m}Pa).

Enrichment/Depletion

In the past, depletion was usually carried out from the initial 0.711% ^{235}U down to a remaining content of about 0.2% ^{235}U. Given today's low prices for uranium, it is

TABLE 13.1
Half-Lives, Specific Activities, and Relative Abundance of Uranium Isotopes in Natural Uranium and in DU

			Relative Isotopic Abundance [%]			
			Natural Uranium		DU	
Isotope	Half-Life (a)	Specific Activity [Bq/mg]	By Mass	By Activity	By Mass	By Activity
^{238}U	4.51×10^9	12.44	99.28	48.2	99.8	87.5
^{235}U	7.1×10^8	80	0.72	2.2	0.2	1.1
^{234}U	2.47×10^5	230 700	0.0055	49.5	0.0007	11.4

TABLE 13.2
Depleted Uranium, DU (^{235}U 0.2%)

Isotope	Relative Isotopic Abundance	Specific Activity [Bq/mg DU]
^{238}U	99.7990%	12.38
^{235}U	0.2000%	0.16
^{234}U	0.0010%	2.29
^{234}Th	Traces (decay product)	12.27
^{234m}Pa	Traces (decay product)	12.27
^{231}Th	Traces (decay product)	0.16
	Total	**39.42**

more economical to deplete to approximately 0.3%. According to federal specifications, the U.S. Department of Defence (DoD) is allowed to use DU for its purpose only when its ^{235}U content is less than 0.3 % [7]. However, typically the percentage concentration by mass of ^{235}U in DU used for military purposes is 0.2% [5]. Figure 13.1 illustrates the simpler of two cases, where DU is produced by isotope separation using natural uranium as feed material.

Besides fission products and transuranic elements like plutonium and neptunium, spent fuel rods from a nuclear power plant still contain a substantial amount of ^{235}U. When the fuel rods are reprocessed, the uranium is separated from all other elements. In the U.S., for decades this mixture of recycled uranium contaminated by traces of other isotopes has been enriched and subsequently depleted again.

This reprocessed uranium (REPU), which is fed into enrichment and thus into the fuel cycle, may contain one plutonium atom in 100 million uranium atoms, at most [10,12]. Therefore, traces of plutonium and other specific isotopes may be detected in the produced DU and can be found in the corresponding products.

Figure 13.2 shows enrichment with a mixture of both natural uranium and recycled uranium (REPU) as feed material. Unlike the case where pure natural uranium is used, here the uranium isotopes ^{236}U and ^{232}U as well as traces of plutonium and other transuranic elements may be found in the DU [18].

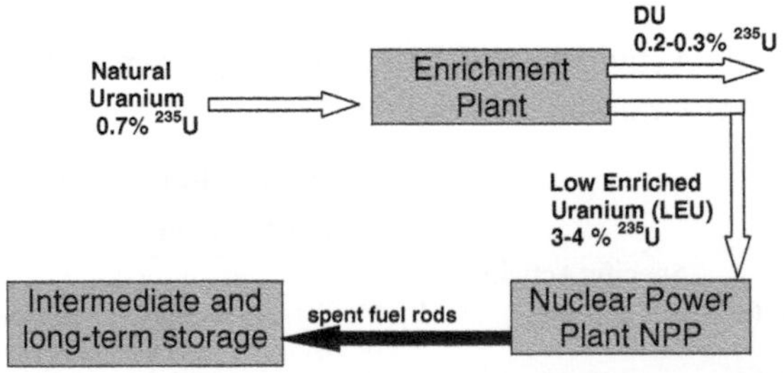

FIGURE 13.1 An illustration of the simpler of two means of producing DU, in which isotope separation is used with natural uranium as feed material.

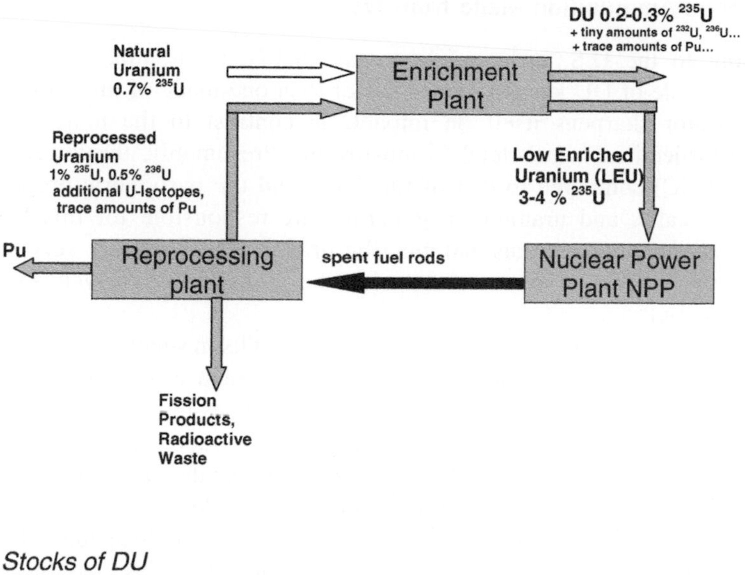

FIGURE 13.2 Enrichment with a mixture of both natural uranium and recycled uranium (REPU) as feed material. Unlike the case where pure natural uranium is used, here the uranium isotopes ^{236}U and ^{232}U, as well as traces of plutonium and other transuranic elements, may be found in the DU. (From Ramachandra, K.B., 2000. Tank-Automotive and Armaments Command (TACOM) and Army Materiel Command (AMC) Review of Transuranics (TRU) in Depleted Uranium (DU) Armor. U.S. Department of the Army, AMSAM-TMD-SB.)

In June 1998, the U.S. Department of Energy (DoE) stored about 500,000 t of depleted uranium in chemical form as uranium hexafluoride. No figures are published for stockpiles in other countries with enrichment facilities. Together, they certainly store at least again as much DU as the U.S. In 1995 as well as in 1996, the worldwide production was about 35,000 t of uranium. From this fact, it can be estimated, that an additional approximately 30,000 t are every year piled up onto this already huge stock of DU [7].

The often-heard claim that the wartime use of DU was a cheap way to solve a waste problem is certainly not true. The total quantity of DU in ammunition that was used in Iraq and the Balkans corresponds to barely a couple of days of DU-production worldwide.

Application of DU

The DU has had a wide range of peaceful applications, such as the provision of radiation shielding for medical sources or as counterweights in planes. In military applications DU is used for heavy tank armor and, owing to its high density (19.07 kg/dm^3), its property of becoming "sharper" as it penetrates any armour plating, in antitank munitions. The use of DU in missiles, formed as penetrators combined with charges of high explosives to hit targets below the surface, is not confirmed yet.

Specifics on Ammunition Made from DU

According to the U.S. technical literature [9,17], upon impact on armor, a projectile made of DU keeps its form better than one made of tungsten or steel; the penetrator sharpens itself on impact, in contrast to the more expensive tungsten projectiles, which tend to mushroom. Presumably, the lower melting point (1132°C) compared to that of tungsten, and the pyrophoric properties of DU (specifically, and uranium in general), are responsible for this behavior. (*Pyrophoric properties* means that metallic uranium is chemically very reactive. In the form of powder it can react with the atmospheric oxygen and ignite spontaneously.)

In any case, with DU ammunition the impact results in smaller but deeper holes in the armor than with conventional ammunition of the same caliber. After penetrating the armor, and as soon as the DU projectile again comes into contact with air, the part of DU which is now in the form of a liquid or powder starts burning, thereby increasing its destructive effect. Often this leads to setting the fuel tank on fire and/or detonating the ammunition stored in the armoured vehicle.

Two types of ammunition with two different uses can be distinguished. On one hand, there is the armor-piercing ammunition used in tank battles. Into this category belong the armor-piercing rounds, which have been fired from 105 mm and 120 mm guns by the multinational forces in Iraq. These penetrators contain approx 3.9 kg and 4.9 kg of DU, respectively [19].

On the other hand one must mention the air-to-ground rounds fired from 25 mm and 30 mm guns, containing 148 g and 299 g, respectively, of DU [19]. Only a small fraction of these per salvo hit the target.

Because of the superiority of this type of ammunition, it has already been introduced by the military forces of several countries. "Depleted uranium weapons have been acquired by 17 countries including Britain, France, Russia, Greece, Turkey, Israel, Saudi Arabia, Bahrain, Egypt, Kuwait, Pakistan, Thailand, South Korea, Taiwan, and other countries which the Pentagon will not disclose for national security reasons" [24].

Natural Uranium

Natural uranium is found in trace amounts in all rocks and soil, in water and air, and in materials made from natural substances. It is a reactive metal, and therefore it is not present as free uranium in the environment. In addition to the uranium naturally present in minerals, the uranium metal and compounds produced by industrial activities can also be released back to the environment.

Uranium can combine with other elements in the environment to form uranium compounds. The solubility of these uranium compounds varies greatly. Uranium in the environment is dominated by uranium oxides such as UO_2, which is an anoxic insoluble compound found in minerals, and UO_3, a moderately soluble compound found in surface waters. The chemical form of the uranium compound determines how easily the compound can move through the environment, as well as how chemically toxic it might be.

TABLE 13.3
Summary of the Activity Concentrations of ^{238}U and ^{235}U of Natural Origin in Some Environmental Materials

	Activity Concentration			
	^{238}U		^{235}U	
Material	Reference Value	Range	Reference Value	Range
Soil (Bq/kg)	35	1–690	—	—
Air (μBq/m^3)	1	0.02–18	0.05	—
Drinking water (Bq/kg)	0.001	0.00009–150	0.00004	0.0004–0.5
Leafy vegetables (Bq/kg)	0.02	0.006–2.2	0.001	0.0007–0.0012
Root vegetables (Bq/kg)	0.003	0.0004–2.9	0.0001	0.00005–0.0006
Milk products (Bq/kg)	0.001	0.0001–0.017	0.00005	0.00005–0.0006
Meat products (Bq/kg)	0.002	0.0008–0.02	0.00005	0.00002–0.0005

Source: IAEA, 2003. Radiological Conditions in Areas of Kuwait with Residues of Depleted Uranium. International Atomic Energy Agency, Vienna, August 2003.

EXPOSURE TO NATURAL URANIUM

Uranium is incorporated into the human body mainly through the ingestion of food and water and the inhalation of air. The U.N. Scientific Committee on the Effects of Atomic Radiation (UNSCEAR) has estimated that the average person ingests 1.3 μg of uranium per day, corresponding to an annual intake of 0.46 mg, or 11.6 Bq, primarily through the consumption of drinking water [5]. Typically, the average person receives an annual dose of less than 0.6 μSv from the ingestion of uranium; in addition, the average individual receives an annual dose of about 110 μSv from the ingestion of the decay products of uranium [5].

Uranium in air is associated with particles of dust. UNSCEAR has estimated that the average person inhales 0.6 μg of uranium (15 mBq) each year [5]. This results in an annual dose of 0.048 μSv; the average total annual dose from the inhalation of all radionuclides of natural origin has been estimated to be 5.8 (μSv) [5]. The size of the uranium aerosols and the solubility of the uranium compounds in the lungs and gut influence the transport of uranium in the human body. Most of the uranium ingested is excreted in feces within a few days and never reaches the bloodstream. The remaining fraction will be transferred into the bloodstream. Most of the uranium in the bloodstream is excreted in urine within a few days, but a small fraction remains in the kidneys and other soft tissue, as well as in bones.

Exposure to DU

The radiation emitted from DU is predominantly alpha particles. Alpha particles have a very limited range in tissue; they can barely penetrate the external layer of

TABLE 13.4
Transfers of Uranium into the Biocycle

Amount Transferred from Soil to Biological Material (Daily)		Amount Transferred from Biological Material to Animal (Daily)	
Soil to Grazing Crops	Soil to Vegetables	Grazing Crops to Milk	Grazing Crops to Meat
(Bq/kg) / (Bq/kg) or (mg/kg) / (mg/kg)	(Bq/kg) / (Bq/kg) or (mg/kg) / (mg/kg)	Days/kg	Days/kg
0.05	0.005	0.0005	0.0004

the skin, and hence do not pose a hazard in terms of external irradiation. However, alpha particles are very energetic, and if emitted inside the body can damage nearby cells. Consequently, internal irradiation is an important consideration. DU inhaled or ingested may have a detrimental effect (tissue damage and increased probability of cancer) on lungs and the gut. If DU is absorbed into blood and retained in other organs, particularly in the skeleton, there is an increased probability of cancer in these organs.

Uranium is not generally transferred effectively along food chains, the transfer factors as shown in Table 13.4 may be applied [15].

As an example: A contamination of 10 mg DU/kg surface soil would result in a contamination of food plants of 0.5 mg DU/kg food plant. With a mean consumption of 65 kg food plant by a cow, this would result in a daily intake in that area of about 30 mg DU, which in turn could yield for the following day's milk a maximum of 0.0163 mg DU/kg.

So in environmental assessments, inhalation is usually the exposure pathway that merits primary attention. Processes such as migration through the soil, deposition of resuspended material onto crops, and transfer to groundwater might be of greater interest in the longer term.

In a combat situation the main radiological hazard associated with DU munitions is the inhalation of the aerosols created when DU munitions hit an armored target. Studies carried out at test ranges show that most of the DU aerosols created by the impact of penetrators against an armored target settle within a short time of the impact and in close proximity to the site, although smaller particles may be carried a distance of several hundred meters by the wind [22].

A possible exposure pathway for those visiting or living in DU-affected areas after the aerosols have settled is the inhalation of the DU particles that are resuspended through the action of the wind or human activities.

One possible pathway of exposure that merits consideration is the inadvertent or deliberate ingestion of soil. For example, farmers working in a field in which DU munitions were fired could inadvertently ingest small quantities of soil, and sometimes children deliberately eat soil. Doses from this exposure pathway were, however, found to be much lower than doses associated with other pathways.

Generally, a large proportion of DU munitions fired from an aircraft miss their intended target. The physical state of these munitions once fired will vary from small fragments to whole intact penetrators, either totally or partially encased in their aluminium jackets. Individuals who find and handle such munitions could be exposed via external irradiation due to the beta particles and gamma rays emitted by the DU. However, the dose received would be significant only if a person were in contact with DU projectiles over a considerable period of time, as the contact dose to the skin from DU is about 2.3 mSv/h [13]. It is therefore unlikely that even prolonged contact with DU would lead to skin burns or any other acute radiation effect.

Penetrators that do not hit the target corrode with time, forming fragments and particles containing a variety of oxides of DU and combinations with carbonates, which may range from several millimeters to less than a micrometer in size [11]. Uranium corrosion products formed through this weathering process could be resuspended and ingested or inhaled by people staying in the zone. As a geochemical effect these corrosion products will have a different mobility in the affected soil than the original form brought into the field. A higher mobility could affect ground water in the zone earlier than a decrease in mobility.

INTERNATIONAL SAFETY STANDARDS

The International Atomic Energy Agency (IAEA), together with other relevant international organizations, has established the basic requirements for protection against the risks associated with exposure to ionizing radiation, which are published in the Basic Safety Standards [14].The standards are based primarily on the recommendations of the ICRP [16] and on the assessments of the health effects of radiation of UNSCEAR [20]. The standards do not apply to nonionizing radiation or to the control of the nonradiological aspects of health and safety, such as chemical toxicity.

The Basic Safety Standards cover a wide range of situations that give rise to or could give rise to exposure to radiation and are applicable to exposures from any combination of uranium isotopes, including those found in DU. They do not, however, include dose criteria directly applicable for aiding decision making on remedial actions for DU. The IAEA Safety Standards [6] recommend a generic reference level for aiding decisions on remediation: an individual existing annual effective dose of 10 mSv from all sources, including natural background radiation. In addition, an upper value is recommended at which intervention is justified under almost any circumstances: an existing annual equivalent dose of 100 mSv to any organ. For perspective, the worldwide average annual effective dose from natural background radiation is 2.4 mSv, with a typical range of 1 to 20 mSv [20]. The most significant contribution to the worldwide average annual effective dose comes from exposure to radon and its decay products (1.15 mSv); exposure to terrestrial gamma rays and cosmic rays accounts for 0.48 mSv and 0.38 mSv, respectively. The contribution of the intake of natural radionuclides in air, food, and water to the average dose is 0.31 mSv, mainly due to ^{40}K (0.17 mSv), ^{210}Po (0.086 mSv), ^{210}Pb (0.032 mSv), and ^{228}Ra (0.021 mSv); uranium isotopes contribute little to the dose (0.0006 mSv) [20] [5].

Chemical Toxicity

The chemical toxicity of a substance is defined by a threshold concentration in the human body, especially in the different organs, below which no health damage can be observed in most individuals. By compliance with legal norms for the threshold values of contaminants in air, soil, water, and food, it can be ensured that even after a long term exposure there is no risk of adverse health effects. In terms of chemical toxicity, heavy metal uranium can cause kidney damage, as nodular changes to the surface of the kidney, lesions to the tubular epithelium, and increased levels of glucose and protein in the urine.

The Canadian MAC for uranium in drinking water of 0.1 (mg/l) (100 μg/l) is under revision. The proposed new value is 0.01 mg/l (10 μg/l). The WHO guideline value for drinking water since the year 2004 has been revised to 15 mg/10 (previously in 2nd edition, 2 μg/l), a value considered to be protective for subclinical renal effects reported in epidemiological studies [8], as it is expected that the U.S. Environmental Protection Agency will establish a guideline value of 20 mg/l. These examples show the continuous and dynamic development of threshold values, reflecting the complexity of effects of uranium on human health.

MASSES USED IN CONFLICTS

DU Assessments in the Balkans: Findings/Results

Results are based on the UNEP DU assessments that took place in the Balkans [2] [3] [4] and the joint IAEA-UNEP mission to Kuwait [5]. It has to be kept in

TABLE 13.5
Threshold Values for Uranium in Air and Drinking Water

	Threshold Values for Uranium		
	Air		Drinking Water
Norm	Uranium (Insoluble Compounds, as U)	Uranium (Soluble Compounds, as U)	Uranium
NIOSH REL	0.25 mg/m^3	0.05 mg/m^3	
STEL	0.6 mg/m^3		
IDLH	10 mg/m^3	10 mg/m^3	
MAC			0.1 mg/l
EPA			0.020 mg/l
WHO (guideline value)			0.015 mg/l

Note: NIOSH: National Institute for Occupational Safety and Health, U.S.; REL: Recommended Exposure Limit; STEL: Short-Term Exposure Limit; IDLH: Immediate Danger to Life or Health (30 min exposure time); MAC: Maximum Acceptable Concentration; EPA: Environmental Protection Agency; WHO: World Health Organization.

TABLE 13.6
DU Used in Conflicts

Country	Offensive Use/Amount (metric tons) Ground–Ground (Tank Battles)	Air–Ground (Usually A10, Other Aircraft/Helicopters)
Iraq (Gulf War, 1991)[1]	40	250
Bosnia and Herzegovina (1994–1995)	—	3.3
Kosovo, Serbia, and Montenegro (1999)	—	10
Afghanistan (19xx/2001)	—	Use confirmed, Total: unknown
Iraq 2003[2]	Total: unknown US: unknown UK: 1.9	Total: unknown US: unknown UK: none

[2] Period end of war until end 2005 occasional use of DU expected.

mind that the time between the military conflict and the corresponding mission to Kosovo was 1–2 years, to Serbia-Montenegro, 2–3 years; to Bosnia-Herzegovina, 7–8 years; and 13 years to Kuwait. It was UNEP's interest to receive the best possible picture on DU in the environment. This included the studies of basic data on DU taken in the environments and the study of possible changes of qualities of DU spread in the environment.

UNEP's approach has always been to work with an internationally balanced team of experts, actively working in the field of DU. Due to fact that UNEP itself has no laboratory capacity, experts from the internationally best-recognized laboratories were already involved in the DU assessments — the sampling campaign — and samples were analyzed in these institutes. The laboratories involved were the Swedish Radiation Protection Institute (SSI), the Swiss SPIEZ Laboratory, the Italian Environmental Protection Agency and Technical Services (APAT), and the University of Bristol, U.K.

(a) Pentrators, Fragments, and Jackets

Many areas that were investigated by the UNEP team usually had been investigated before by local experts and/or multinational troop. At some sites these investigators found and removed penetrators.

At the sites investigated by the UNEP DU Assessments, where about 10,000 penetrators were fired, more than 200 penetrators, fragments, and jackets (casings) were found on surfaces, and more than 500 penetrators buried in the ground were indicated.

TABLE 13.7
Summary of DU Munitions Used Specifically during the Gulf War

DU Munition Type	Rounds Used in the Gulf War	Weight of a DU Round (kg)	Total Weight of DU (t)
	U.S. Army		
M900 (105 mm)	504	3.83	1.93
M829 and M829A1 (120 mm)	9048	3.94/4.64	37.3
Total	**9552**	—	**39.2**
	U.S. Air Force		
API (30 mm)	**783 514**	**0.302**	**237**
	U.S. Navy		
20 mm from Phalanx CIWS	**4 – 5**	**~ 0.1**	**~ 0.0005**
	U.S. Marine Corps		
PGU/20 (25 mm)	**67 436**	**0.148**	**10**
	British Army		
APFSDS (120 mm)	**<100**	**4.85**	**< 0.5**
Total	**~860 600**	—	**~286**

Isotopic Composition of DU Ammunition Used

Uranium Isoptopes

In the range of 37 penetrators and jackets, and several dozens of fragments taken in the missions were analyzed in detail in the nominated laboratories.

UNEP found that the composition of the DU used for 30 mm ammunition has been constant: ^{235}U is present with 0.200 ± 0.005% by mass. The similar value was found on Type 20/25mm-penetrators analyzed by the IAEA from the joint IAEA mission to Kuwait.

^{234}U was measured between 0.00055 and 0.00071% by mass; IAEA measured 1.39–1.57 ± 0.08 10^6 Bq/kg.

The presence of ^{236}U — a resultant product of reprocessing — has been confirmed with a rather constant presence of 0.0028 ± 0.0002 % by mass.

Actinides/Transuranics

There have been rumors that in the DU ammunitions actinides, especially plutonium in high concentrations, are present. UNEP confirmed the presence of plutonium in penetrators. The concentrations have been found to be extremely low: ranging between less than 0.8–88 Bq/kg for the sum of ^{239}Pu and ^{240}Pu, corresponding to a maximum of 38.2 $\cdot 10^9$ g/kg calculated as ^{239}Pu. From the IAEA Kuwait mission $^{239+240}Pu$ was found in the range 0.6–6.2 Bq/kg in the DU penetrators.

The values found by UNEP and IAEA were lower compared to the values published by the U.S. Army in January 2000 on DU used as tank armor ranging from 85–130 Bq/kg [23].

The radiochemical analysis in UNEP investigations also showed that the concentration of Neptunium-237 in the DU penetrators from Bosnia and Herzegovina was from less than 4–16.2 Bq/kg, which corresponds to a maximum of $6.2 \cdot 10^{-7}$ g/kg.

Corrosion of Penetrators

Studies of the corrosion on the DU penetrators began by weighing and measuring them in the same state as when they were collected in the field, and again after removing the surface layer of soil and uranium oxide by both mechanical and mild chemical cleaning.

After lying in the ground for over 7 years, the penetrators were heavily corroded and intensive pitting (corrosion attack producing small holes) of the DU surface had taken place. According to U.S. military literature, a penetrator's original weight is ~292 g. Thus, the penetrators studied had lost 66–93 g due to corrosion.

These findings differed from UNEP results of penetrator studies from Kosovo, which were only slightly corroded after 1.5 years in the ground. These new findings showed that the level of corrosion increased dramatically with time. Once corrosion starts, the exposed surface tends to increase and the pitting effect will thus accelerate. Penetrators laying on the ground surface were much less corroded than those buried below the ground surface. Bearing in mind the state of the penetrators when they were found, UNEP established the losses to be 2–5 g in Kosovo after 1.5 years; 11–38 g in Serbia and Montenegro after 2.5 years; and 66-93 g in BiH over 7 years after the conflict (not corrected for loss of weight due to formation of DU dust during impact). In conclusion, based on these findings, no more penetrators consisting of metallic DU will be found in the Balkans grounds after 25–35 years. Instead of metallic penetrators, contaminated spots in the ground will be found containing DU decomposition products.

It has been recognized during the IAEA Kuwait mission that penetrators and fragments showed the similar corrosion pictures under desert conditions as in the Balkans. However, a complete study has not been undertaken.

Jackets/Casings

Jackets/casings made of aluminium are the slightly DU contaminated parts of the bullet once separated from the penetrators. Not more than a few milligrams of DU can be found in the center hole where the penetrator was held before entering an object.

Specific Comments

One has to keep in mind that the penetrators and the corrosion products currently hidden in the ground may be dug up during construction works in the future and as a result the corresponding risks as described above might occur; for example, without personal protection equipment, contamination of the construction workers and, as a consequence, internal and external doses, poisoning may result.

Penetrators, fragments, and jackets laying on surfaces and accessible to the public are UNEP's main concern and pose the main risk from DU ammunition once dropped in the environment.

(b) Contamination Points

More than 500 points of ground contamination by DU were found over the missions led by UNEP. For UNEP a contamination point is defined as the point where a DU projectile hits the ground surface, keeping in mind that this is the case for most of the penetrators in an air to ground attack.

Analyses showed that the concentration of DU at these points can vary from 0.01 to 100 g DU/kg soil. The dispersion of DU into the ground occurs mainly in the first 20 cm depth at present and the concentration decreases with depth (1–2 orders of magnitude for each 10 cm).

Measured from the center of the impact, an area of no more than 20 × 20 cm is significantly affected by the impact. Soft surfaces were found resulting in higher ground contamination than hard surfaces.

(c) Localized Ground Contamination

In the areas of air to ground attacks with DU ammunition, the presence of DU could be measured starting already with a value clearly below the natural level of uranium decreasingly up to distances not more than 150 m from the center. Based on this finding the DU contamination can be defined as a *localized ground contamination.*

The risks related to contamination points are the dispersion/resuspension of DU, and the direct, none protected, contact of such points which will result in contamination. Therefore, it is advisable to remove, cover, or mark the points.

(d) Widespread Contamination

A contamination of the ground surface may be either *localized (see [C])* or *widespread* over a large area, depending on the properties of the aerosols and the prevailing meteorological conditions.

Widespread contamination exists in the case where the contamination can be found over a couple of hundred meters from the source of contamination resulting in an affected large area.

Widespread contamination might occur as a result of initial dispersion in air during the military attack and by dispersion/resuspension of DU from, e.g., contamination points and/or corroded penetrators and fragments laying on surfaces by the wind.

UNEP found no signs of widespread contamination.

(e) Contamination of Vehicles

At the sites investigated in Kosovo and Bosnia-Herzegovina no vehicle hit by DU could be found, all vehicles had been removed. The history of the vehicles remained unsolved.

One single armored personnel carrier (APC) could be studied in the Serbia-Montenegro Mission: The APC was hit by one 30mm DU penetrator. The smear samples collected showed that contamination by DU dust occurred inside the APC. Expressed as total uranium, the deposition of uranium ranged from 0.053 to 0.261 $\mu g/cm^2$ and around the impact hole, 0.155 $\mu g/cm^2$ was measured.

A calculation with these data showed that within the APC, 200–2000 mg uranium were deposited directly after the penetration with an initial concentration in the air of 50–500 mg uranium/m^3. Of course, this theoretical initial concentration was present only for a short time (some minutes) before deposition on the inner surface took place. Based on a daily inhalation volume of 23 m^3 for adult males (WHO), and assuming a stay of 1 min after the attack, a rough estimate of the respiration intake for the crew members would result in a range of 1–10 mg DU. Such an intake could lead to a temporary weak impairment of renal function and to a committed effective radiation dose of the order of 1 [mSv].

A very comprehensive study on the effects of DU has been published by the U.S. Army Center for Health Promotion and Preventive Medicine in 2004 (Depleted Uranium Aerosol Doses and Risks: Summary of U.S. Assessment, Capstone, Prepared for U.S. Army Center for Health Promotion and Preventive Medicine and the U.S. Army Heavy Metals Office) [21].

(f) Contamination of Buildings

In the DU assessments of Kosovo and Serbia-Montenegro, no DU contaminated buildings were recognized.

In Bosnia-Herzegovina at two sites which have been heavily attacked by A10 aircrafts penetrators, fragments and DU dust were found in buildings which were still in use by the local population or army.

At one site a building (storage barn) was studied in detail. Both scratch and smear samples were collected. Scratch samples most likely represented primary deposition of debris and dust from the initial impact of DU penetrators on the concrete floor inside the building. A DU contamination inside the building ranging between 11 and 1070 [mg/m^2] was found. Smear samples taken from surfaces of artillery guns and grenade boxes — understood as secondary contamination — showed a surface contamination of 59–270 $\mu g/m^2$. The resuspension factor for this building was calculated and found to be 5.2 10 9 m^{-1}.

Even though calculations showed a minor radiological risk by entering or staying for a certain period of time in the building, UNEP recommended avoiding unnecessary contamination with a number of cleanup measures easy to realize.

(g) Contamination of Water

UNEP in its assessments always included the sampling and analysis of water. In Kosovo and Serbia-Montenegro, no contamination by DU of water could be found. Also water samples taken during the IAEA Kuwait mission showed no indication of the presence of DU.

However, in Bosnia and Herzegovina, DU contamination of groundwater was found at 1 of the 11 sites investigated. The concentration of DU in the water was low from the radiological and chemical-toxicological point of view and compared to the natural level of uranium present in the water, but was indicative of possible water contamination in the future.

The finding justifies — as a precautionary action — the monitoring for DU, respectively the total uranium concentration in water used as drinking water taken inside attacked areas and 200–300 m along the direction of the groundwater flow from the attacked areas.

(h) Contamination of Air

Measurements on possible contamination of air by resuspended DU from contaminated ground were carried out in Serbia-Montenegro and in Bosnia-Herzegovina. A resuspension experiment was included in the IAEA Kuwait assessment.

In the UNEP Balkans investigations on air at heavily attacked sites, some samples showed clear indications of DU in the air. The levels found in the air were low, of some nanograms per cubic meter (ng/m3) and have been discussed in the UNEP reports in detail.

Normal uranium concentration in the air as mentioned in Table 3 is 1 μBq m^{-3} (8×10^{-5} μg m^{-3}, 0.08 ng m^{-3}, respectively) ± factor of 10 upward and downward.

Resulting effective doses by inhalation are 0.02–0.03 μSv per year from ^{238}U and ^{234}U, respectively.

From the radiation dose viewpoint, it is concluded that in normal conditions, such as when air sampling has taken place the DU intake by inhalation and effective doses from airborne uranium are very small, even if the concentration is several hundred times higher than normal. Heavy metal risks are also insignificant in these cases.

Calculations showed that for the worst but rather unrealistic case for the general public, assuming desert conditions with vehicular movement, etc., the DU concentration in air can approach the limits for chronic occupational exposure and exceed the limit for short-term exposure of the general public. In the same worst-case scenario assuming normal living conditions in the contaminated area, DU concentration can be higher than a Minimal Risk Level (MRL) for chronic inhalation exposure of uranium as a toxic metal.

Two experiments to evaluate the resuspension of residues of DU have been performed during the IAEA Kuwait mission in desert conditions. The two experiments were performed with explosives where rather low-DU-contaminated sand was blown up. It clearly came out that results from these experiments should be interpreted with caution, and any extrapolation to different conditions may not be justified as they were conducted with a specific DU-contaminated soil/sand and followed a protocol that may only approximately represent real conditions. These experiment conditions showed that DU concentrations in the air never reached alarming levels.

(i) Contamination of Botanical Material

UNEP in its assessments to the Balkans was including sampling and analysis of bio indicators, focused on lichen. Rarely samples of vegetables were included.

A result was that some lichen showed traces of DU that indicated earlier or ongoing contamination through the air caused by the initial phase, the attack and/or as a result of steady resuspension from contaminated ground. No vegetables contaminated by DU were found.

The farming areas in Kuwait were investigated in detail in the IAEA Kuwait assessment.

There was no evidence of the presence of DU in any of the soil samples taken in the farms. In some crop samples lowest concentrations of DU were found. These results did not fit into the overall picture of the site. Neither the water taken from the site nor the soil or other vegetables showed indications of the presence of DU. It was therefore concluded that a possible explanation was cross contamination at the time of packaging the samples at the end of the campaign.

(j) Awareness Raising

As one result from the DU assessments in the Balkans, UNEP published the brochure "Depleted Uranium Awareness," answering the questions "What is DU?", "What does it look like? ", "When was it used?", Where can you find it?" and discussing health risks and the risk of DU exposure in targeted areas, finally giving precautionary steps.

This brochure found wide distribution and has been an element in the UNEP capacity-building process for local experts in different countries. It also has been used as a basis for local versions of DU awareness flyers published in the local language.

(k) Iraq

UNEP DU assessments in Iraq are planned once the situation in the country allows. UNEP has received from the U.K. Ministry of Defence the complete details of DU used by the U.K. troops during the 2003 Iraq war. UNEP activities on DU exposures in Iraq in general are supported by the U.K. for environmental assessments and training, including assessments of sites containing DU, and training on monitoring, sampling techniques and clean up. A seminar, different workshops, and training courses were already held outside Iraq for recognized Iraqi experts from the Ministry of Environment. In addition, a number of selected field-work related equipment and instruments were provided with training in detail for the Iraqi Radiation Protection Centre. The process of building up the capacity of recognized Iraqi institutes continues. The close cooperation with the Iraqi Ministry of Environment providing information on DU targeted areas, in combination with in depth analysis of other sources including satellite imagery, allows UNEP to have the corresponding DU databases ready to undertake the DU Assessments in Iraq.

CONCLUSIONS

Following the DU assessment in the Balkans and the IAEA Mission to Kuwait, the collective information from these reports can be used to minimize any health and environmental risks from depleted uranium.

These studies confirm that the behavior of DU in the environment is a complex issue, and that DU can be found in soil, vegetation, water, and air in certain conditions many years after the conflict.

It is a fact that, regarding the issue of DU in the environment from the scientific point of view, there are still open questions, especially on the behavior of residues of DU in postconflict desert conditions. Answers are also being sought about the mobility of the DU degradation products in the ground.

It is a fact that not all countries in a postconflict situation with a history of attack by DU ammunition have been assessed. Many residents of the DU targeted areas, without clear guidance from their governments based on scientific facts live permanently in fear and continue to repeat rumors concerning health consequences from the presence of DU in their environments. It has, additionally, to be kept in mind that in the former regime of Iraq, DU was even used in propaganda against the multinational troops. UNEP came to the conclusion that the only way to avoid ongoing uncertainties and rumors, etc., was to have DU assessments performed by highly professional, international teams. Their analyses, informed by international standards, may bring further clarification to the issue of DU.

Based on UNEP's work to date, the residue of DU munitions does not present a significiant risk to human health at national levels. On a site-specific basis, the main risks are toxological based on exposure to a heavy metal. The radiological risks are insignificiant, and less than or equal to background natural radiation. However, based on the precautionary principle and on existing scientific uncertainty regarding the environmental behavior of DU, risk-reduction measures such as access restrictions and clean-up should be adopted. In addition, long-term monitoring of groundwater should be employed. The authorities should also be active by informing the local residents and workers at such sites of the hazards associated with collecting DU residues and other possible actions.

A new situation in the history of DU was presented in the 2003 Iraq war; the offensive use of DU in the air-to-ground attack changed. DU has been used against buildings in towns and villages. It remains open what consequences the presence of a significant number of DU penetrators and residues has in an urban area in the attacked building itself, in the surroundings of the building, and for the vicinity of the town or the village in general.

REFERENCES

1. UNEP/UNCHS Balkans Task Force (BTF), 1999. The Potentials Effects on Human Health and the Environment Arising from Possible Use of Depleted Uranium during the 1999 Kosovo Conflict: A Preliminary Assessment. Geneva, October 1999.
2. UNEP, 2001. Depleted Uranium in Kosovo, Post-Conflict Environmental Assessment. UNEP Scientific Team Mission to Kosovo (November 5–19, 2000). United Nations Environment Programme, Geneva, March 2001.
3. UNEP, 2002. Depleted Uranium in Serbia and Montenegro. Post-Conflict Environmental assessment in FRY. United Nations Environment Programme, Geneva, March 2002.
4. UNEP, 2003. Depleted Uranium in Bosnia and Herzegovina. Post-Conflict Environmental Assessment. United Nations Environmental Programme, Geneva, March 2003.
5. IAEA, 2003. Radiological Conditions in Areas of Kuwait with Residues of Depleted Uranium. International Atomic Energy Agency, Vienna, August 2003.

6. IAEA, 2003. Remediation of Areas Contaminated by Past Activities and Accidents, Safety Standards Series No. WS-R-3, International Atomic Energy Agency, Vienna, November 2003.
7. EU, 2001. Depleted Uranium: Environmental and Health Effects in the Gulf War, Bosnia and Kosovo. Keller, M., Anet, B., Burger, M., Schmid, E., Wicki, A., Wirz, Ch., Spiez Laboratory in cooperation with Chambers G., European Parliament, Directorate-General for Research, Working Paper, Scientific and Technological Options Assessment Series, STOA 100 EN 05-2001, April 2001. Web: www. europarl.eu.int.
8. WHO, 2004. Uranium in Drinking-water. Summary statement. Extract from Chapter 12 — Chemical fact sheets of WHO, *Guidelines for Drinking-Water Quality*, 3rd ed. World Health Organization, 2004.
9. Andrew, S.P., Caliguri, R.D., Eiselstein, L.E., 1992. Relationship between Dynamic Properties and Penetration Mechanisms of Tungsten and Depleted Uranium Penetrators, 13th International Symposium on Ballistics. Stockholm.
10. Benedict, M., Pigford, T.H., Levi, H.W., 1981. *Nuclear Chemical Engineering*, p. 487.
11. DANESI, P.R. et al., 2003. Depleted uranium particles in selected Kosovo samples. *J. Environ. Radioact.* 64, 143–154.
12. DOE, 1963. Plutonium Content of Depleted Uranium. Department of Energy. 1963. Web: http://www.oakridge.doe.gov/Foia/GDP-0002.awd.
13. Fetter, S., Von Hippel, F.N., 1999. The hazard posed by depleted uranium munitions. *Sci. Global Security* 8, 125–161.
14. Food and Agriculture Organization of the United Nations, International Atomic Energy Agency, International Labour Organisation, OECD Nuclear Energy Agency, Pan American Health Organization, World Health Organization, International Basic Safety Standards for Protection against Ionizing Radiation and for the Safety of Radiation Sources, Safety Series No. 115, IAEA, Vienna, 1996.
15. HSK (Hauptabteilung für die Sicherheit der Kernanlagen), 1997. Berechnung der Strahlenexposition in der Umgebung aufgrund von Emissionen radioaktiver Stoffe aus Kernanlagen, HSK-R-41/d.
16. ICRP, 1990. Recommendations of the International Commission on Radiological Protection, Publication 60. Pergamon Press, Oxford and New York (1991). International Commission on Radiological Protection, 1990.
17. Magness, L.S., Farrand, T.G., 1990. Deformation Behavior and Its Relationship to the Penetration Performance of High-Density KE Penetrator Materials. Ballistic Research Laboratory, Aberdeen Proving Ground, MD. 21005-5066.
18. Ramachandra, K.B., 2000. Tank-Automotive and Armaments Command (TACOM) and Army Materiel Command (AMC) Review of Transuranics (TRU) in Depleted Uranium (DU) Armor. U.S. Department of the Army, AMSAM-TMD-SB.
19. Rostker, B., 2000. Environmental Exposure Report, Depleted Uranium in the Gulf (II). Web: http://www.gulflink.osd.mil/du_ii/.
20. UN, 2000. Sources and Effects of Ionizing Radiation (Report to the General Assembly), Scientific Committee on the Effects of Atomic Radiation (UNSCEAR), United Nations, 2000.
21. U.S. Army Center for Health Promotion and Preventive Medicine, 2004. Depleted Uranium Aerosol Doses and Risks: Summary of U.S. Assessment, Capstone, Prepared for U.S. Army Center for Health Promotion and Preventive Medicine and the U.S. Army Heavy Metals Office, October 2004. Web: http://www.deploymentlink.osd.mil/du_library/du_capstone/index.pdf.

22. U.S. Army Center for Health Promotion and Preventive Medicine, Depleted Uranium, Human Exposure Assessment and Health Risk Characterization in Support of the Environmental Exposure Report "Depleted Uranium in the Gulf" of the Office of the Special Assistant to the Secretary of Defense for Gulf War Illnesses, Medical Readiness and Military Deployments (OSAGWI), Health Risk Assessment Consultation No. 26-MF 7555-00D, USACHPPM, Washington, DC, 2000.
23. U.S. Army Material Command, 2000. Tank-Automotive and Armaments Command (TACOM) and Army Material Command (AMC) Review of Transuranics (TRU) in Depleted Uranium Armor. January 19, 2000. Memorandum. ATTN: AMCSF (Mr. Pittenger), 5001 Eisenhower Avenue, Alexandria, VA 22333-001.
24. Zajic, V.S. Review of Radioactivity, Military Use, and Health Effects of Depleted Uranium. 1999. Web: http://members.tripod.com/vzajic/.

Index

C

D

E

F

G

H

I

K

L

M

N

O

P

Q

R

S

T

U

V

W

X

Printed and bound by CPI Group (UK) Ltd, Croydon, CR0 4YY
23/10/2024
01778263-0012